Birkhäuser

Operator Theory: Advances and Applications

Volume 240

Founded in 1979 by Israel Gohberg

Joseph A. Ball • Michael A. Dritschel • A.F.M. ter Elst
Pierre Portal • Denis Potapov
Editors

Operator Theory in Harmonic and Non-commutative Analysis

23rd International Workshop in Operator Theory and its Applications, Sydney, July 2012

Birkhäuser

Editors
Joseph A. Ball
Department of Mathematics
Virginia Polytechnic Institute
Blacksburg, VA, USA

Michael A. Dritschel
Department of Mathematics
University of Newcastle upon Tyne
Newcastle upon Tyne, UK

A.F.M. ter Elst
Department of Mathematics
University of Auckland
Auckland, New Zealand

Pierre Portal
Mathematical Sciences Institute
Université Lille 1
Villeneuve d'Ascq, France
and
The Australian National University
Canberra, ACT, Australia

Denis Potapov
School of Mathematics and Statistics
University of New South Wales
Sydney, NSW, Australia

ISSN 0255-0156 ISSN 2296-4878 (electronic)
ISBN 978-3-319-06265-5 ISBN 978-3-319-06266-2 (eBook)
DOI 10.1007/978-3-319-06266-2
Springer Cham Heidelberg New York Dordrecht London

Library of Congress Control Number: 2014942552

Mathematics Subject Classification (2010): 30E05, 30H20, 34B24, 34C25, 34K13, 34L05, 34L40, 35Q58, 42A45, 42B37, 46L53, 47A10, 47A13, 47A20, 47A40, 47A48, 47A55, 47A60, 47A75, 47B07, 47B10, 47B20, 47B33, 47B35, 47B38, 47B40, 47D06, 47L20, 60H15

Printed on acid-free paper

Springer Basel is part of Springer Science+Business Media (www.birkhauser-science.com)

Contents

Introduction

The XXIII International Workshop on Operator Theory and its Applications (IWOTA 2012) was held at the University of New South Wales (Sydney, Australia) from 16 July to 20 July 2012. With 140 participants from all parts of the world, and 22 plenary speakers representing many different branches of operator theory, the meeting was a great success. Building on the strengths of Australian mathematical analysis, the meeting focused on the role of operator theory in harmonic and non-commutative analysis. Other themes were also well represented, from pure operator theory in Banach spaces through to engineering applications. The meeting certainly demonstrated the unity within the diversity of the field with discussions highlighting many connections between different branches of operator theory. It was financially supported by the Australian Mathematical Sciences Institute (AMSI), the University of New South Wales, the Centre for Mathematics and its Applications of the Australian National University, and the National Science Foundation of the USA. This support was vital and is gratefully acknowledged.

This volume contains the proceedings of the conference. It reflects the quality and the diversity of the research presented at IWOTA 2012. Each paper has been carefully refereed and has only been accepted if it meets the standards of the journal Integral Equations and Operator Theory. We are very thankful to the authors and the referees for their contributions.

The editors:

Joe Ball, Michael Dritschel, Tom ter Elst,
Pierre Portal, and Denis Potapov.

Operator Theory:
Advances and Applications, Vol. 240, 1–29

Tent Spaces over Metric Measure Spaces under Doubling and Related Assumptions

Alex Amenta

Abstract. In this article, we define the Coifman–Meyer–Stein tent spaces $T^{p,q,\alpha}(X)$ associated with an arbitrary metric measure space (X, d, μ) under minimal geometric assumptions. While gradually strengthening our geometric assumptions, we prove duality, interpolation, and change of aperture theorems for the tent spaces. Because of the inherent technicalities in dealing with abstract metric measure spaces, most proofs are presented in full detail.

Mathematics Subject Classification (2010). 42B35.

Keywords. Duality, change of aperture, complex interpolation, volume doubling, Hardy–Littlewood maximal operator.

1. Introduction

The purpose of this article is to indicate how the theory of tent spaces, as developed by Coifman, Meyer, and Stein for Euclidean space in [7], can be extended to more general metric measure spaces. Let X denote the metric measure space under consideration. If X is doubling, then the methods of [7] seem at first to carry over without much modification. However, there are some technicalities to be considered, even in this context. This is already apparent in the proof of the atomic decomposition given in [17].

Further still, there is an issue with the proof of the main interpolation result of [7] (see Remark 3.20 below). Alternate proofs of the interpolation result have since appeared in the literature – see for example [12], [4], [6], and [14] – but these proofs are given in the Euclidean context, and no indication is given of their general applicability. In fact, the methods of [12] and [4] can be used to obtain a partial interpolation result under weaker assumptions than doubling. This result relies on some tent space duality; we show in Section 3.2 that this holds once we assume that the uncentred Hardy–Littlewood maximal operator is of strong type (r, r) for all $r > 1$.[1]

Supported by the Australian Research Council Discovery grant DP120103692.

[1]This fact is already implicit in [7].

Finally, we consider the problem of proving the change of aperture result when X is doubling. The proof in [7] implicitly uses a geometric property of X which we term (NI), or 'nice intersections'. This property is independent of doubling, but holds for many doubling spaces which appear in applications – in particular, all complete Riemannian manifolds have 'nice intersections'. We provide a proof which does not require this assumption.

2. Spatial assumptions

Throughout this article, we implicitly assume that (X, d, μ) is a metric measure space; that is, (X, d) is a metric space and μ is a Borel measure on X. The ball centred at $x \in X$ of radius $r > 0$ is the set

$$B(x, r) := \{y \in X : d(x, y) < r\},$$

and we write $V(x, r) := \mu(B(x, r))$ for the volume of this set. We assume that the volume function $V(x, r)$ is finite[2] and positive; one can show that V is automatically measurable on $X \times \mathbb{R}_+$.

There are four geometric assumptions which we isolate for future reference:

(Proper): a subset $S \subset X$ is compact if and only if it is both closed and bounded, and the volume function $V(x, r)$ is lower semicontinuous as a function of (x, r);[3]

(HL): the uncentred Hardy–Littlewood maximal operator $\mathcal{M}$, defined for measurable functions f on X by

$$\mathcal{M}(f)(x) := \sup_{B \ni x} \frac{1}{\mu(B)} \int_B |f(y)| \, d\mu(y) \tag{1}$$

where the supremum is taken over all balls B containing x, is of strong type (r, r) for all $r > 1$;

(Doubling): there exists a constant $C > 0$ such that for all $x \in X$ and $r > 0$,

$$V(x, 2r) \leq CV(x, r);$$

(NI): for all $\alpha, \beta > 0$ there exists a positive constant $c_{\alpha,\beta} > 0$ such that for all $r > 0$ and for all $x, y \in X$ with $d(x, y) < \alpha r$,

$$\frac{\mu(B(x, \alpha r) \cap B(y, \beta r))}{V(x, \alpha r)} \geq c_{\alpha,\beta}.$$

We do not assume that X satisfies any of these assumptions unless mentioned otherwise. However, readers are advised to take (X, d, μ) to be a complete Riemannian manifold with its geodesic distance and Riemannian volume if they are not interested in such technicalities.

---

[2]Since X is a metric space, this implies that μ is σ-finite.

[3]Note that this is a strengthening of the usual definition of a proper metric space, as the usual definition does not involve a measure. We have abused notation by using the word 'proper' in this way, as it is convenient in this context.

It is well known that doubling implies (HL). However, the converse is not true. See for example [10] and [18], where it is shown that (HL) is true for $\mathbb{R}^2$ with the Gaussian measure. We will only consider (NI) along with doubling, so we remark that doubling does not imply (NI): one can see this by taking $\mathbb{R}^2$ (now with Lebesgue measure) and removing an open strip.[4] One can show that all complete doubling length spaces – in particular, all complete doubling Riemannian manifolds – satisfy (NI).

3. The basic tent space theory

3.1. Initial definitions and consequences

Let X^+ denote the 'upper half-space' $X \times \mathbb{R}_+$, equipped with the product measure $d\mu(y)\,dt/t$ and the product topology. Since X and $\mathbb{R}_+$ are metric spaces, with $\mathbb{R}_+$ separable, the Borel σ-algebra on X^+ is equal to the product of the Borel σ-algebras on X and $\mathbb{R}_+$, and so the product measure on X^+ is Borel (see [5, Lemma 6.4.2(i)]).

We say that a subset $C \subset X^+$ is *cylindrical* if it is contained in a cylinder: that is, if there exists $x \in X$ and $a, b, r > 0$ such that $C \subset B(x,r) \times (a,b)$. Note that cylindricity is equivalent to boundedness when X^+ is equipped with an appropriate metric, and that compact subsets of X^+ are cylindrical.

Cones and tents are defined as usual: for each $x \in X$ and $\alpha > 0$, the *cone of aperture α with vertex x* is the set

$$\Gamma^\alpha(x) := \{(y,t) \in X^+ : y \in B(x, \alpha t)\}.$$

For any subset $F \subset X$ we write

$$\Gamma^\alpha(F) := \bigcup_{x \in F} \Gamma^\alpha(x).$$

For any subset $O \subset X$, the *tent of aperture α over O* is defined to be the set

$$T^\alpha(O) := (\Gamma^\alpha(O^c))^c.$$

Writing

$$F_O(y,t) := \frac{\operatorname{dist}(y, O^c)}{t} = t^{-1} \inf_{x \in O^c} d(y,x),$$

one can check that $T^\alpha(O) = F_O^{-1}((\alpha, \infty))$. Since F_O is continuous (due to the continuity of $\operatorname{dist}(\cdot, O^c)$), we find that tents over open sets are measurable, and so it follows that cones over closed sets are also measurable. We remark that tents (resp. cones) over non-open (resp. non-closed) sets may not be measurable.

Let $F \subset X$ be such that $O := F^c$ has finite measure. Given $\gamma \in (0,1)$, we say that a point $x \in X$ has *global γ-density with respect to F* if for all balls B

[4] One could instead remove an open bounded region with sufficiently regular boundary, for example an open square. This yields a connected example.

containing x,

$$\frac{\mu(B \cap F)}{\mu(B)} \geq \gamma.$$

We denote the set of all such points by F_γ^*, and define $O_\gamma^* := (F_\gamma^*)^c$. An important fact here is the equality

$$O_\gamma^* = \{x \in X : \mathcal{M}(\mathbf{1}_O)(x) > 1 - \gamma\},$$

where $\mathbf{1}_O$ is the indicator function of O. We emphasise that $\mathcal{M}$ denotes the *uncentred* maximal operator. When O is open (i.e., when F is closed), this shows that $O \subset O_\gamma^*$ and hence that $F_\gamma^* \subset F$. Furthermore, the function $\mathcal{M}(\mathbf{1}_O)$ is lower semicontinuous whenever $\mathbf{1}_O$ is locally integrable (which is always true, since we assumed O has finite measure), which implies that F_γ^* is closed (hence measurable) and that O_γ^* is open (hence also measurable). Note that if X is doubling, then since $\mathcal{M}$ is of weak-type $(1,1)$, we have that

$$\mu(O_\gamma^*) \lesssim_{\gamma,X} \mu(O).$$

Remark 3.1. In our definition of points of γ-density, we used balls containing x rather than balls centred at x (as is usually done). This is done in order to avoid using the centred maximal function, which may not be measurable without assuming continuity of the volume function $V(x,r)$.

Here we find it convenient to introduce the notion of the α-*shadow* of a subset of X^+. For a subset $C \subset X^+$, we define the α-shadow of C to be the set

$$S^\alpha(C) := \{x \in X : \Gamma^\alpha(x) \cap C \neq \varnothing\}.$$

Shadows are always open, for if $A \subset X^+$ is any subset, and if $x \in S^\alpha(A)$, then there exists a point $(z,t_z) \in \Gamma^\alpha(x) \cap A$, and one can easily show that $B(x, \alpha t_z - d(x,z))$ is contained in $S^\alpha(A)$.

The starting point of the tent space theory is the definition of the operators $\mathcal{A}_q^\alpha$ and $\mathcal{C}_q^\alpha$. For $q \in (0,\infty)$, the former is usually defined for measurable functions f on $\mathbb{R}_+^{n+1}$ (with values in $\mathbb{R}$ or $\mathbb{C}$, depending on context) by

$$\mathcal{A}_q^\alpha(f)(x)^q := \iint_{\Gamma^\alpha(x)} |f(y,t)|^q \, \frac{d\lambda(y)\, dt}{t^{n+1}}$$

where $x \in \mathbb{R}^n$ and λ is the Lebesgue measure. There are four reasonable ways to generalise this definition to our possibly non-doubling metric measure space X:[5] these take the form

$$\mathcal{A}_q^\alpha(f)(x)^q := \iint_{\Gamma^\alpha(x)} |f(y,t)|^q \, \frac{d\mu(y)}{V(\mathbf{a},\mathbf{b}t)} \, \frac{dt}{t}$$

where $\mathbf{a} \in \{x,y\}$ and $\mathbf{b} \in \{1,\alpha\}$. In all of these definitions, if a function f on X^+ is supported on a subset $C \subset X^+$, then $\mathcal{A}_q^\alpha(f)$ is supported on $S^\alpha(C)$; we will use this fact repeatedly in what follows. Measurability of $\mathcal{A}_q^\alpha(f)(x)$ in x when $\mathbf{a} = y$ follows from Lemma A.6 in the Appendix; the choice $\mathbf{a} = x$ can be taken care

[5] We do not claim that these are the only reasonable generalisations.

of with a straightforward modification of this lemma. The choice $\mathbf{a} = x$, $\mathbf{b} = 1$ appears in [2, 17], and the choice $\mathbf{a} = y$, $\mathbf{b} = 1$ appears in [15, §3]. These definitions all lead to equivalent tent spaces when X is doubling. We will take $\mathbf{a} = y$, $\mathbf{b} = \alpha$ in our definition, as it leads to the following fundamental technique, which works with no geometric assumptions on X.

Lemma 3.2 (Averaging trick). *Let $\alpha > 0$, and suppose Φ is a nonnegative measurable function on X^+. Then*

$$\int_X \iint_{\Gamma^\alpha(x)} \Phi(y,t) \frac{d\mu(y)}{V(y,\alpha t)} \frac{dt}{t} \, d\mu(x) = \iint_{X^+} \Phi(y,t) \, d\mu(y) \frac{dt}{t}.$$

Proof. This is a straightforward application of the Fubini–Tonelli theorem, which we present explicitly due to its importance in what follows:

$$\begin{aligned}
&\int_X \iint_{\Gamma^\alpha(x)} \Phi(y,t) \frac{d\mu(y)}{V(y,\alpha t)} \frac{dt}{t} \, d\mu(x) \\
&= \int_X \int_0^\infty \int_X \mathbf{1}_{B(x,\alpha t)}(y) \Phi(y,t) \frac{d\mu(y)}{V(y,\alpha t)} \frac{dt}{t} \, d\mu(x) \\
&= \int_0^\infty \int_X \int_X \mathbf{1}_{B(y,\alpha t)}(x) \, d\mu(x) \, \Phi(y,t) \frac{d\mu(y)}{V(y,\alpha t)} \frac{dt}{t} \\
&= \int_0^\infty \int_X \frac{V(y,\alpha t)}{V(y,\alpha t)} \Phi(y,t) \, d\mu(y) \frac{dt}{t} \\
&= \iint_{X^+} \Phi(y,t) \, d\mu(y) \frac{dt}{t}.
\end{aligned}$$

□

We will also need the following lemma in order to prove that our tent spaces are complete. Here we need to make some geometric assumptions.

Lemma 3.3. *Let X be proper or doubling. Let $p, q, \alpha > 0$, let $K \subset X^+$ be cylindrical, and suppose f is a measurable function on X^+. Then*

$$\left\| \mathcal{A}_q^\alpha(\mathbf{1}_K f) \right\|_{L^p(X)} \lesssim \|f\|_{L^q(K)} \lesssim \left\| \mathcal{A}_q^\alpha(f) \right\|_{L^p(X)}, \tag{2}$$

with implicit constants depending on p, q, α, and K.

Proof. Write

$$K \subset B(x,r) \times (a,b) =: C$$

for some $x \in X$ and $a, b, r > 0$. We claim that there exist constants $c_0, c_1 > 0$ such that for all $(y,t) \in C$,

$$c_0 \leq V(y,\alpha t) \leq c_1.$$

If X is proper, this is an immediate consequence of the lower semicontinuity of the ball volume function (recall that we are assuming this whenever we assume X is proper) and the compactness of the closed cylinder $\overline{B(x,r)} \times [a,b]$. If X is doubling, then we argue as follows. Since $V(y,\alpha t)$ is increasing in t, we have that

$$\min_{(y,t) \in C} V(y,\alpha t) \geq \min_{y \in B(x,r)} V(y,\alpha a)$$

and

$$\max_{(y,t)\in C} V(y,\alpha t) \leq \max_{y\in B(x,r)} V(y,\alpha b).$$

By the argument in the proof of Lemma A.4 (in particular, by (16)), there exists $c_0 > 0$ such that

$$\min_{y\in B(x,r)} V(y,\alpha a) \geq c_0.$$

Furthermore, since

$$V(y,\alpha b) \leq V(x,\alpha b + r)$$

for all $y \in B(x,r)$, we have that

$$\max_{y\in B(x,r)} V(y,\alpha b) \leq V(x,\alpha b + r) =: c_1,$$

proving the claim.

To prove the first estimate of (2), write

$$\begin{aligned}
&\left\| \mathcal{A}_q^\alpha(\mathbf{1}_K f) \right\|_{L^p(X)} \\
&= \left(\int_{S^\alpha(K)} \left(\iint_{\Gamma^\alpha(x)} \mathbf{1}_K(y,t) |f(y,t)|^q \, \frac{d\mu(y)}{V(y,\alpha t)} \, \frac{dt}{t} \right)^{\frac{p}{q}} d\mu(x) \right)^{\frac{1}{p}} \\
&\lesssim_{c_0,q} \left(\int_{S^\alpha(K)} \left(\iint_K |f(y,t)|^q \, d\mu(y) \, \frac{dt}{t} \right)^{\frac{p}{q}} d\mu(x) \right)^{\frac{1}{p}} \\
&\lesssim_K \|f\|_{L^q(K)}.
\end{aligned}$$

To prove the second estimate, first choose finitely many points $(x_n)_{n=1}^N$ such that

$$\overline{B(x,r)} \subset \bigcup_{n=1}^N B(x_n, \alpha a/2)$$

using either compactness of $\overline{B(x,r)}$ (in the proper case) or doubling.[6] We then have

$$\begin{aligned}
&\left(\iint_K |f(y,t)|^q \, d\mu(y) \, \frac{dt}{t} \right)^{\frac{1}{q}} \\
&\lesssim_{c_1} \left(\iint_K \sum_{n=1}^N \mathbf{1}_{B(x_n,\alpha a/2)}(y) |f(y,t)|^q \, \frac{d\mu(y)}{V(y,\alpha t)} \, \frac{dt}{t} \right)^{\frac{1}{q}} \\
&\lesssim_{X,q} \sum_{n=1}^N \left(\iint_K \mathbf{1}_{B(x_n,\alpha a/2)}(y) |f(y,t)|^q \, \frac{d\mu(y)}{V(y,\alpha t)} \, \frac{dt}{t} \right)^{\frac{1}{q}}.
\end{aligned}$$

[6] In the doubling case, this is a consequence of what is usually called 'geometric doubling'. A proof that this follows from the doubling condition can be found in [8, §III.1].

If $x, y \in B(x_n, \alpha a/2)$, then $d(x,y) < \alpha a < \alpha t$ (since $t > a$), and so

$$\iint_K \mathbf{1}_{B(x_n,\alpha a/2)}(y)|f(y,t)|^q \frac{d\mu(y)}{V(y,\alpha t)} \frac{dt}{t} \leq \iint_{\Gamma^\alpha(x)} |f(y,t)|^q \frac{d\mu(y)}{V(y,\alpha t)} \frac{dt}{t}. \qquad (3)$$

When $p \geq q$, we use Hölder's inequality along with (3) to write

$$\begin{aligned}
&\sum_{n=1}^N \left(\iint_K \mathbf{1}_{B(x_n,\alpha a/2)}(y)|f(y,t)|^q \frac{d\mu(y)}{V(y,\alpha t)} \frac{dt}{t}\right)^{\frac{1}{q}} \\
&= \sum_{n=1}^N \Big(\frac{1}{V(x_n,\alpha a/2)} \int_{B(x_n,\alpha a/2)} \\
&\qquad \iint_K \mathbf{1}_{B(x_n,\alpha a/2)}(y)|f(y,t)|^q \frac{d\mu(y)}{V(y,\alpha t)} \frac{dt}{t}\, d\mu(x)\Big)^{\frac{1}{q}} \\
&\leq \sum_{n=1}^N \Big(\frac{1}{V(x_n,\alpha a/2)} \int_{B(x_n,\alpha a/2)} \\
&\qquad \left(\iint_K \mathbf{1}_{B(x_n,\alpha a/2)}(y)|f(y,t)|^q \frac{d\mu(y)}{V(y,\alpha t)} \frac{dt}{t}\right)^{\frac{p}{q}} d\mu(x)\Big)^{\frac{1}{p}} \\
&\leq \sum_{n=1}^N \left(\frac{1}{V(x_n,\alpha a/2)} \int_{B(x_n,\alpha a/2)} \left(\iint_{\Gamma^\alpha(x)} |f(y,t)|^q \frac{d\mu(y)}{V(y,\alpha t)} \frac{dt}{t}\right)^{\frac{p}{q}} d\mu(x)\right)^{\frac{1}{p}} \\
&\lesssim_{K,p} \left\|\mathcal{A}_q^\alpha(f)\right\|_{L^p(X)},
\end{aligned}$$

completing the proof in this case. When $p < q$, the situation can be handled using Minkowski's inequality as follows. Using $p/q < 1$, we have

$$\begin{aligned}
&\left(\frac{1}{V(x_n,\alpha a/2)} \int_{B(x_n,\alpha a/2)} \iint_K \mathbf{1}_{B(x_n,\alpha a/2)}(y)|f(y,t)|^q \frac{d\mu(y)}{V(y,\alpha t)} \frac{dt}{t}\, d\mu(x)\right)^{\frac{1}{q}} \\
&\leq \Bigg(\frac{1}{V(x_n,\alpha a/2)} \Bigg(\int_{B(x_n,\alpha a/2)} \\
&\qquad \left(\iint_K \mathbf{1}_{B(x_n,\alpha a/2)}(y)|f(y,t)|^q \frac{d\mu(y)}{V(y,\alpha t)} \frac{dt}{t}\right)^{\frac{p}{q}} d\mu(x)\Bigg)^{\frac{q}{p}}\Bigg)^{\frac{1}{q}} \\
&\leq C\Bigg(\frac{1}{V(x_n,\alpha a/2)} \int_{B(x_n,\alpha a/2)} \\
&\qquad \left(\iint_K \mathbf{1}_{B(x_n,\alpha a/2)}(y)|f(y,t)|^q \frac{d\mu(y)}{V(y,\alpha t)} \frac{dt}{t}\right)^{\frac{p}{q}} d\mu(x)\Bigg)^{\frac{1}{p}},
\end{aligned}$$

where

$$C = C(p,q,\alpha,K) = \max_n \left(V(x_n, \alpha a/2)^{\frac{1}{p}-\frac{1}{q}} \right).$$

We can then proceed as in the case where $p \geq q$. □

As usual, with $\alpha > 0$ and $p, q \in (0,\infty)$, we define the tent space (quasi-)norm of a measurable function f on X^+ by

$$\|f\|_{T^{p,q,\alpha}(X)} := \left\| \mathcal{A}_q^\alpha(f) \right\|_{L^p(X)},$$

and the tent space $T^{p,q,\alpha}(X)$ to be the (quasi-)normed vector space consisting of all such f (defined almost everywhere) for which this quantity is finite.

Remark 3.4. One can define the tent space as either a real or complex vector space, according to one's own preference. We will implicitly work in the complex setting (so our functions will always be $\mathbb{C}$-valued). Apart from complex interpolation, which demands that we consider complex Banach spaces, the difference is immaterial.

Proposition 3.5. *Let X be proper or doubling. For all $p, q, \alpha \in (0,\infty)$, the tent space $T^{p,q,\alpha}(X)$ is complete and contains $L^q_c(X^+)$ (the space of functions $f \in L^q(X^+)$ with cylindrical support) as a dense subspace.*

Proof. Let $(f_n)_{n\in\mathbb{N}}$ be a Cauchy sequence in $T^{p,q,\alpha}(X)$. Then by Lemma 3.3, for every cylindrical subset $K \subset X^+$ the sequence $(\mathbf{1}_K f_n)_{n\in\mathbb{N}}$ is Cauchy in $L^q(K)$. We thus obtain a limit

$$f_K := \lim_{n\to\infty} \mathbf{1}_K f_n \in L^q(K)$$

for each K. If K_1 and K_2 are two cylindrical subsets of X^+, then $f_{K_1}|_{K_1\cap K_2} = f_{K_2}|_{K_1\cap K_2}$, so by making use of an increasing sequence $\{K_m\}_{m\in\mathbb{N}}$ of cylindrical subsets of X^+ whose union is X^+ (for example, we could take $K_m := B(x,m) \times (1/m, m)$ for some $x \in X$) we obtain a function $f \in L^q_{\mathrm{loc}}(X^+)$ with $f|_{K_m} = f_{K_m}$ for each $m \in \mathbb{N}$.[7] This is our candidate limit for the sequence $(f_n)_{n\in\mathbb{N}}$. To see that f lies in $T^{p,q,\alpha}(X)$, write for any $m, n \in \mathbb{N}$

$$\begin{aligned} \|\mathbf{1}_{K_m} f\|_{T^{p,q,\alpha}(X)} &\lesssim_{p,q} \|\mathbf{1}_{K_m}(f - f_n)\|_{T^{p,q,\alpha}(X)} + \|\mathbf{1}_{K_m} f_n\|_{T^{p,q,\alpha}(X)} \\ &\leq C_{p,q,\alpha,X,m} \|f - f_n\|_{L^q(K_m)} + \|f_n\|_{T^{p,q,\alpha}(X)}, \end{aligned}$$

the (p,q)-dependence in the first estimate being relevant only for $p < 1$ or $q < 1$, and the second estimate coming from Lemma 3.3. Since the sequence $(f_n)_{n\in\mathbb{N}}$ converges to $\mathbf{1}_{K_m} f$ in $L^q(K_m)$ and is Cauchy in $T^{p,q,\alpha}(X)$, we have that

$$\|\mathbf{1}_{K_m} f\|_{T^{p,q,\alpha}(X)} \lesssim \sup_{n\in\mathbb{N}} \|f_n\|_{T^{p,q,\alpha}(X)}$$

uniformly in m. Hence $\|f\|_{T^{p,q,\alpha}(X)}$ is finite. We now claim that for all $\varepsilon > 0$ there exists $m \in \mathbb{N}$ such that for all sufficiently large $n \in \mathbb{N}$, we have

$$\left\|\mathbf{1}_{K_m^c}(f_n - f)\right\|_{T^{p,q,\alpha}(X)} \leq \varepsilon.$$

[7] We interpret 'locally integrable on X^+' as meaning 'integrable on all cylinders', rather than 'integrable on all compact sets'.

Indeed, since the sequence $(f_n)_{n\in\mathbb{N}}$ is Cauchy in $T^{p,q,\alpha}(X)$, there exists $N \in \mathbb{N}$ such that for all $n, n' \geq N$ we have $\|f_n - f_{n'}\|_{T^{p,q,\alpha}(X)} < \varepsilon/2$. Furthermore, since

$$\lim_{m\to\infty} \left\|\mathbf{1}_{K_m^c}(f_N - f)\right\|_{T^{p,q,\alpha}(X)} = 0$$

by the Dominated Convergence Theorem, we can choose m such that

$$\left\|\mathbf{1}_{K_m^c}(f_N - f)\right\|_{T^{p,q,\alpha}(X)} < \varepsilon/2.$$

Then for all $n \geq N$,

$$\begin{aligned}
&\left\|\mathbf{1}_{K_m^c}(f_n - f)\right\|_{T^{p,q,\alpha}(X)} \\
&\quad \lesssim_{p,q} \left\|\mathbf{1}_{K_m^c}(f_n - f_N)\right\|_{T^{p,q,\alpha}(X)} + \left\|\mathbf{1}_{K_m^c}(f_N - f)\right\|_{T^{p,q,\alpha}(X)} \\
&\quad \leq \|f_n - f_N\|_{T^{p,q,\alpha}(X)} + \left\|\mathbf{1}_{K_m^c}(f_N - f)\right\|_{T^{p,q,\alpha}(X)} \\
&\quad < \varepsilon,
\end{aligned}$$

proving the claim. Finally, by the previous remark, for all $\varepsilon > 0$ we can find m such that for all sufficiently large $n \in \mathbb{N}$ we have

$$\begin{aligned}
\|f_n - f\|_{T^{p,q,\alpha}(X)} &\lesssim_{p,q} \|\mathbf{1}_{K_m}(f_n - f)\|_{T^{p,q,\alpha}(X)} + \left\|\mathbf{1}_{K_m^c}(f_n - f)\right\|_{T^{p,q,\alpha}(X)} \\
&< \|\mathbf{1}_{K_m}(f_n - f)\|_{T^{p,q,\alpha}(X)} + \varepsilon \\
&\leq C(p, q, \alpha, X, m) \|f_n - f\|_{L^q(K_m)} + \varepsilon.
\end{aligned}$$

Taking the limit of both sides as $n \to \infty$, we find that $\lim_{n\to\infty} f_n = f$ in $T^{p,q,\alpha}(X)$, and therefore $T^{p,q,\alpha}(X)$ is complete. To see that $L_c^q(X^+)$ is dense in $T^{p,q,\alpha}(X)$, simply write $f \in T^{p,q,\alpha}(X)$ as the pointwise limit

$$f = \lim_{n\to\infty} \mathbf{1}_{K_n} f.$$

By the Dominated Convergence Theorem, this convergence holds in $T^{p,q,\alpha}(X)$. □

We note that Lemma 3.2 implies that in the case where $p = q$, we have $T^{p,p,\alpha}(X) = L^p(X^+)$ for all $\alpha > 0$.

In the same way as Lemma 3.2, we can prove the analogue of [7, Lemma 1].

Lemma 3.6 (First integration lemma). *For any nonnegative measurable function Φ on X^+, with F a measurable subset of X and $\alpha > 0$,*

$$\int_F \iint_{\Gamma^\alpha(x)} \Phi(y, t)\, d\mu(y)\, dt\, d\mu(x) \leq \iint_{\Gamma^\alpha(F)} \Phi(y, t) V(y, \alpha t)\, d\mu(y)\, dt.$$

Remark 3.7. There is one clear disadvantage of our choice of tent space norm: it is no longer clear that

$$\|\cdot\|_{T^{p,q,\alpha}(X)} \leq \|\cdot\|_{T^{p,q,\beta}(X)} \tag{4}$$

when $\alpha < \beta$. In fact, this may not even be true for general non-doubling spaces. This is no great loss, since for doubling spaces we can revert to the 'original' tent space norm (with $\mathbf{a} = x$ and $\mathbf{b} = 1$) at the cost of a constant depending only on X, and for this choice of norm (4) is immediate.

In order to define the tent spaces $T^{\infty,q,\alpha}(X)$, we need to introduce the operator $\mathcal{C}_q^\alpha$. For measurable functions f on X^+, we define

$$\mathcal{C}_q^\alpha(f)(x) := \sup_{B \ni x} \left(\frac{1}{\mu(B)} \iint_{T^\alpha(B)} |f(y,t)|^q \, d\mu(y) \, \frac{dt}{t} \right)^{\frac{1}{q}},$$

where the supremum is taken over all balls containing x. Since $\mathcal{C}_q^\alpha(f)$ is lower semicontinuous (see Lemma A.7), $\mathcal{C}_q^\alpha(f)$ is measurable. We define the (quasi-)norm $\|\cdot\|_{T^{\infty,q,\alpha}(X)}$ for functions f on X^+ by

$$\|f\|_{T^{\infty,q,\alpha}(X)} := \left\|\mathcal{C}_q^\alpha(f)\right\|_{L^\infty(X)},$$

and the tent space $T^{\infty,q,\alpha}(X)$ as the (quasi-)normed vector space of measurable functions f on X^+, defined almost everywhere, for which $\|f\|_{T^{\infty,q,\alpha}(X)}$ is finite. The proof that $T^{\infty,q,\alpha}(X)$ is a (quasi-)Banach space is similar to that of Proposition 3.5 once we have established the following analogue of Lemma 3.3.

Lemma 3.8. *Let $q, \alpha > 0$, let $K \subset X^+$ be cylindrical, and suppose f is a measurable function on X^+. Then*

$$\|f\|_{L^q(K)} \lesssim \|f\|_{T^{\infty,q,\alpha}(X)}, \tag{5}$$

with implicit constant depending only on α, q, and K (but not otherwise on X). Furthermore, if X is proper or doubling, then we also have

$$\|\mathbf{1}_K f\|_{T^{\infty,q,\alpha}(X)} \lesssim \|f\|_{L^q(K)},$$

again with implicit constant depending only on α, q, and K.

Proof. We use Lemma A.4. To prove the first estimate, for each $\varepsilon > 0$ we can choose a ball B_ε such that $T^\alpha(B_\varepsilon) \supset K$ and $\mu(B_\varepsilon) < \beta_1(K) + \varepsilon$. Then

$$\begin{aligned}
\|f\|_{L^q(K)} &\leq \left\|\mathbf{1}_{T^\alpha(B_\varepsilon)} f\right\|_{L^q(X^+)} \\
&= \mu(B_\varepsilon)^{\frac{1}{q}} \mu(B_\varepsilon)^{-\frac{1}{q}} \left\|\mathbf{1}_{T^\alpha(B_\varepsilon)} f\right\|_{L^q(X^+)} \\
&\leq (\beta_1(K) + \varepsilon)^{\frac{1}{q}} \|f\|_{T^{\infty,q,\alpha}(X)}.
\end{aligned}$$

In the final line we used that $\mu(B_\varepsilon) > 0$ to conclude that $\mu(B_\varepsilon)^{-1/q} \left\|\mathbf{1}_{T^\alpha(B_\varepsilon)} f\right\|_{L^q(X^+)}$ is less than the *essential* supremum of $\mathcal{C}_q^\alpha(f)$. Since $\varepsilon > 0$ was arbitrary, we have the first estimate. For the second estimate, assuming that X is proper or doubling, observe that

$$\begin{aligned}
\|\mathbf{1}_K f\|_{T^{\infty,q,\alpha}(X)} &\leq \sup_{B \subset X} \left(\frac{1}{\mu(B)} \iint_{T^\alpha(B) \cap K} |f(y,t)|^q \, d\mu(y) \, \frac{dt}{t} \right)^{\frac{1}{q}} \\
&\leq \left(\frac{1}{\beta_0(K)} \iint_K |f(y,t)|^q \, d\mu(y) \frac{dt}{t} \right)^{\frac{1}{q}} \\
&= \beta_0(K)^{-\frac{1}{q}} \|f\|_{L^q(K)},
\end{aligned}$$

completing the proof. □

Remark 3.9. In this section we did not impose any geometric conditions on our space X besides our standing assumptions on the measure μ and the properness assumption (in the absence of doubling). Thus we have defined the tent space $T^{p,q,\alpha}(X)$ in considerable generality. However, what we have defined is a *global* tent space, and so this concept may not be inherently useful when X is non-doubling. Instead, our interest is to determine precisely where geometric assumptions are needed in the tent space theory.

3.2. Duality, the vector-valued approach, and complex interpolation

3.2.1. Midpoint results. The geometric assumption (HL) from Section 2 now comes into play. For $r \geq 1$, we denote the Hölder conjugate of r by $r' := r/(r-1)$ with $r' = \infty$ when $r = 1$.

Proposition 3.10. *Suppose that X is either proper or doubling, and satisfies assumption* (HL). *Then for $p, q \in (1,\infty)$ and $\alpha > 0$, the pairing*

$$\langle f, g\rangle := \iint_{X^+} f(y,t)\overline{g(y,t)}\, d\mu(y)\, \frac{dt}{t} \qquad (f \in T^{p,q,\alpha}(X), g \in T^{p',q',\alpha}(X))$$

realises $T^{p',q',\alpha}(X)$ as the Banach space dual of $T^{p,q,\alpha}(X)$, up to equivalence of norms.

This is proved in the same way as in [7]. We provide the details in the interest of self-containment.

Proof. We first remark that if $p = q$, the duality statement is a trivial consequence of the equality $T^{p,p,\alpha}(X) = L^p(X^+)$. In general, suppose $f \in T^{p,q,\alpha}(X)$ and $g \in T^{p',q',\alpha}(X)$. Then by the averaging trick and Hölder's inequality, we have

$$\begin{aligned} |\langle f, g\rangle| &\leq \int_X \iint_{\Gamma^\alpha(x)} |f(y,t)\overline{g(y,t)}| \frac{d\mu(y)}{V(y,\alpha t)} \frac{dt}{t}\, d\mu(x) \\ &\leq \int_X \mathcal{A}^\alpha_q(f)(x)\mathcal{A}^\alpha_{q'}(g)(x)\, d\mu(x) \\ &\leq \|f\|_{T^{p,q,\alpha}(X)} \|g\|_{T^{p',q',\alpha}(X)}. \end{aligned} \tag{6}$$

Thus every $g \in T^{p',q',\alpha}(X)$ induces a bounded linear functional on $T^{p,q,\alpha}(X)$ via the pairing $\langle\cdot,\cdot\rangle$, and so $T^{p',q',\alpha}(X) \subset (T^{p,q,\alpha}(X))^*$. Conversely, suppose $\ell \in (T^{p,q,\alpha}(X))^*$. If $K \subset X^+$ is cylindrical, then by the properness or doubling assumption, we can invoke Lemma 3.3 to show that ℓ induces a bounded linear functional $\ell_K \in (L^q(K))^*$, which can in turn be identified with a function $g_K \in L^{q'}(K)$. By covering X^+ with an increasing sequence of cylindrical subsets, we thus obtain a function $g \in L^{q'}_{\text{loc}}(X^+)$ such that $g|_K = g_K$ for all cylindrical $K \subset X^+$. If $f \in L^q(X^+)$ is cylindrically supported, then we have

$$\begin{aligned} \iint_{X^+} f(y,t)\overline{g(y,t)}\, d\mu(y)\, \frac{dt}{t} &= \iint_{\operatorname{supp} f} f(y,t)\overline{g_{\operatorname{supp} f}(y,t)}\, d\mu(y)\, \frac{dt}{t} \\ &= \ell_{\operatorname{supp} f}(f) = \ell(f), \end{aligned} \tag{7}$$

recalling that $f \in T^{p,q,\alpha}(X)$ by Lemma 3.3. Since the cylindrically supported $L^q(X^+)$ functions are dense in $T^{p,q,\alpha}(X)$, the representation (7) of $\ell(f)$ in terms of g is valid for all $f \in T^{p,q,\alpha}(X)$ by dominated convergence and the inequality (6), provided we show that g is in $T^{p',q',\alpha}(X)$. Now suppose $p < q$. We will show that g lies in $T^{p',q',\alpha}(X)$, thus showing directly that $(T^{p,q,\alpha}(X))^*$ is contained in $T^{p',q',\alpha}(X)$. It suffices to show this for g_K, where $K \subset X^+$ is an arbitrary cylindrical subset, provided we obtain an estimate which is uniform in K. We estimate

$$\|g_K\|_{T^{p',q',\alpha}(X)}^{q'} = \left\| \mathcal{A}_{q'}^{\alpha}(g_K)^{q'} \right\|_{L^{p'/q'}(X)}$$

by duality. Let $\psi \in L^{(p'/q')'}(X)$ be nonnegative, with $\|\psi\|_{L^{(p'/q')'}(X)} \leq 1$. Then by the Fubini–Tonelli theorem,

$$\begin{aligned}
&\int_X \mathcal{A}_{q'}^{\alpha}(g_K)(x)^{q'} \psi(x)\, d\mu(x) \\
&= \int_X \iint_{X^+} \mathbf{1}_{B(y,\alpha t)}(x) |g_K(y,t)|^{q'} \frac{d\mu(y)}{V(y,\alpha t)} \frac{dt}{t} \psi(x)\, d\mu(x) \\
&= \int_0^\infty \int_X \frac{1}{V(y,\alpha t)} \int_{B(y,\alpha t)} \psi(x)\, d\mu(x)\, |g_K(y,t)|^{q'}\, d\mu(y) \frac{dt}{t} \\
&= \iint_{X^+} M_{\alpha t}\psi(y) |g_K(y,t)|^{q'}\, d\mu(y) \frac{dt}{t},
\end{aligned}$$

where M_s is the averaging operator defined for $y \in X$ and $s > 0$ by

$$M_s\psi(y) := \frac{1}{V(y,s)} \int_{B(y,s)} \psi(x)\, d\mu(x).$$

Thus we can write formally

$$\int_X \mathcal{A}_{q'}^{\alpha}(g_K)(x)^{q'} \psi(x)\, d\mu(x) = \langle f_\psi, g \rangle, \tag{8}$$

where we define

$$f_\psi(y,t) := \begin{cases} M_{\alpha t}\psi(y) \overline{g_K(y,t)}^{q'/2} g_K(y,t)^{(q'/2)-1} & \text{when } g_K(y,t) \neq 0, \\ 0 & \text{when } g_K(y,t) = 0, \end{cases}$$

noting that $g_K(y,t)^{(q'/2)-1}$ is not defined when $g_K(y,t) = 0$ and $q' < 2$. However, the equality (8) is not valid until we show that f_ψ lies in $T^{p,q,\alpha}(X)$. To this end, estimate

$$\begin{aligned}
\mathcal{A}_q^{\alpha}(f_\psi) &\leq \left(\iint_{\Gamma^\alpha(x)} M_{\alpha t}\psi(y)^q |g_K(y,t)|^{q(q'-1)} \frac{d\mu(y)}{V(y,\alpha t)} \frac{dt}{t} \right)^{\frac{1}{q}} \\
&\leq \left(\iint_{\Gamma^\alpha(x)} \mathcal{M}\psi(x)^q |g_K(y,t)|^{q'} \frac{d\mu(y)}{V(y,\alpha t)} \frac{dt}{t} \right)^{\frac{1}{q}} \\
&= \mathcal{M}\psi(x) \mathcal{A}_{q'}^{\alpha}(g_K)(x)^{q'/q}.
\end{aligned}$$

Taking r such that $1/p = 1/r + 1/(p'/q')'$ and using (HL), we then have

$$\begin{aligned}\left\|\mathcal{A}_q^\alpha(f_\psi)\right\|_{L^p(X)} &\leq \left\|(\mathcal{M}\psi)\mathcal{A}_{q'}^\alpha(g_K)^{q'/q}\right\|_{L^p(X)} \\ &\leq \|\mathcal{M}\psi\|_{L^{(p'/q')'}(X)} \left\|\mathcal{A}_{q'}^\alpha(g_K)^{q'/q}\right\|_{L^r(X)} \\ &\lesssim_X \|\psi\|_{L^{(p'/q')'}(X)} \left\|\mathcal{A}_{q'}^\alpha(g_K)\right\|_{L^{rq'/q}(X)}^{q'/q} \\ &\leq \left\|\mathcal{A}_{q'}^\alpha(g_K)\right\|_{L^{rq'/q}(X)}^{q'/q}.\end{aligned}$$

One can show that $rq'/q = p'$, and so f_ψ is in $T^{p,q,\alpha}(X)$ by Lemma 3.3. By (8), taking the supremum over all ψ under consideration, we can write

$$\begin{aligned}\|g_K\|_{T^{p',q',\alpha}(X)}^{q'} &\leq \|\ell\| \, \|f_\psi\|_{T^{p,q,\alpha}(X)} \\ &\lesssim_X \|\ell\| \, \|g_K\|_{T^{p',q',\alpha}(X)}^{q'/q},\end{aligned}$$

and consequently, using that $\|g_K\|_{T^{p',q',\alpha}(X)} < \infty$,

$$\|g_K\|_{T^{p',q',\alpha}(X)} \lesssim_X \|\ell\|.$$

Since this estimate is independent of K, we have shown that $g \in T^{p',q',\alpha}(X)$, and therefore that $(T^{p,q,\alpha}(X))^*$ is contained in $T^{p',q',\alpha}(X)$. This completes the proof when $p < q$. To prove the statement for $p > q$, it suffices to show that the tent space $T^{p',q',\alpha}(X)$ is reflexive. Thanks to the Eberlein–Šmulian theorem (see [1, Corollary 1.6.4]), this is equivalent to showing that every bounded sequence in $T^{p',q',\alpha}(X)$ has a weakly convergent subsequence. Let $\{f_n\}_{n\in\mathbb{N}}$ be a sequence in $T^{p',q',\alpha}(X)$ with $\|f_n\|_{T^{p',q',\alpha}(X)} \leq 1$ for all $n \in \mathbb{N}$. Then by Lemma 3.3, for all cylindrical $K \subset X^+$ the sequence $\{f_n\}_{n\in\mathbb{N}}$ is bounded in $L^{q'}(K)$, and so by reflexivity of $L^{q'}(K)$ we can find a subsequence $\{f_{n_j}\}_{j\in\mathbb{N}}$ which converges weakly in $L^{q'}(K)$. We will show that this subsequence also converges weakly in $T^{p',q',\alpha}(X)$. Let $\ell \in (T^{p',q',\alpha}(X))^*$. Since $p' < q'$, we have already shown that there exists a function $g \in T^{p,q,\alpha}(X)$ such that $\ell(f) = \langle f, g\rangle$. For every $\varepsilon > 0$, we can find a cylindrical set $K_\varepsilon \subset X^+$ such that

$$\|g - \mathbf{1}_{K_\varepsilon} g\|_{T^{p,q,\alpha}(X)} \leq \varepsilon.$$

Thus for all $i, j \in \mathbb{N}$ and for all $\varepsilon > 0$ we have

$$\begin{aligned}&\ell(f_{n_i}) - \ell(f_{n_j}) \\ &\quad = \langle f_{n_i} - f_{n_j}, \mathbf{1}_{K_\varepsilon} g\rangle + \langle f_{n_i} - f_{n_j}, g - \mathbf{1}_{K_\varepsilon} g\rangle \\ &\quad \leq \langle f_{n_i} - f_{n_j}, \mathbf{1}_{K_\varepsilon} g\rangle + \left(\|f_{n_i}\|_{T^{p',q',\alpha}(X)} + \left\|f_{n_j}\right\|_{T^{p',q',\alpha}(X)}\right) \|g - \mathbf{1}_{K_\varepsilon} g\|_{T^{p,q,\alpha}} \\ &\quad \leq \langle f_{n_i} - f_{n_j}, \mathbf{1}_{K_\varepsilon} g\rangle + 2\varepsilon.\end{aligned}$$

As $i, j \to \infty$, the first term on the right-hand side above tends to 0, and so we conclude that $\{f_{n_j}\}_{n\in\mathbb{N}}$ converges weakly in $T^{p',q',\alpha}(X)$. This completes the proof. □

Remark 3.11. As mentioned earlier, property (HL) is weaker than doubling, but this is still a strong assumption. We note that for Proposition 3.10 to hold for a given pair (p,q), the uncentred Hardy–Littlewood maximal operator need only be of strong type $((p'/q')', (p'/q')')$. Since $(p'/q')'$ is increasing in p and decreasing in q, the condition required on X is stronger as $p \to 1$ and $q \to \infty$.

Given Proposition 3.10, we can set up the vector-valued approach to tent spaces (first considered in [12]) using the method of [4]. Fix $p \in (0,\infty)$, $q \in (1,\infty)$, and $\alpha > 0$. For simplicity of notation, write

$$L^q_\alpha(X^+) := L^q\left(X^+; \frac{d\mu(y)}{V(y,\alpha t)}\,\frac{dt}{t}\right).$$

We define an operator $T_\alpha \colon T^{p,q,\alpha}(X) \to L^p(X; L^q_\alpha(X^+))$ from the tent space into the $L^q_\alpha(X^+)$-valued L^p space on X (see [9, §2] for vector-valued Lebesgue spaces) by setting

$$T_\alpha f(x)(y,t) := f(y,t)\mathbf{1}_{\Gamma^\alpha(x)}(y,t).$$

One can easily check that

$$\|T_\alpha f\|_{L^p(X;L^q_\alpha(X^+))} = \|f\|_{T^{p,q,\alpha}(X)},$$

and so the tent space $T^{p,q,\alpha}(X)$ can be identified with its image under T_α in $L^p(X; L^q_\alpha(X^+))$, provided that $T_\alpha f$ is indeed a strongly measurable function of $x \in X$. This can be shown for $q \in (1,\infty)$ by recourse to Pettis' measurability theorem [9, §2.1, Theorem 2], which reduces the question to that of weak measurability of $T_\alpha f$. To prove weak measurability, suppose $g \in L^{q'}_\alpha(X)$; then

$$\langle T_\alpha f(x), g\rangle = \iint_{\Gamma^\alpha(x)} f(y,t)\overline{g(y,t)}\,\frac{d\mu(y)}{V(y,\alpha t)}\,\frac{dt}{t},$$

which is measurable in x by Lemma A.6. Thus $T_\alpha f$ is weakly measurable, and therefore $T_\alpha f$ is strongly measurable as claimed.

Now assume $p,q \in (1,\infty)$ and consider the operator Π_α, sending X^+-valued functions on X to $\mathbb{C}$-valued functions on X^+, given by

$$(\Pi_\alpha F)(y,t) := \frac{1}{V(y,\alpha t)} \int_{B(y,\alpha t)} F(x)(y,t)\,d\mu(x)$$

whenever this expression is defined. Using the duality pairing from Proposition 3.10 and the duality pairing $\langle\langle\cdot,\cdot\rangle\rangle$ for vector-valued L^p spaces, for $f \in T^{p,q,\alpha}(X)$ and $G \in L^{p'}(X; L^{q'}_\alpha(X^+))$ we have

$$\begin{aligned}
\langle\langle T_\alpha f, G\rangle\rangle &= \int_X \iint_{X^+} T_\alpha f(x)(y,t)\overline{G(x)(y,t)}\,\frac{d\mu(y)}{V(y,\alpha t)}\,\frac{dt}{t}\,d\mu(x) \\
&= \iint_{X^+} \frac{f(y,t)}{V(y,\alpha t)} \int_X \mathbf{1}_{B(y,\alpha t)}(x)\overline{G(x)(y,t)}\,d\mu(x)\,d\mu(y)\,\frac{dt}{t} \\
&= \iint_{X^+} f(y,t)\overline{(\Pi_\alpha G)(y,t)}\,d\mu(y)\,\frac{dt}{t} \\
&= \langle\langle f, \Pi_\alpha G\rangle\rangle.
\end{aligned}$$

Thus Π_α maps $L^{p'}(X; L^{q'}_\alpha(X^+))$ to $T^{p',q',\alpha}(X)$, by virtue of being the adjoint of T_α. Consequently, the operator $P_\alpha := T_\alpha \Pi_\alpha$ is bounded from $L^p(X; L^q_\alpha(X^+))$ to itself for $p, q \in (1, \infty)$. A quick computation shows that $\Pi_\alpha T_\alpha = I$, so that P_α projects $L^p(X; L^q_\alpha(X^+))$ onto $T_\alpha(T^{p,q,\alpha}(X))$. This shows that $T_\alpha(T^{p,q,\alpha}(X))$ is a complemented subspace of $L^p(X; L^q_\alpha(X^+))$. This observation leads to the basic interpolation result for tent spaces. Here $[\cdot, \cdot]_\theta$ denotes the complex interpolation functor (see [3, Chapter 4]).

Proposition 3.12. *Suppose that X is either proper or doubling, and satisfies assumption* (HL). *Then for p_0, p_1, q_0, and q_1 in $(1, \infty)$, $\theta \in [0, 1]$, and $\alpha > 0$, we have (up to equivalence of norms)*

$$[T^{p_0,q_0,\alpha}(X), T^{p_1,q_1,\alpha}(X)]_\theta = T^{p,q,\alpha}(X),$$

where $1/p = (1-\theta)/p_0 + \theta/p_1$ and $1/q = (1-\theta)/q_0 + \theta/q_1$.

Proof. Recall the identification

$$T^{r,s,\alpha}(X) \cong T_\alpha T^{r,s,\alpha}(X) \subset L^r(X; L^s_\alpha(X^+))$$

for all $r \in (0, \infty)$ and $s \in (1, \infty)$. Since

$$\begin{aligned}[L^{p_0}(X; L^{q_0}_\alpha(X^+)), L^{p_1}(X; L^{q_1}_\alpha(X^+))]_\theta &= L^p(X; [L^{q_0}_\alpha(X^+), L^{q_1}_\alpha(X^+)]_\theta) \\ &= L^p(X; L^q_\alpha(X^+))\end{aligned}$$

applying the standard result on interpolation of complemented subspaces with common projections (see [19, Theorem 1.17.1.1]) yields

$$\begin{aligned}[T^{p_0,q_0,\alpha}(X), T^{p_1,q_1,\alpha}(X)]_\theta &= L^p(X; L^q_\alpha(X^+)) \cap (T^{p_0,q_0,\alpha}(X) + T^{p_1,q_1,\alpha}(X)) \\ &= T^{p,q,\alpha}(X).\end{aligned}$$

□

Remark 3.13. Since [19, Theorem 1.17.1.1] is true for any interpolation functor (not just complex interpolation), analogues of Proposition 3.12 hold for any interpolation functor F for which the spaces $L^p(X; L^q_\alpha(X^+))$ form an appropriate interpolation scale. In particular, Proposition 3.12 (appropriately modified) holds for real interpolation.

Remark 3.14. Following the first submission of this article, the anonymous referee suggested a more direct proof of Proposition 3.12, which avoids interpolation of complemented subspaces.

Since T_α acts as an isometry both from $T^{p_0,q_0,\alpha}(X)$ to $L^{p_0}(X; L^{q_0}_\alpha(X^+))$ and from $T^{p_1,q_1,\alpha}(X)$ to $L^{p_1}(X; L^{q_1}_\alpha(X^+))$, if $f \in [T^{p_0,q_0,\alpha}(X), T^{p_1,q_1,\alpha}(X)]_\theta$, then

$$\|f\|_{T^{p,q,\alpha}(X)} = \|T_\alpha f\|_{L^p(X; L^q_\alpha(X^+))} \leq \|f\|_{[T^{p_0,q_0,\alpha}(X), T^{p_1,q_1,\alpha}(X)]_\theta}$$

due to the exactness of the complex interpolation functor (and similarly for the real interpolation functor). Hence $[T^{p_0,q_0,\alpha}(X), T^{p_1,q_1,\alpha}(X)]_\theta \subset T^{p,q,\alpha}(X)$, and the reverse containment follows by duality. We have chosen to include both proofs for their own intrinsic interest.

3.2.2. Endpoint results. We now consider the tent spaces

$$T^{1,q,\alpha}(X) \quad \text{and} \quad T^{\infty,q,\alpha}(X),$$

and their relation to the rest of the tent space scale. In this section, we prove the following duality result using the method of [7].

Proposition 3.15. *Suppose X is doubling, and let $\alpha > 0$ and $q \in (1, \infty)$. Then the pairing $\langle \cdot, \cdot \rangle$ of Proposition 3.10 realises $T^{\infty,q,\alpha}(X)$ as the Banach space dual of $T^{1,q,\alpha}(X)$, up to equivalence of norms.*

As in [7], we require a small series of definitions and lemmas to prove this result. We define *truncated cones* for $x \in X$, $\alpha, h > 0$ by

$$\Gamma_h^\alpha(x) := \Gamma^\alpha(x) \cap \{(y,t) \in X^+ : t < h\},$$

and corresponding Lusin operators for $q > 0$ by

$$\mathcal{A}_q^\alpha(f|h)(x) := \left(\iint_{\Gamma_h^\alpha(x)} |f(y,t)|^q \frac{d\mu(y)}{V(y,\alpha t)} \frac{dt}{t} \right)^{\frac{1}{q}}.$$

One can show that $\mathcal{A}_q^\alpha(f|h)$ is measurable in the same way as for $\mathcal{A}_q^\alpha(f)$.

Lemma 3.16. *For each measurable function g on X^+, each $q \in [1, \infty)$, and each $M > 0$, define*

$$h_{g,q,M}^\alpha(x) := \sup\{h > 0 : \mathcal{A}_q^\alpha(g|h)(x) \leq M\mathcal{C}_q^\alpha(g)(x)\}$$

for $x \in X$. If X is doubling, then for sufficiently large M (depending on X, q, and α), whenever $B \subset X$ is a ball of radius r,

$$\mu\{x \in B : h_{g,q,M}^\alpha(x) \geq r\} \gtrsim_{X,\alpha} \mu(B).$$

Proof. Let $B \subset X$ be a ball of radius r. Applying Lemmas A.5 and 3.6, the definition of $\mathcal{C}_q^\alpha$, and doubling, we have

$$\begin{aligned}
&\int_B \mathcal{A}_q^\alpha(g|r)(x)^q \, d\mu(x) \\
&= \int_B \iint_{\Gamma_r^\alpha(x)} \mathbf{1}_{T^\alpha((2\alpha+1)B)}(y,t) |g(y,t)|^q \frac{d\mu(y)}{V(y,\alpha t)} \frac{dt}{t} \, d\mu(x) \\
&\leq \int_B \iint_{\Gamma^\alpha(x)} \mathbf{1}_{T^\alpha((2\alpha+1)B)}(y,t) |g(y,t)|^q \frac{d\mu(y)}{V(y,\alpha t)} \frac{dt}{t} \, d\mu(x) \\
&\leq \iint_{T^\alpha((2\alpha+1)B)} |g(y,t)|^q \, d\mu(y) \frac{dt}{t} \\
&\leq \mu((2\alpha+1)B) \inf_{x \in B} \mathcal{C}_q^\alpha(g)(x)^q \\
&\lesssim_{X,\alpha} \mu(B) \inf_{x \in B} \mathcal{C}_q^\alpha(g)(x)^q.
\end{aligned}$$

We can estimate

$$\begin{aligned}\int_B &\mathcal{A}_q^\alpha(g|r)(x)^q\,d\mu(x)\\ &\geq (M\inf_{x\in B}\mathcal{C}_q^\alpha(g)(x))^q\mu\{x\in B:\mathcal{A}_q^\alpha(g|r)(x) > M\inf_{x\in B}\mathcal{C}_q^\alpha(g)(x)\},\end{aligned}$$

and after rearranging and combining with the previous estimate we get

$$M^q\left(\mu(B)-\mu\{x\in B:\mathcal{A}_q^\alpha(g|r)(x)\leq M\inf_{x\in B}\mathcal{C}_q^\alpha(g)(x)\}\right)\lesssim_{X,\alpha}\mu(B).$$

More rearranging and straightforward estimating yields

$$\mu\{x\in B:\mathcal{A}_q^\alpha(g|r)(x)\leq M\mathcal{C}_q^\alpha(g)(x)\}\geq(1-M^{-q}C_{X,\alpha})\mu(B).$$

Since $h_{g,q,M}^\alpha(x)\geq r$ if and only if $\mathcal{A}_q^\alpha(g|r)(x)\leq M\mathcal{C}_q^\alpha(g)(x)$ as $\mathcal{A}_q^\alpha(g|h)$ is increasing in h, we can rewrite this as

$$\mu\{x\in B:h_{g,q,M}^\alpha(x)\geq r\}\geq(1-M^{-q}C_{X,\alpha})\mu(B).$$

Choosing $M > C_{X,\alpha}^{1/q}$ completes the proof. □

Corollary 3.17. *With X, g, q, and α as in the statement of the previous lemma, there exists $M = M(X,q,\alpha)$ such that whenever Φ is a nonnegative measurable function on X^+, we have*

$$\iint_{X^+}\Phi(y,t)V(y,\alpha t)\,d\mu(y)\,dt\lesssim_{X,\alpha}\int_X\iint_{\Gamma^\alpha_{h^\alpha_{g,q,M}(x)/\alpha}(x)}\Phi(y,t)\,d\mu(y)\,dt\,d\mu(x).$$

Proof. This is a straightforward application of the Fubini–Tonelli theorem along with the previous lemma. Taking M sufficiently large, Lemma 3.16 gives

$$\begin{aligned}&\iint_{X^+}\Phi(y,t)V(y,\alpha t)\,d\mu(y)\,dt\\ &\lesssim_{X,\alpha}\iint_{X^+}\Phi(y,t)\int_{\{x\in B(y,\alpha t):h^\alpha_{g,q,\mathrm{M}}(x)\geq\alpha t\}}d\mu(x)\,d\mu(y)\,dt\\ &\quad=\int_X\int_0^{h^\alpha_{g,q,M}(x)/\alpha}\int_{B(x,\alpha t)}\Phi(y,t)\,d\mu(y)\,dt\,d\mu(x)\\ &\quad=\int_X\iint_{\Gamma^\alpha_{h^\alpha_{g,q,M}(x)/\alpha}(x)}\Phi(y,t)\,d\mu(y)\,dt\,d\mu(x)\end{aligned}$$

as required. □

We are now ready for the proof of the main duality result.

Proof of Proposition 3.15. First suppose $f\in T^{1,q,\alpha}(X)$ and $g\in T^{\infty,q',\alpha}(X)$. By Corollary 3.17, there exists $M = M(X,q,\alpha) > 0$ such that

$$\iint_{X^+}|f(y,t)||g(y,t)|\,d\mu(y)\,\frac{dt}{t}\lesssim_{X,\alpha}\int_X\iint_{\Gamma^\alpha_{h(x)}(x)}|f(y,t)||g(y,t)|\,\frac{d\mu(y)}{V(y,\alpha t)}\,\frac{dt}{t}\,d\mu(x),$$

where $h(x) := h^\alpha_{g,q',M}(x)/\alpha$. Using Hölder's inequality and the definition of $h(x)$, we find that

$$\begin{aligned}
&\int_X \left(\iint_{\Gamma^\alpha_{h(x)}(x)} |f(y,t)||g(y,t)| \frac{d\mu(y)}{V(y,\alpha t)} \frac{dt}{t} \right) d\mu(x) \\
&\leq \int_X \mathcal{A}^\alpha_q(f|h(x))(x) \mathcal{A}^\alpha_{q'}(g|h(x))(x)\, d\mu(x) \\
&\leq M \int_X \mathcal{A}^\alpha_q(f)(x) \mathcal{C}^\alpha_{q'}(g)(x)\, d\mu(x) \\
&\lesssim_{X,q,\alpha} \|f\|_{T^{1,q,\alpha}(X)} \|g\|_{T^{\infty,q,\alpha}(X)} .
\end{aligned}$$

Hence every $g \in T^{\infty,q',\alpha}(X)$ induces a bounded linear functional on $T^{1,q,\alpha}(X)$ via the pairing $\langle f, g\rangle$ above, and so $T^{\infty,q',\alpha}(X) \subset (T^{1,q,\alpha}(X))^*$. Conversely, suppose $\ell \in (T^{1,q,\alpha}(X))^*$. Then as in the proof of Proposition 3.10, from ℓ we construct a function $g \in L^{q'}_{\text{loc}}(X^+)$ such that

$$\iint_{X^+} f(y,t)\overline{g(y,t)}\, d\mu(y) \frac{dt}{t} = \ell(f)$$

for all $f \in T^{1,q,\alpha}(X)$ with cylindrical support. We just need to show that g is in $T^{\infty,q',\alpha}(X)$. By the definition of the $T^{\infty,q',\alpha}(X)$ norm, it suffices to estimate

$$\left(\frac{1}{\mu(B)} \iint_{T^\alpha(B)} |g(y,t)|^{q'}\, d\mu(y) \frac{dt}{t} \right)^{\frac{1}{q'}},$$

where $B \subset X$ is an arbitrary ball. For all nonnegative $\psi \in L^q(T^\alpha(B))$ with $\|\psi\|_{L^q(T^\alpha(B))} \leq 1$, using that $S^\alpha(T^\alpha(B)) = B$ we have that

$$\begin{aligned}
\|\psi\|_{T^{1,q,\alpha}(X)} &= \int_B \mathcal{A}^\alpha_q(\psi)(x)\, d\mu(x) \leq \mu(B)^{1/q'} \|\psi\|_{T^{q,q,\alpha}(X)} \\
&= \mu(B)^{1/q'} \|\psi\|_{L^q(X^+)} \leq \mu(B)^{1/q'}.
\end{aligned}$$

In particular, ψ is in $T^{1,q,\alpha}(X)$, so we can write

$$\iint_{T^\alpha(B)} \overline{g}\psi\, d\mu \frac{dt}{t} = \ell(\psi).$$

Arguing by duality and using the above computation, we then have

$$\begin{aligned}
\left(\frac{1}{\mu(B)} \iint_{T^\alpha(B)} |g(y,t)|^{q'}\, d\mu(y) \frac{dt}{t} \right)^{1/q'} &= \mu(B)^{-1/q'} \sup_\psi \iint_{T^\alpha(B)} \overline{g}\psi\, d\mu \frac{dt}{t} \\
&= \mu(B)^{-1/q'} \sup_\psi \ell(\psi) \\
&\leq \mu(B)^{-1/q'} \|\ell\| \|\psi\|_{T^{1,q,\alpha}(X)} \\
&\leq \|\ell\| ,
\end{aligned}$$

where the supremum is taken over all ψ described above. Now taking the supremum over all balls $B \subset X$, we find that

$$\|g\|_{T^{\infty,q',\alpha}(X)} \leq \|\ell\|,$$

which completes the proof that $(T^{1,q,\alpha}(X))^* \subset T^{\infty,q',\alpha}(X)$. □

Once Proposition 3.15 is established, we can obtain the full scale of interpolation using the 'convex reduction' argument of [4, Theorem 3] and Wolff's reiteration theorem (see [20] and [13]).

Proposition 3.18. *Suppose that X is doubling. Then for $p_0, p_1 \in [1,\infty]$ (not both equal to ∞), q_0 and q_1 in $(1,\infty)$, $\theta \in [0,1]$, and $\alpha > 0$, we have (up to equivalence of norms)*

$$[T^{p_0,q_0,\alpha}(X), T^{p_1,q_1,\alpha}(X)]_\theta = T^{p,q,\alpha}(X),$$

where $1/p = (1-\theta)/p_0 + \theta/p_1$ and $1/q = (1-\theta)/q_0 + \theta/q_1$.

Proof. First we will show that

$$[T^{1,q_0,\alpha}(X), T^{p_1,q_1,\alpha}(X)]_\theta \supset T^{p,q,\alpha}(X). \tag{9}$$

Suppose $f \in T^{p,q,\alpha}(X)$ is a cylindrically supported simple function. Then there exists another cylindrically supported simple function g such that $f = g^2$. Then

$$\|f\|_{T^{p,q,\alpha}(X)} = \|g\|^2_{T^{2p,2q,\alpha}(X)},$$

and so g is in $T^{2p,2q,\alpha}(X)$. By Proposition 3.12 we have the identification

$$T^{2p,2q,\alpha}(X) = [T^{2,2q_0,\alpha}(X), T^{2p_1,2q_1,\alpha}(X)]_\theta \tag{10}$$

up to equivalence of norms, and so by the definition of the complex interpolation functor (see Section A.3), there exists for each $\varepsilon > 0$ a function

$$G_\varepsilon \in \mathcal{F}(T^{2,2q_0,\alpha}(X), T^{2p_1,2q_1,\alpha}(X))$$

such that $G_\varepsilon(\theta) = g$ and

$$\begin{aligned}\|G_\varepsilon\|_{\mathcal{F}(T^{2,2q_0,\alpha}(X),T^{2p_1,2q_1,\alpha}(X)} &\leq (1+\varepsilon)\|g\|_{[T^{2,2q_0,\alpha}(X),T^{2p_1,2q_1,\alpha}(X)]_\theta} \\ &\simeq (1+\varepsilon)\|g\|_{T^{2p,2q,\alpha}(X)},\end{aligned}$$

the implicit constant coming from the norm equivalence (10). Define $F_\varepsilon := G_\varepsilon^2$. Then we have

$$F_\varepsilon \in \mathcal{F}(T^{1,q_0,\alpha}(X), T^{p_1,q_1,\alpha}(X)),$$

with

$$\begin{aligned}\|F_\varepsilon\|_{\mathcal{F}(T^{1,q_0,\alpha}(X),T^{p_1,q_1,\alpha}(X))} &= \|G_\varepsilon\|^2_{\mathcal{F}(T^{2,2q_0,\alpha}(X),T^{2p_1,2q_1,\alpha}(X))} \\ &\lesssim (1+\varepsilon)^2\|g\|^2_{T^{2p,2q,\alpha}(X)} \\ &= (1+\varepsilon)^2\|f\|_{T^{p,q,\alpha}(X)}.\end{aligned}$$

Therefore

$$\|f\|_{[T^{1,q_0,\alpha}(X),T^{p_1,q_1,\alpha}(X)]_\theta} \lesssim \|f\|_{T^{p,q,\alpha}(X)},$$

and so the inclusion (9) follows from the fact that cylindrically supported simple functions are dense in $T^{p,q,\alpha}(X)$. By the duality theorem [3, Corollary 4.5.2] for interpolation (using that $T^{p_1,q_1,\alpha}(X)$ is reflexive, the inclusion (9), and Propositions 3.10 and 3.15, we have

$$[T^{p_1',q_1',\alpha}(X), T^{\infty,q_0',\alpha}(X)]_{1-\theta} \subset T^{p',q',\alpha}(X).$$

Therefore we have the containment

$$[T^{p_0,q_0,\alpha}(X), T^{\infty,q_1,\alpha}(X)]_\theta \subset T^{p,q,\alpha}(X). \tag{11}$$

The reverse containment can be obtained from

$$[T^{1,q_0,\alpha}(X), T^{p_1,q_1,\alpha}(X)]_\theta \subset T^{p,q,\alpha}(X) \tag{12}$$

(for p_1, q_0, $q_1 \in (1,\infty)$) by duality. The containment (12) can be obtained as in Remark 3.14, with $p_0 = 1$ not changing the validity of this method.[8] Finally, it remains to consider the case when $p_0 = 1$ and $p_1 = \infty$. This is covered by Wolff reiteration. Set $A_1 = T^{1,q_0,\alpha}(X)$, $A_2 = T^{p,q,\alpha}(X)$, $A_3 = T^{p+1,q_3,\alpha}(X)$, and $A_4 = T^{\infty,q_1,\alpha}(X)$ for an approprate choice of q_3.[9] Then for an appropriate index η, we have $[A_1, A_3]_{\theta/\eta} = A_2$ and $[A_2, A_4]_{(\eta-\theta)/(1-\theta)} = A_3$. Therefore by Wolff reiteration, we have $[A_1, A_4]_\theta = A_2$; that is, $[T^{1,q_0,\alpha}(X), T^{\infty,q_1,\alpha}(X)]_\theta = T^{p,q,\alpha}(X)$. This completes the proof. □

Remark 3.19. Note that doubling is not explicitly used in the above proof; it is only required to the extent that it is needed to prove Propositions 3.10 and 3.15 (as Proposition 3.12 follows from 3.10). If these propositions could be proven under some assumptions other than doubling, then it would follow that Proposition 3.18 holds under these assumptions.

Remark 3.20. The proof of [7, Lemma 5], which amounts to proving the containment (9), contains a mistake which is seemingly irrepairable without resorting to more advanced techniques. This mistake appears on page 323, line -3, when it is stated that "$A(f_k)$ is supported in $O_k^* - O_{k+1}$" (and in particular, that $A(f_k)$ is supported in O_{k+1}^c). However (reverting to our notation), since $f_k := \mathbf{1}_{T((O_k)_\gamma^*) \setminus T((O_{k+1})_\gamma^*)} f$, $\mathcal{A}_2^1(f_k)$ is supported on

$$S^1(T((O_k)_\gamma^*) \setminus T((O_{k+1})_\gamma^*)) = (O_k)_\gamma^*$$

and we cannot conclude that $\mathcal{A}_2^1(f_k)$ is supported away from O_{k+1}. Simple one-dimensional examples can be constructed which show that this is false in general. Hence the containment (9) is not fully proven in [7]; the first valid proof in the Euclidean case that we know of is in [4] (the full range of interpolation is not obtained in [12]).

---

[8] We thank the anonymous referee once more for this suggestion.

[9] More precisely, we need to take $1/q_3 = (1 - 1/p')/q_0 + (1/p')/q_1$.

3.3. Change of aperture

Under the doubling assumption, the change of aperture result can be proven without assuming (NI) by means of the vector-valued method. The proof is a combination of the techniques of [12] and [4].

Proposition 3.21. *Suppose X is doubling. For $\alpha, \beta \in (0,\infty)$ and $p, q \in (0,\infty)$, the tent space (quasi-)norms $\|\cdot\|_{T^{p,q,\alpha}(X)}$ and $\|\cdot\|_{T^{p,q,\beta}(X)}$ are equivalent.*

Proof. First suppose $p, q \in (1,\infty)$. Since X is doubling, we can replace our definition of $\mathcal{A}_q^\alpha$ with the definition

$$\mathcal{A}_q^\alpha(f)(x)^q := \iint_{\Gamma^\alpha(x)} |f(y,t)|^q \frac{d\mu(y)}{V(y,t)} \frac{dt}{t};$$

using the notation of Section 3.1, this is the definition with $\mathbf{a} = y$ and $\mathbf{b} = 1$. Having made this change, the vector-valued approach to tent spaces (see Section 3.2) transforms as follows. The tent space $T^{p,q,\alpha}(X)$ now embeds isometrically into $L^p(X; L_1^q(X^+))$ via the operator T_α defined, as before, by

$$T_\alpha f(x)(y,t) := f(y,t)\mathbf{1}_{\Gamma^\alpha(x)}(y,t)$$

for $f \in T^{p,q,\alpha}(X)$. The adjoint of T_α is the operator Π_α, now defined by

$$(\Pi_\alpha G)(y,t) := \frac{1}{V(y,t)} \int_{B(y,\alpha t)} G(z)(y,t)\, d\mu(z)$$

for $G \in L^p(X; L_1^q(X^+))$. The composition $P_\alpha := T_\alpha \Pi_\alpha$ is then a bounded projection from $L^p(X; L_1^q(X^+))$ onto $T_\alpha T^{p,q,\alpha}(X)$, and can be written in the form

$$P_\alpha G(x)(y,t) = \frac{\mathbf{1}_{\Gamma^\alpha(x)}(y,t)}{V(y,t)} \int_{B(y,\alpha t)} G(z)(y,t)\, d\mu(z).$$

For $f \in T^{p,q,\alpha}(X)$, we can easily compute

$$P_\beta T_\alpha f(x)(y,t) = T_\beta f(x)(y,t) \frac{V(y, \min(\alpha,\beta)t)}{V(y,t)}. \tag{13}$$

Without loss of generality, suppose $\beta > \alpha$. Then we obviously have

$$\|\cdot\|_{T^{p,q,\alpha}(X)} \lesssim_{q,\alpha,\beta,X} \|\cdot\|_{T^{p,q,\beta}(X)}$$

by Remark 3.7. It remains to show that

$$\|\cdot\|_{T^{p,q,\beta}(X)} \lesssim_{p,q,\alpha,\beta,X} \|\cdot\|_{T^{p,q,\alpha}(X)}. \tag{14}$$

From (13) and doubling, for $f \in T^{p,q,\alpha}(X)$ we have that

$$T_\beta f(x)(y,t) \lesssim_{X,\alpha} P_\beta T_\alpha f(x)(y,t),$$

and so we can write

$$\begin{aligned} \|f\|_{T^{p,q,\beta}(X)} &= \|T_\beta f\|_{L^p(X;L_1^q(X^+))} \lesssim_{X,\alpha} \|P_\beta T_\alpha f\|_{L^p(X;L_1^q(X^+))} \\ &\le \|P_\beta\|_{\mathcal{L}(L^p(X;L_1^q(X^+)))} \|T_\alpha f\|_{L^p(X;L_1^q(X^+))} \lesssim_{p,q,\beta,X} \|f\|_{T^{p,q,\alpha}(X)} \end{aligned}$$

since P_β is a bounded operator on $L^p(X; L_1^q(X^+))$. This shows (14), and completes the proof for $p, q \in (1, \infty)$. Now suppose that at least one of p and q is not in $(1, \infty)$, and suppose $f \in T^{p,q,\alpha}(X)$ is a cylindrically supported simple function. Choose an integer M such that both Mp and Mq are in $(1, \infty)$. Then there exists a cylindrically supported simple function g with $g^M = f$. We then have

$$\begin{aligned} \|f\|_{T^{p,q,\alpha}(X)}^{1/M} &= \|g^M\|_{T^{p,q,\alpha}(X)}^{1/M} = \|g\|_{T^{Mp,Mq,\alpha}(X)} \\ &\simeq_{p,q,\alpha,\beta,X} \|g\|_{T^{Mp,Mq,\beta}(X)} = \|f\|_{T^{p,q,\beta}(X)}^{1/M}, \end{aligned}$$

and so the result is true for cylindrically supported simple functions, with an implicit constant which does not depend on the support of such a function. Since the cylindrically supported simple functions are dense in $T^{p,q,\alpha}(X)$, the proof is complete. □

Remark 3.22. Written more precisely, with $p, q \in (0, \infty)$ and $\beta < 1$, the inequality (14) is of the form

$$\|\cdot\|_{T^{p,q,1}(X)} \lesssim_{p,q,X} \sup_{(y,t)\in X^+} \left(\frac{V(y,t)}{V(y,\beta t)}\right)^M \|\cdot\|_{T^{p,q,\beta}(X)}.$$

where M is such that $Mp, Mq \in (1, \infty)$.

3.4. Relations between $\mathcal{A}$ and $\mathcal{C}$

Again, this proposition follows from the methods of [7].

Proposition 3.23. *Suppose X satisfies (HL), and suppose $0 < q < p < \infty$ and $\alpha > 0$. Then*

$$\left\|\mathcal{C}_q^\alpha(f)\right\|_{L^p(X)} \lesssim_{p,q,X} \left\|\mathcal{A}_q^\alpha(f)\right\|_{L^p(X)}.$$

Proof. Let $B \subset X$ be a ball. Then by the Fubini–Tonelli theorem, using the fact that $S^\alpha(T^\alpha(B)) = B$,

$$\begin{aligned} &\frac{1}{\mu(B)} \iint_{T^\alpha(B)} |f(y,t)|^q \, d\mu(y) \, \frac{dt}{t} \\ &= \frac{1}{\mu(B)} \iint_{T^\alpha(B)} \frac{|f(y,t)|^q}{V(y,\alpha t)} \int_{B(y,\alpha t)} d\mu(x) \, d\mu(y) \, \frac{dt}{t} \\ &= \frac{1}{\mu(B)} \int_X \iint_{T^\alpha(B)} \mathbf{1}_{B(y,\alpha t)}(x) |f(y,t)|^q \, \frac{d\mu(y)}{V(y,\alpha t)} \, \frac{dt}{t} \, d\mu(x) \\ &= \frac{1}{\mu(B)} \int_B \iint_{T^\alpha(B)} \mathbf{1}_{B(x,\alpha t)}(y) |f(y,t)|^q \, \frac{d\mu(y)}{V(y,\alpha t)} \, \frac{dt}{t} \, d\mu(x) \\ &\leq \frac{1}{\mu(B)} \int_B \iint_{X^+} \mathbf{1}_{B(x,\alpha t)}(y) |f(y,t)|^q \, \frac{d\mu(y)}{V(y,\alpha t)} \, \frac{dt}{t} \, d\mu(x) \\ &= \frac{1}{\mu(B)} \int_B \mathcal{A}_q^\alpha(f)(x)^q \, d\mu(x). \end{aligned}$$

Now fix $x \in X$ and take the supremum of both sides of this inequality over all balls B containing x. We find that

$$\mathcal{C}_q^\alpha(f)(x)^q \leq \mathcal{M}(\mathcal{A}_q^\alpha(f)^q)(x).$$

Since $p/q > 1$, we can apply (HL) to get

$$\begin{aligned}\left\|\mathcal{C}_q^\alpha(f)\right\|_{L^p(X)} &\leq \left\|\mathcal{M}(\mathcal{A}_q^\alpha(f)^q)^{1/q}\right\|_{L^p(X)} \\ &= \left\|\mathcal{M}(\mathcal{A}_q^\alpha(f)^q)\right\|_{L^{p/q}(X)}^{1/q} \\ &\lesssim_{p,q,X} \left\|\mathcal{A}_q^\alpha(f)^q\right\|_{L^{p/q}(X)}^{1/q} \\ &= \left\|\mathcal{A}_q^\alpha(f)\right\|_{L^p(X)}\end{aligned}$$

as desired. □

Remark 3.24. If X is doubling, and if $p, q \in (0, \infty)$, then for $\alpha > 0$ we also have that

$$\left\|\mathcal{A}_q^\alpha(f)\right\|_{L^p(X)} \lesssim_{p,q,X} \left\|\mathcal{C}_q^\alpha(f)\right\|_{L^p(X)}.$$

This can be proven as in [7, §6], completely analogously to the proofs above.

Appendix: Assorted lemmas and notation

A.1. Tents, cones, and shadows

Lemma A.1. *Suppose A and B are subsets of X, with A open, and suppose $T^\alpha(A) \subset T^\alpha(B)$. Then $A \subset B$.*

Proof. Suppose $x \in A$. Then $\operatorname{dist}(x, A^c) > 0$ since A is open, and so $\operatorname{dist}(x, A^c) > \alpha t$ for some $t > 0$. Hence $(x, t) \in T^\alpha(A) \subset T^\alpha(B)$, so that $\operatorname{dist}(x, B^c) > \alpha t > 0$. Therefore $x \in B$. □

Lemma A.2. *Let $C \subset X^+$ be cylindrical, and suppose $\alpha > 0$. Then $S^\alpha(C)$ is bounded.*

Proof. Write $C \subset B(x, r) \times (a, b)$ for some $x \in X$ and $r, a, b > 0$. Then $S^\alpha(C) \subset S^\alpha(B(x, r) \times (a, b))$, and one can easily show that

$$S^\alpha(B(x, r) \times (a, b)) \subset B(x, r + \alpha b),$$

showing the boundedness of $S^\alpha(C)$. □

Lemma A.3. *Let $C \subset X^+$, and suppose $\alpha > 0$. Then $T^\alpha(S^\alpha(C))$ is the minimal α-tent containing C, in the sense that $T^\alpha(S) \supset C$ for some $S \subset X$ implies that $T^\alpha(S^\alpha(C)) \subset T^\alpha(S)$.*

Proof. A straightforward set-theoretic manipulation shows that C is contained in $T^\alpha(S^\alpha(C))$. We need to show that $S^\alpha(C)$ is minimal with respect to this property. Suppose that $S \subset X$ is such that $C \subset T^\alpha(S)$, and suppose (w, t_w) is in $T^\alpha(S^\alpha(C))$. With the aim of showing that $\operatorname{dist}(w, S^c) > \alpha t_w$, suppose that $y \in S^c$. Then $\Gamma^\alpha(y) \cap T^\alpha(S) = \varnothing$, and so $\Gamma^\alpha(y) \cap C = \varnothing$ since $T^\alpha(S)$ contains C. Thus $y \in S^\alpha(C)^c$, and so

$$d(w, y) \geq \operatorname{dist}(w, S^\alpha(C)^c) > \alpha t_w$$

since $(w, t_w) \in T^\alpha(S^\alpha(C))$. Taking an infimum over $y \in S^c$, we get that

$$\operatorname{dist}(w, S^c) > \alpha t_w,$$

which says precisely that (w, t_w) is in $T^\alpha(S)$. Therefore $T^\alpha(S^\alpha(C)) \subset T^\alpha(S)$ as desired. □

Lemma A.4. *For a cylindrical subset $K \subset X^+$, define*

$$\beta_0(K) := \inf_{B \subset X} \{\mu(B) : T^\alpha(B) \cap K \neq \varnothing\} \text{ and } \beta_1(K) := \inf_{B \subset X} \{\mu(B) : T^\alpha(B) \supset K\},$$

with both infima taken over the set of balls B in X. Then $\beta_1(K)$ is positive, and if X is proper or doubling, then $\beta_0(K)$ is also positive.

Proof. We first prove that $\beta_0 := \beta_0(K)$ is positive, assuming that X is proper or doubling. Write

$$K \subset \overline{C} := \overline{B(x_0, r_0)} \times [a_0, b_0]$$

for some $x_0 \in X$ and $a_0, b_0, r_0 > 0$. If B is a ball such that $T^\alpha(B) \cap K \neq \varnothing$, then we must have $T^\alpha(B) \cap \overline{C} \neq \varnothing$, and so we can estimate

$$\beta_0 \geq \inf_{B \subset X} \{\mu(B) : T^\alpha(B) \cap \overline{C} \neq \varnothing\}.$$

Note that if $B = B(c(B), r(B))$ is a ball with $c(B) \in \overline{B(x_0, r_0)}$, then $T^\alpha(B) \cap \overline{C} \neq \varnothing$ if and only if $r(B) \geq \alpha a_0$. Defining

$$I(x) := \inf\{V(x, r) : r > 0, T^\alpha(B(x, r)) \cap \overline{C} \neq \varnothing\}$$

for $x \in X$, we thus see that $I(x) = V(x, \alpha a_0)$ when $x \in \overline{B(x_0, r_0)}$, and so $I|_{\overline{B(x_0, r_0)}}$ is lower semicontinuous as long as the volume function is lower semicontinuous. Now suppose $B = B(y, \rho)$ is any ball with $T^\alpha(B) \cap \overline{C} \neq \varnothing$. Let (z, t_z) be a point in $T^\alpha(B) \cap \overline{C}$. We claim that the ball

$$\widetilde{B} := B\left(z, \frac{1}{2}(\rho - d(z, y) + \alpha t_z)\right)$$

is contained in B, centred in $\overline{B(x_0, r_0)}$, and is such that $T^\alpha(\widetilde{B}) \cap \overline{C} \neq \varnothing$. The second fact is obvious: $(z, t_z) \in \overline{C}$ implies $z \in \overline{B(x_0, r_0)}$. For the first fact, observe that

$$\begin{aligned}\widetilde{B} &\subset B(y, d(z,y) + (\rho - d(z,y) + \alpha t_z)/2)\\ &= B(y, (\rho + d(z,y) + \alpha t_z)/2)\\ &\subset B(y, (\rho + (\rho - \alpha t_z) + \alpha t_z)/2)\\ &= B(y,\rho),\end{aligned}$$

since $(z,t_z) \in T^\alpha(B)$ implies that $d(z,y) < \rho - \alpha t_z$. Finally, we have $(z,t_z) \in T^\alpha(\widetilde{B})$: since $c(\widetilde{B}) = z$, we just need to show that $t_z < r(\widetilde{B})/\alpha$. Indeed, we have

$$\frac{r(\widetilde{B})}{\alpha} = \frac{1}{2}\left(\frac{\rho - d(z,y)}{\alpha} + t_z\right),$$

and $t_z < (\rho - d(z,y))/\alpha$ as above. The previous paragraph shows that

$$\inf_{x\in X} I(x) \geq \inf_{x\in\overline{B(x_0,r_0)}} I(x),$$

and so we are reduced to showing that the right-hand side of this inequality is positive, since $\beta_0 \geq \inf_{x\in X} I(x)$.

If X is proper: Since $\overline{B(x_0,r_0)}$ is compact and $I|_{\overline{B(x_0,r_0)}}$ is lower semicontinuous, $I|_{\overline{B(x_0,r_0)}}$ attains its infimum on $\overline{B(x_0,r_0)}$. That is,

$$\inf_{x\in\overline{B(x_0,r_0)}} I(x) = \min_{x\in\overline{B(x_0,r_0)}} I_x > 0, \tag{15}$$

by positivity of the ball volume function.

If X is doubling: Since $I(x) = V(x,\alpha a_0)$ when $x \in \overline{B(x_0,r_0)}$, we can write

$$\inf_{x\in\overline{B(x_0,r_0)}} I(x) \geq \inf_{x\in\overline{B(x_0,r_0)}} V(x,\varepsilon),$$

where $\varepsilon = \min(\alpha a_0, 3r_0)$. If $x \in \overline{B(x_0,r_0)}$, then $\overline{B(x_0,r_0)} \subset \overline{B(x,2r_0)} \subset B(x,3r_0)$, and so since $3r_0/\varepsilon \geq 1$,

$$\begin{aligned}V(x_0,r_0) &\leq V(x,3r_0) = V(x,\varepsilon(3r_0/\varepsilon))\\ &\lesssim_X V(x,\varepsilon).\end{aligned}$$

Hence $V(x,\varepsilon) \gtrsim_X V(x_0,r_0)$, and therefore

$$\inf_{x\in\overline{B(x_0,r_0)}} V(x,\varepsilon) \gtrsim V(x_0,r_0) > 0 \tag{16}$$

as desired.

We now prove that $\beta_1 = \beta_1(K)$ is positive. Recall from Lemma A.3 that if $T^\alpha(B) \supset K$, then $T^\alpha(B) \supset T^\alpha(S^\alpha(K))$. Since shadows are open, Lemma A.1 tells us that $B \supset S^\alpha(K)$. Hence $\mu(B) \geq \mu(S^\alpha(K))$, and so

$$\beta_1 \geq \mu(S^\alpha(K)) > 0$$

by positivity of the ball volume function.[10] □

[10]If $S^\alpha(K)$ is a ball, then $\beta_1(K) = \mu(S^\alpha(K))$.

Lemma A.5. *Let B be an open ball in X of radius r. Then for all $x \in B$, the truncated cone $\Gamma_r^\alpha(x)$ is contained in $T^\alpha((2\alpha+1)B)$.*

Proof. Suppose $(y,t) \in \Gamma_r^\alpha(x)$ and $z \in ((2\alpha+1)B)^c$, so that $d(y,x) < \alpha t < \alpha r$ and $d(c(B),z) \geq (2\alpha+1)r$. Then by the triangle inequality

$$\begin{aligned} d(y,z) &\geq d(c(B),z) - d(c(B),x) - d(x,y) \\ &> (2\alpha+1)r - r - \alpha r \\ &= \alpha r \\ &> \alpha t, \end{aligned}$$

so that $\operatorname{dist}(y,((2\alpha+1)B)^c) > \alpha t$, which yields $(y,t) \in T^\alpha((2\alpha+1)B)$. □

A.2. Measurability

We assume (X,d,μ) has the implicit assumptions from Section 2.

Lemma A.6. *Let $\alpha > 0$, and suppose Φ is a non-negative measurable function on X^+. Then the function*

$$g \colon x \mapsto \iint_{\Gamma^\alpha(x)} \Phi(y,t)\, d\mu(y)\, \frac{dt}{t}$$

is μ-measurable.

We present two proofs of this lemma: one uses an abstract measurability result, while the other is elementary (and in fact stronger, proving that g is not only measurable but lower semicontinuous).

First proof. By [16, Theorem 3.1], it suffices to show that the function

$$F(x,(y,t)) := \mathbf{1}_{B(y,\alpha t)}(x)\Phi(y,t)$$

is measurable on $X \times X^+$. For $\varepsilon > 0$, define

$$f_\varepsilon(x,(y,t)) := \frac{\operatorname{dist}(x, \overline{B(y,\alpha t)})}{\operatorname{dist}(x,\overline{B(y,\alpha t)}) + \operatorname{dist}(x, B(y,\alpha t+\varepsilon)^c)}.$$

Then $f_\varepsilon(x,(y,t))$ is continuous in x, and converges pointwise to $\mathbf{1}_{B(y,\alpha t)}(x)$ as $\varepsilon \to 0$. Hence

$$F(x,(y,t)) = \lim_{\varepsilon\to 0} f_\varepsilon(x,(y,t))\Phi(y,t) =: \lim_{\varepsilon \to 0} F_\varepsilon(x,(y,t)),$$

and therefore it suffices to show that each $F_\varepsilon(x,(y,t))$ is measurable on $X \times X^+$. Since F_ε is continuous in x and measurable in (y,t), F_ε is measurable on $X \times X^+$,[11] and the proof is complete. □

[11] See [11, Theorem 1], which tells us that F_ε is Lusin measurable; this implies Borel measurability on $X \times X^+$.

Second proof. For all $x \in X$ and $\varepsilon > 0$, define the vertically translated cone

$$\Gamma_\varepsilon^\alpha(x) := \{(y,t) \in X^+ : (y, t-\varepsilon) \in \Gamma^\alpha(x)\} \subset \Gamma^\alpha(x).$$

If $y \in B(x, \alpha\varepsilon)$, then is it easy to show that $\Gamma_\varepsilon^\alpha(x) \subset \Gamma^\alpha(y)$: indeed, if $(z,t) \in \Gamma_\varepsilon^\alpha(x)$, then $d(z,x) < \alpha(t-\varepsilon)$, and so

$$d(z,y) \leq d(z,x) + d(x,y) < \alpha(t-\varepsilon) + \alpha\varepsilon = \alpha t.$$

For all $x \in X$ and $\varepsilon > 0$, define

$$g_\varepsilon(x) := \iint_{\Gamma_\varepsilon^\alpha(x)} \Phi(y,t)\, d\mu(y)\, \frac{dt}{t}.$$

For each $x \in X$, as $\varepsilon \searrow 0$, we have $g_\varepsilon(x) \nearrow g(x)$ by monotone convergence. Fix $\lambda > 0$, and suppose that $g(x) > \lambda$. Then there exists $\varepsilon(x)$ such that $g_{\varepsilon(x)}(x) > \lambda$. If $y \in B(x, \alpha\varepsilon(x))$, then by the previous paragraph we have

$$g(y) \geq g_{\varepsilon(x)}(x) > \lambda.$$

Therefore g is lower semicontinuous, and thus measurable. □

Lemma A.7. *Let f be a measurable function on X^+, $q \in (0,\infty)$, and $\alpha > 0$. Then $\mathcal{C}_q^\alpha(f)$ is lower semicontinuous.*

Proof. Let $\lambda > 0$, and suppose $x \in X$ is such that $\mathcal{C}_q^\alpha(f)(x) > \lambda$. Then there exists a ball $B \ni x$ such that

$$\frac{1}{\mu(B)} \iint_{T^\alpha(B)} |f(y,t)|^q \, d\mu(y)\, \frac{dt}{t} > \lambda^q.$$

Hence for any $z \in B$, we have $\mathcal{C}_q^\alpha(f)(z) > \lambda$, and so the set $\{x \in X : \mathcal{C}_q^\alpha(f)(x) > \lambda\}$ is open. □

A.3. Interpolation

Here we fix some notation involving complex interpolation.

An *interpolation pair* is a pair (B_0, B_1) of complex Banach spaces which admit embeddings into a single complex Hausdorff topological vector space. To such a pair we can associate the Banach space $B_0 + B_1$, endowed with the norm

$$\|x\|_{B_0+B_1} := \inf\{\|x_0\|_{B_0} + \|x_1\|_{B_1} : x_0 \in B_0, x_1 \in B_1, x = x_0 + x_1\}.$$

We can then consider the space $\mathcal{F}(B_0, B_1)$ of functions f from the closed strip

$$\overline{S} = \{z \in \mathbb{C} : 0 \leq \mathrm{Re}(z) \leq 1\}$$

into the Banach space $B_0 + B_1$, such that

- f is analytic on the interior of S and continuous on $\overline{S}$,
- $f(z) \in B_j$ whenever $\mathrm{Re}(z) = j$ ($j \in \{0,1\}$), and
- the traces $f_j := f|_{\mathrm{Re}\, z = j}$ ($j \in \{0,1\}$) are continuous maps into B_j which vanish at infinity.

The space $\mathcal{F}(B_0, B_1)$ is a Banach space when endowed with the norm

$$\|f\|_{\mathcal{F}(B_0,B_1)} := \max\left(\sup_{\operatorname{Re} z=0} \|f(z)\|_{B_0}, \sup_{\operatorname{Re} z=1} \|f(z)\|_{B_1} \right).$$

We define the *complex interpolation space* $[B_0, B_1]_\theta$ for $\theta \in [0,1]$ to be the subspace of $B_0 + B_1$ defined by

$$[B_0, B_1]_\theta := \{f(\theta) : f \in \mathcal{F}(B_0, B_1)\}$$

endowed with the norm

$$\|x\|_{[B_0,B_1]_\theta} := \inf_{f(\theta)=x} \|f\|_{\mathcal{F}(B_0,B_1)}.$$

Acknowledgements

We thank Pierre Portal and Pascal Auscher for their comments and suggestions, particularly regarding the proofs of Lemmas 3.3 and A.6. We further thank Lashi Bandara, Li Chen, Mikko Kemppainen and Yi Huang for discussions on this work, as well as the participants of the Workshop in Harmonic Analysis and Geometry at the Australian National University for their interest and suggestions. Finally, we thank the referee for their detailed comments.

References

[1] F. Albiac and N.J. Kalton, *Topics in Banach space theory*, Graduate Texts in Mathematics, vol. 233, Springer, New York, 2006.

[2] P. Auscher, A. McIntosh, and E. Russ, *Hardy spaces of differential forms on Riemannian manifolds*, J. Geom. Anal. **18** (2008), 192–248.

[3] J. Bergh and J. Löfström, *Interpolation spaces*, Grundlehren der mathematischen Wissenschaften, vol. 223, Springer-Verlag, Berlin, 1976.

[4] A. Bernal, *Some results on complex interpolation of T^p_q spaces*, Interpolation spaces and related topics (Ramat-Gan), Israel Mathematical Conference Proceedings, vol. 5, 1992, pp. 1–10.

[5] V.I. Bogachev, *Measure theory*, vol. 2, Springer-Verlag, 2007.

[6] W.S. Cohn and I.E. Verbitsky, *Factorization of tent spaces and Hankel operators*, J. Funct. Anal. **175** (2000), 308–329.

[7] R.R. Coifman, Y. Meyer, and E.M. Stein, *Some new function spaces and their applications to harmonic analysis*, J. Funct. Anal. **62** (1985), 304–335.

[8] R.R. Coifman and G. Weiss, *Analyse harmonique non-commutative sur certains espaces homogenes*, Lecture Notes in Mathematics, vol. 242, Springer-Verlag, Berlin, 1971.

[9] J. Diestel and J.J. Uhl, Jr., *Vector measures*, Mathematical Surveys, vol. 15, American Mathematical Society, Providence, 1977.

[10] L. Forzani, R. Scotto, P. Sjögren, and W. Urbina, *On the L^p-boundedness of the non-centred Gaussian Hardy–Littlewood maximal function*, Proc. Amer. Math. Soc. **130** (2002), no. 1, 73–79.

[11] K. Gowrisankaran, *Measurability of functions in product spaces*, Proc. Amer. Math. Soc. **31** (1972), no. 2, 485–488.

[12] E. Harboure, J. Torrea, and B. Viviani, *A vector-valued approach to tent spaces*, J. Anal. Math. **56** (1991), 125–140.

[13] S. Janson, P. Nilsson, and J. Peetre, *Notes on Wolff's note on interpolation spaces*, Proc. London Math. Soc. **s3-48** (1984), no. 2, 283–299, with appendix by Misha Zafran.

[14] M. Kemppainen, *The vector-valued tent spaces T^1 and T^∞*, J. Austral. Math. Soc. (to appear), arxiv:1105.0261.

[15] J. Maas, J. van Neerven, and P. Portal, *Conical square functions and non-tangential maximal functions with respect to the Gaussian measure*, Publ. Mat. **55** (2011), 313–341.

[16] L. Mattner, *Product measurability, parameter integrals, and a Fubini counterexample*, Enseign. Math. **45** (1999), 271–279.

[17] E. Russ, *The atomic decomposition for tent spaces on spaces of homogeneous type*, CMA/AMSI research symposium "Asymptotic geometric analysis, harmonic analysis and related topics" (Canberra) (A. McIntosh and P. Portal, eds.), Proceedings of the Centre for Mathematics and its Applications, vol. 42, 2007, pp. 125–135.

[18] P. Sjögren, *A remark on the maximal function for measures in $\mathbb{R}^n$*, Amer. J. Math. **105** (1983), no. 5, 1231–1233.

[19] H. Triebel, *Interpolation theory, function spaces, differential operators*, North-Holland Mathematical Library, vol. 18, North-Holland Publishing Company, Amsterdam, 1978.

[20] T.H. Wolff, *A note on interpolation spaces*, Harmonic Analysis (Minneapolis, Minn., 1981), Lecture Notes in Mathematics, vol. 908, Springer, Berlin, 1982, pp. 199–204.

Alex Amenta
Mathematical Sciences Institute
Australian National University
Acton ACT 0200
e-mail: amenta@fastmail.fm

Operator Theory:
Advances and Applications, Vol. 240, 31–43

Remarks on Functional Calculus for Perturbed First-order Dirac Operators

Pascal Auscher and Sebastian Stahlhut

Abstract. We make some remarks on earlier works on R-bisectoriality in L^p of perturbed first-order differential operators by Hytönen, McIntosh and Portal. They have shown that this is equivalent to bounded holomorphic functional calculus in L^p for p in any open interval when suitable hypotheses are made. Hytönen and McIntosh then showed that R-bisectoriality in L^p at one value of p can be extrapolated in a neighborhood of p. We give a different proof of this extrapolation and observe that the Hytönen-McIntosh proof has impact on the splitting of the space into kernel and range.

Mathematics Subject Classification (2010). Primary 47A60; Secondary 42B37, 47F05.

Keywords. Differential operators with bounded measurable coefficients, extrapolation of norm inequalities, R-bisectorial operators, coercivity conditions, kernel/range decomposition.

1. Introduction

Recall that an unbounded operator $\mathcal{A}$ on a Banach space $\mathcal{X}$ is called *bisectorial* of angle $\omega \in [0, \pi/2)$ if it is closed, its spectrum is contained in the closure of

$$S_\omega := \{z \in \mathbb{C}; |\arg(\pm z)| < \omega\},$$

and one has the resolvent estimate

$$\|(I + \lambda\mathcal{A})^{-1}\|_{\mathcal{L}(\mathcal{X})} \leq C_{\omega'} \qquad \forall\, \lambda \notin S_{\omega'}, \quad \forall\, \omega' > \omega.$$

Assuming reflexivity of $\mathcal{X}$, this implies that the domain is dense and also the fact that the null space and the closure of the range split. More precisely, we say that the operator $\mathcal{A}$ *kernel/range decomposes* if $\mathcal{X} = \mathsf{N}(\mathcal{A}) \oplus \overline{\mathsf{R}(\mathcal{A})}$ ($\oplus$ means that the sum is topological). Here $\mathsf{N}(\mathcal{A})$ denotes the kernel or null space and $\mathsf{R}(\mathcal{A})$ its range, while the domain is denoted by $\mathsf{D}(\mathcal{A})$. Bisectoriality in a reflexive space is stable under taking adjoints.

For any bisectorial operator, one can define a calculus of bounded operators by the Cauchy integral formula,

$$\psi(\mathcal{A}) := \frac{1}{2\pi i}\int_{\partial S_{\omega'}} \psi(\lambda)\left(I - \frac{1}{\lambda}\mathcal{A}\right)^{-1}\frac{d\lambda}{\lambda},$$
$$\psi \in \Psi(S_{\omega''}) := \left\{\phi \in H^\infty(S_{\omega''}) : \phi \in O\big(\inf(|z|, |z^{-1}|)^\alpha\big), \alpha > 0\right\},$$

with $\omega'' > \omega' > \omega$ and where $H^\infty(S_{\omega''})$ is the space of bounded holomorphic functions in $S_{\omega''}$. If this calculus may be boundedly extended to all $\psi \in H^\infty(S_{\omega''})$ for all $\omega'' > \omega$, then $\mathcal{A}$ is said to have an H^∞-*calculus* of angle ω.

Assume $\mathcal{X} = L^q$ of some σ-finite measure space and $q \in (1, \infty)$. A closed operator $\mathcal{A}$ is called R-*bisectorial* of angle ω if its spectrum is contained in S_ω and for all $\omega' > \omega$, there exists a constant $C > 0$ such that

$$\left\| \left(\sum_{j=1}^k |(I + \lambda_j \mathcal{A})^{-1} u_j|^2 \right)^{1/2} \right\|_q \le C \left\| \left(\sum_{j=1}^k |u_j|^2 \right)^{1/2} \right\|_q$$

for all $k \in \mathbb{N}$, $\lambda_1, \ldots, \lambda_k \notin S_{\omega'}$ and $u_1, \ldots, u_k \in L^q$. This is the so-called R-boundedness criterion applied to the resolvent family. Note that the definition implies that $\mathcal{A}$ is bisectorial. This notion can be defined on any Banach space but we do not need this here.

In [10] and [11], the equivalence between bounded H^∞-calculus and R-bisectoriality is studied for some perturbed first-order Hodge–Dirac and Dirac type bisectorial operators in L^p spaces (earlier work on such operators appears in [1]). For general bisectorial operators in subspaces of L^p, $p \in (1, \infty)$, it is known that the former implies the latter [14, Theorem 5.3], but the converse is not known. For these first-order operators, the converse holds provided the R-bisectoriality in L^p is assumed for all p in a given open subinterval of $(1, \infty)$, not just for one fixed value of $p \in (1, \infty)$. Subsequently, in [9], the R-bisectoriality in L^p for these first-order operators is shown to be stable under perturbation of p, allowing to apply [10] and [11] and complete the study. The main part of the argument uses an extrapolation "à la" Calderón and Zygmund, by real methods. Here, we wish to observe that this part can be obtained with an extrapolation "à la" Šneĭberg using complex function theory. Nevertheless, the argument in [9] is useful to obtain further characterizations of R-bisectoriality in L^p in terms of kernel/range decomposition. Indeed, we shall see that for the first-order operators in L^p considered in [9], this property remains true by perturbation of p in the same interval as for perturbation of R-bisectoriality.

Our plan is to first review properties of perturbed Dirac type operators at some abstract level of generality. Then we consider the first-order differential operators of [11, 9]. We next show the Šneĭberg extrapolation for (R-)bisectoriality of such operators and conclude for the equivalence of R-bisectoriality and H^∞-calculus. We then show that H^∞-calculus, R-bisectoriality, bisectoriality hold simultaneously to kernel/range decomposition on a certain open interval. We interpret this with the motivating example coming from a second-order differential operator in divergence form, showing that this interval agrees with an interval studied in [2].

2. Abstract results

In this section, we assume without mentioning the following: $\mathcal{X}$ is a reflexive complex Banach space. The duality between $\mathcal{X}$ and its dual $\mathcal{X}^*$ is denoted $\langle u^*, u\rangle$ and is anti-linear in u^* and linear in u. Next, D is a closed, densely defined operator on $\mathcal{X}$ and B is a bounded operator on $\mathcal{X}$. We state a first proposition on properties of BD, DB and their duals under various hypotheses.

Proposition 2.1.

1. *BD with $\mathsf{D}(BD) = \mathsf{D}(D)$ is densely defined. Its adjoint $(BD)^*$ is closed, and $\mathsf{D}((BD)^*) = \{u \in \mathcal{X};\ B^*u \in \mathsf{D}(D^*)\} = \mathsf{D}(D^*B^*)$ with $(BD)^* = D^*B^*$.*
2. *Assume that $\|Bu\| \gtrsim \|u\|$ for all $u \in \overline{\mathsf{R}(D)}$. Then,*
   (i) *$B|_{\overline{\mathsf{R}(D)}} \colon \overline{\mathsf{R}(D)} \to \overline{\mathsf{R}(BD)}$ is an isomorphism.*
   (ii) *BD and D^*B^* are both densely defined and closed.*
   (iii) *$DB|_{\overline{\mathsf{R}(D)}}$ and $BD|_{\overline{\mathsf{R}(BD)}}$ are similar under conjugation by $B|_{\overline{\mathsf{R}(D)}}$.*
3. *Assume that $\|Bu\| \gtrsim \|u\|$ for all $u \in \overline{\mathsf{R}(D)}$ and $\mathcal{X} = \mathsf{N}(D) \oplus \overline{\mathsf{R}(D)}$. Then $\mathsf{N}(D) = \mathsf{N}(BD)$.*
4. *Assume that $\|Bu\| \gtrsim \|u\|$ for all $u \in \overline{\mathsf{R}(D)}$ and $\mathcal{X} = \mathsf{N}(D) \oplus \overline{\mathsf{R}(BD)}$. Then,*
   (i) *$\mathcal{X} = \mathsf{N}(DB) \oplus \overline{\mathsf{R}(D)}$.*
   (ii) *$\mathsf{R}(DB) = \mathsf{R}(D)$.*
5. *Assume that $\|Bu\| \gtrsim \|u\|$ for all $u \in \overline{\mathsf{R}(D)}$ and $\mathcal{X} = \mathsf{N}(D) \oplus \overline{\mathsf{R}(BD)}$. Then,*
   (i) *$\mathcal{X} = \mathsf{N}(D^*B^*) \oplus \overline{\mathsf{R}(D^*)}$.*
   (ii) *$\big(\overline{\mathsf{R}(BD)}\big)^* = \overline{\mathsf{R}(D^*)}$ in the duality $\langle\ ,\ \rangle$, with comparable norms.*
   (iii) *$\|B^*u^*\| \gtrsim \|u^*\|$ for all $u^* \in \overline{\mathsf{R}(D^*)}$, hence $B^*|_{\overline{\mathsf{R}(D^*)}} \colon \overline{\mathsf{R}(D^*)} \to \overline{\mathsf{R}(B^*D^*)}$ is an isomorphism.*
   (iv) *$(DB)^* = B^*D^*$.*
   (v) *$D^*B^*|_{\overline{\mathsf{R}(D^*)}}$ and $B^*D^*|_{\overline{\mathsf{R}(B^*D^*)}}$ are similar under conjugation by $B^*|_{\overline{\mathsf{R}(D^*)}}$.*
   (vi) *$\overline{\mathsf{R}(B^*D^*)} = \big(\overline{\mathsf{R}(D)}\big)^*$ in the duality $\langle\ ,\ \rangle$, with comparable norms.*
   (vii) *$D^*B^*|_{\overline{\mathsf{R}(D^*)}}$ is the adjoint of $BD|_{\overline{\mathsf{R}(BD)}}$ in the duality $\langle\ ,\ \rangle$.*
   (viii) *$B^*D^*|_{\overline{\mathsf{R}(B^*D^*)}}$ is the adjoint to $DB|_{\overline{\mathsf{R}(D)}}$ in the duality $\langle\ ,\ \rangle$.*

Proof. We skip the elementary proofs of (1) and (2) except for (2iii). See the proof of [8, Lemma 4.1] where this is explicitly stated on a Hilbert space. The reflexivity of $\mathcal{X}$ is used to deduce that $D^*B^* = (BD)^*$ is densely defined. We next show (2iii). Note that $\overline{\mathsf{R}(D)}$ is an invariant subspace for DB. Let $\beta = B|_{\overline{\mathsf{R}(D)}}$. If $u \in \mathsf{D}(BD|_{\overline{\mathsf{R}(BD)}}) = \overline{\mathsf{R}(BD)} \cap \mathsf{D}(BD) = \overline{\mathsf{R}(BD)} \cap \mathsf{D}(D)$, then $\beta^{-1}u \in \overline{\mathsf{R}(D)} \cap \mathsf{D}(D\beta) = \overline{\mathsf{R}(D)} \cap \mathsf{D}(DB) = \mathsf{D}(DB|_{\overline{\mathsf{R}(D)}})$ and

$$BDu = \beta Du = \beta(DB)(\beta^{-1}u).$$

We now prove (3). Clearly $\mathsf{N}(D) \subset \mathsf{N}(BD)$. Conversely, let $u \in \mathsf{N}(BD)$. From $\mathcal{X} = \mathsf{N}(D) \oplus \overline{\mathsf{R}(D)}$ write $u = v + w$ with $v \in \mathsf{N}(D)$ and $w \in \overline{\mathsf{R}(D)}$. It follows that $Du = Dw$ and $0 = BDu = BDw$. As $B|_{\overline{\mathsf{R}(D)}} \colon \overline{\mathsf{R}(D)} \to \overline{\mathsf{R}(BD)}$ is an isomorphism, we have $w = 0$. Hence, $u = v \in \mathsf{N}(D)$.

We next prove (4). We know that DB is closed. Its null space is $\mathsf{N}(DB) = \{u \in \mathcal{X}\,;\ Bu \in \mathsf{N}(D)\}$.

Let us first show (i), namely that $\mathcal{X} = \mathsf{N}(DB) \oplus \overline{\mathsf{R}(D)}$. As $\mathcal{X} = \mathsf{N}(D) \oplus \overline{\mathsf{R}(BD)}$ by assumption, the projection P_1 on $\overline{\mathsf{R}(BD)}$ along $\mathsf{N}(D)$ is bounded on $\mathcal{X}$. Let $u \in \mathcal{X}$. As $P_1Bu \in \overline{\mathsf{R}(BD)}$, there exists $v \in \overline{\mathsf{R}(D)}$ such that $P_1Bu = Bv$ and $\|v\| \lesssim \|Bv\| = \|P_1Bu\| \lesssim \|u\|$. Since $Bu = (I - P_1)Bu + P_1Bu$ and $(I - P_1)Bu \in \mathsf{N}(D)$, we have $B(u - v) \in \mathsf{N}(D)$, that is $u - v \in \mathsf{N}(DB)$. It follows that $u = u - v + v \in \mathsf{N}(DB) + \overline{\mathsf{R}(D)}$ with $\|v\| + \|u - v\| \lesssim \|u\|$.

Next, we see that $\mathsf{R}(DB) = \mathsf{R}(D)$. Indeed, the inclusion $\mathsf{R}(DB) \subseteq \mathsf{R}(D)$ is trivial. For the other direction, if $v \in \mathsf{R}(D)$, then one can find $u \in \mathsf{D}(D)$ such that $v = Du$. Using $\mathcal{X} = \mathsf{N}(D) \oplus \overline{\mathsf{R}(BD)}$, one can select $u \in \overline{\mathsf{R}(BD)} = B\overline{\mathsf{R}(D)}$ and write $u = Bw$ with $w \in \overline{\mathsf{R}(D)}$. Hence $v = DBw \in \mathsf{R}(DB)$.

We turn to the proof of (5). Item (i) is proved as Lemma 6.2 in [11]. To see (ii), we observe that if $u^* \in \overline{\mathsf{R}(D^*)}$, then $\overline{\mathsf{R}(BD)} \ni u \mapsto \langle u^*, u\rangle$ is a continuous linear functional. Conversely, if $\ell \in \big(\overline{\mathsf{R}(BD)}\big)^*$, then by the Hahn–Banach theorem, there is $u^* \in \mathcal{X}^*$ such that $\ell(u) = \langle u^*, u\rangle$ for all $u \in \overline{\mathsf{R}(BD)}$. Write $u^* = v^* + w^*$ with $v^* \in \mathsf{N}(D^*B^*)$ and $w^* \in \overline{\mathsf{R}(D^*)}$ by (i). Since $\langle v^*, u\rangle = 0$ for all $u \in \overline{\mathsf{R}(BD)}$, we have $\ell(u) = \langle w^*, u\rangle$ for all $u \in \overline{\mathsf{R}(BD)}$ with $w^* \in \overline{\mathsf{R}(D^*)}$.

To see (iii), consider again $\beta = B|_{\overline{\mathsf{R}(D)}}$. Let $u^* \in \overline{\mathsf{R}(D^*)}, u \in \overline{\mathsf{R}(D)}$. Then

$$\langle B^*u^*, u\rangle = \langle u^*, Bu\rangle = \langle u^*, \beta u\rangle.$$

Using (ii), we have proved $B^*|_{\overline{\mathsf{R}(D^*)}} = \beta^*$ and the conclusion follows.

To see item (iv), we remark that combining (iii) and item (2) applied to B^*D^*, we have $(B^*D^*)^* = DB$, hence $(DB)^* = B^*D^*$ by reflexivity.

Item (v) follows from item (iii) as for item (2iii).

Item (vi) follows from $\overline{\mathsf{R}(D)} = \beta^{-1}\overline{\mathsf{R}(BD)}$ with β as above: $\big(\overline{\mathsf{R}(D)}\big)^* = \beta^*\big(\overline{\mathsf{R}(BD)}\big)^*$ and we conclude using item (ii) and $B^*|_{\overline{\mathsf{R}(D^*)}} = \beta^*$.

Item (vii) follows from the dualities $(BD)^* = D^*B^*$ and $\big(\overline{\mathsf{R}(BD)}\big)^* = \overline{\mathsf{R}(D^*)}$.

To prove item (viii), we recall that $DB|_{\overline{\mathsf{R}(D)}} = \beta^{-1}BD|_{\overline{\mathsf{R}(BD)}}\beta$. Thus using what precedes,

$$\begin{aligned}(DB|_{\overline{\mathsf{R}(D)}})^* &= \beta^*(BD|_{\overline{\mathsf{R}(BD)}})^*(\beta^*)^{-1} = B^*(D^*B^*|_{\overline{\mathsf{R}(D^*)}})(\beta^*)^{-1}\\ &= B^*D^*|_{\overline{\mathsf{R}(B^*D^*)}}.\end{aligned}$$

□

Remark 2.2. Note that the property $\|Bu\| \gtrsim \|u\|$ for all $u \in \overline{\mathsf{R}(D)}$ alone does not seem to imply $\|B^*u^*\| \gtrsim \|u^*\|$ for all $u^* \in \overline{\mathsf{R}(D^*)}$. Hence the situation for BD and B^*D^* is not completely symmetric without further hypotheses.

Here is an easy way to check the assumptions above from kernel/range decomposition assumptions.

Corollary 2.3. *Assume that $\|Bu\| \gtrsim \|u\|$ for all $u \in \overline{\mathsf{R}(D)}$. If D and BD kernel/range decompose, then $\mathcal{X} = \mathsf{N}(D) \oplus \overline{\mathsf{R}(BD)}$. In particular, this holds if D and BD are bisectorial.*

Proof. By Proposition 2.1, (3), $\mathsf{N}(D) = \mathsf{N}(BD)$. We conclude from $\mathcal{X} = \mathsf{N}(BD) \oplus \overline{\mathsf{R}(BD)}$. □

Corollary 2.4. *Assume that $\|Bu\| \gtrsim \|u\|$ for all $u \in \overline{\mathsf{R}(D)}$ and that D kernel/range decomposes. If BD kernel/range decomposes, so does DB. If BD is bisectorial, so is DB, with the same angle as BD. The same holds if R-bisectorial replaces bisectorial everywhere when $\mathcal{X} = L^p$.*

Proof. The statement about kernel/range decomposition is a consequence of Corollary 2.3 and Proposition 2.1, item (4). Assume next that BD is bisectorial and let us show that DB is bisectorial. By Proposition 2.1, item (2), $DB|_{\overline{\mathsf{R}(D)}}$ and $BD|_{\overline{\mathsf{R}(BD)}}$ are similar, thus $DB|_{\overline{\mathsf{R}(D)}}$ is bisectorial. Trivially $DB|_{\mathsf{N}(DB)} = 0$ is also bisectorial. As $\mathcal{X} = \mathsf{N}(DB) \oplus \overline{\mathsf{R}(D)}$ by Corollary 2.3 and Proposition 2.1, item (4), we conclude that DB is bisectorial in $\mathcal{X}$.

The proof for R-bisectoriality is similar. □

Remark 2.5. The converse, DB (R-)bisectorial implies BD (R-)bisectorial, seems unclear under the above assumptions on B and D, even if $\mathcal{X}$ is reflexive which we assumed. So it appears that the theory is not completely symmetric for BD and for DB under such assumptions.

Corollary 2.6. *Assume that D kernel/range decomposes. The following are equivalent:*

1. *$\|Bu\| \gtrsim \|u\|$ for all $u \in \overline{\mathsf{R}(D)}$ and BD bisectorial in $\mathcal{X}$.*
2. *$\|B^*u^*\| \gtrsim \|u^*\|$ for all $u^* \in \overline{\mathsf{R}(D^*)}$ and B^*D^* bisectorial in $\mathcal{X}^*$.*

Moreover the angles are the same. If either of them holds, then DB and D^*B^* are also bisectorial, with the same angle. The same holds with R-bisectorial replacing bisectorial everywhere if $\mathcal{X}$ is an L^p space with σ-finite measure and $1 < p < \infty$.

Proof. It is enough to assume (1) by symmetry (recall that we assume $\mathcal{X}$ reflexive). That $\|B^*u^*\| \gtrsim \|u^*\|$ for all $u^* \in \overline{\mathsf{R}(D^*)}$ follows from Corollary 2.3 and Proposition 2.1, item (5). Next, as $B^*D^* = (DB)^*$ by Proposition 2.1, item (5), and as DB is bisectorial by Corollary 2.4, B^*D^* is also bisectorial by general theory. This proves the equivalence. Checking details, one sees that the angles are the same. Bisectoriality of DB and D^*B^* are already used in the proofs. The proof is the same for R-bisectoriality, which is stable under taking adjoints on reflexive L^p space with σ-finite measure (see [15, Corollary 2.11]). □

3. First-order constant coefficients differential systems

Assume now that D is a first-order differential operator on $\mathbb{R}^n$ acting on functions valued in $\mathbb{C}^N$ whose symbol satisfies the conditions (D0), (D1) and (D2) in [9]. We do not assume that D is self-adjoint. Let $1 < q < \infty$ and $\mathsf{D}_q(D) = \{u \in L^q\,;\ Du \in L^q\}$ with $L^q := L^q(\mathbb{R}^n;\mathbb{C}^N)$ and $D_q = D$ on $\mathsf{D}_q(D)$. We keep using the notation D instead of D_q for simplicity. The followings properties have been shown in [11].

1. D is a R-bisectorial operator with H^∞-calculus in L^q.
2. $L^q = N_q(D) \oplus \overline{\mathsf{R}_q(D)}$.
3. $N_q(D)$ and $\overline{\mathsf{R}_q(D)}$, $1 < q < \infty$, are complex interpolation families.
4. D has the coercivity condition
$$\|\nabla u\|_q \lesssim \|Du\|_q \quad \text{for all } u \in \mathsf{D}_q(D) \cap \overline{\mathsf{R}_q(D)} \subset W^{1,q}.$$
Here, we use the notation ∇u for $\nabla \otimes u$.
5. The same properties hold for D^*.

Let us add one more property.

Proposition 3.1. *Let $t > 0$. The spaces $\mathsf{D}_q(D)$, $1 < q < \infty$, equipped with the norm $|||f|||_{q,t} := \|f\|_q + t\|Df\|_q$, form a complex interpolation family. The same holds for D^*.*

Proof. Since D is bisectorial in L^q, we have $\|(I+itD)^{-1}u\|_q \le C\|u\|_q$ with C independent of t. [To be precise, we should write D_q for q and use that the resolvents are compatible for different values of q, that is, the resolvents for different q agree on the intersection of the L^q's.] Thus $(I+itD)^{-1}\colon (L^q, \|\ \|_q) \to (\mathsf{D}_q(D), |||\ |||_{q,t})$ is an isomorphism with uniform bounds with respect to t:
$$\|u\|_q \le \|(I+itD)^{-1}u\|_q + t\|D(I+itD)^{-1}u\|_q \le (2C+1)\|u\|_q.$$
The conclusion follows by the fonctoriality of complex interpolation. □

4. Perturbed first-order differential systems

Let $B \in L^\infty(\mathbb{R}^n;\mathcal{L}(\mathbb{C}^N))$. Identified with the operator of multiplication by $B(x)$, $B \in \mathcal{L}(L^q)$ for all q. Its adjoint B^* has the same property. With D as before, introduce the set of coercivity of B

$$\mathcal{I}(BD) = \{q \in (1,\infty)\,;\, \|Bu\|_q \gtrsim \|u\|_q \text{ for all } u \in \mathsf{R}_q(D)\}.$$

By density, we may replace $\mathsf{R}_q(D)$ by its closure. For $q \in \mathcal{I}(BD)$, $B|_{\overline{\mathsf{R}_q(D)}}$: $\overline{\mathsf{R}_q(D)} \to \overline{\mathsf{R}_q(BD)}$ is an isomorphism. Let

$$b_q = \inf\left(\frac{\|Bu\|_q}{\|u\|_q}\,;\, u \in \overline{\mathsf{R}_q(D)}, u \neq 0\right) > 0.$$

Lemma 4.1. *The set $\mathcal{I}(BD)$ of coercivity of B is open.*

Proof. We have for all $1 < q < \infty$, $\|Bu\|_q \leq \|B\|_\infty \|u\|_q$. Thus, the bounded map $B\colon \overline{\mathsf{R}_q(D)} \to L^q$ is bounded below by b_q for each $q \in \mathcal{I}(BD)$. Using that $\overline{\mathsf{R}_q(D)}$ and L^q are complex interpolation families, the result follows from a result of Šneĭberg [16] (see also Kalton–Mitrea [12]). □

Remark 4.2. If B is invertible in $L^\infty(\mathbb{R}^n; \mathcal{L}(\mathbb{C}^N))$, then B is invertible in $\mathcal{L}(L^q)$ and its inverse is the operator of multiplication by B^{-1}. In this case, $\mathcal{I}(BD) = (1,\infty)$.

For further use, let us recall the statement of Šneĭberg (concerning lower bound) and Kalton–Mitrea (concerning invertibility even in the quasi-Banach case).

Proposition 4.3. *Let (X_s) and (Y_s) be two complex interpolation families of Banach spaces for $0 < s < 1$. Let T be an operator with $C = \sup_{0<s<1} \|T\|_{\mathcal{L}(X_s,Y_s)} < \infty$. Assume that for $s_0 \in (0,1)$ and $\delta > 0$, $\|Tu\|_{Y_{s_0}} \geq \delta \|u\|_{X_{s_0}}$ for all $u \in X_{s_0}$. Then, there is an interval J around s_0 whose length is bounded below by a number depending on C, δ, s_0 on which $\|Tu\|_{Y_s} \geq \frac{\delta}{2}\|u\|_{X_s}$ for all $u \in X_s$ and all $s \in J$. If, moreover, T is invertible in $\mathcal{L}(X_{s_0}, Y_{s_0})$ with lower bound δ then there is an interval J whose length is bounded below by a number depending on C, δ, s_0 on which T is invertible with $\|Tu\|_{Y_s} \geq \frac{\delta}{2}\|u\|_{X_s}$ for all $u \in X_s$ and all $s \in J$.*

Our point is that the lower bound on the size of J is universal for complex families. See Proposition 4.5 below. Define two more sets related to the operator BD:

$$\mathcal{B}(BD) = \{q \in \mathcal{I}(BD)\,;\, BD \text{ bisectorial in } L^q\}$$
$$\mathcal{R}(BD) = \{q \in \mathcal{I}(BD)\,;\, BD\ R\text{–bisectorial in } L^q\}$$

Note that these are subsets of $\mathcal{I}(BD)$. We can define the analogous sets for B^*D^*.

Proposition 4.4. *Let $1 < p < \infty$. Then $p \in \mathcal{R}(BD)$ (resp. $\mathcal{B}(BD)$) if and only if $p' \in \mathcal{R}(B^*D^*)$ (resp. $\mathcal{B}(B^*D^*)$).*

Proof. This is Corollary 2.6. □

Our next results are the new observation of this paper, simplifying the approach of [9].

Proposition 4.5. *The sets $\mathcal{B}(BD)$ and $\mathcal{R}(BD)$ are open.*

Proof. Let us consider the openness of $\mathcal{B}(BD)$ first. We know that for all $p \in \mathcal{I}(BD)$, BD is densely defined and closed on L^p from Proposition 2.1, item (2). Fix $q \in \mathcal{B}(BD)$. Let ω be the angle of bisectoriality in L^q and $\omega < \mu < \pi/2$. Let $C_\mu = \sup_{\lambda \notin S_\mu} \|(I+\lambda BD)^{-1}\|_{\mathcal{L}(L^q)}$. Fix $\lambda \notin S_\mu$. Then $\|\lambda BD(I+\lambda BD)^{-1}\|_{\mathcal{L}(L^q)} \le C_\mu + 1$. Thus for all $u \in \mathsf{D}_q(D)$,

$$\begin{aligned}\|(I+\lambda BD)u\|_q &\ge (2C_\mu)^{-1}\|u\|_q + (2C_\mu+2)^{-1}|\lambda|\|BDu\|_q \\ &\ge (2C_\mu)^{-1}\|u\|_q + (2C_\mu+2)^{-1}b_q|\lambda|\|Du\|_q \\ &\ge \delta|||u|||_{q,|\lambda|}\end{aligned}$$

with $\delta = \inf((2C_\mu)^{-1}, (2C_\mu+2)^{-1}b_q) > 0$. Also

$$\|(I+\lambda BD)u\|_q \le C|||u|||_{q,|\lambda|}$$

with $C = \sup(1, \|B\|_\infty)$. Applying Proposition 4.3 thanks to Proposition 3.1, we obtain an open interval J around q contained in $\mathcal{I}(BD)$ such that for all $\lambda \notin S_\mu$ and $p \in J$, $(I+\lambda BD)^{-1}$ is bounded on L^p with bound $2/\delta$.

The proof for perturbation of R-bisectoriality is basically the same, with C_μ being the R-bound of $(I+\lambda BD)^{-1}$, that is the best constant in the inequality

$$\left\| \left(\sum_{j=1}^k |(I+\lambda_j BD)^{-1}u_j|^2 \right)^{1/2} \right\|_q \le C \left\| \left(\sum_{j=1}^k |u_j|^2 \right)^{1/2} \right\|_q$$

for all $k \in \mathbb{N}$, $\lambda_1, \dots, \lambda_k \notin S_\mu$ and $u_1, \dots, u_k \in L^q$. One works in the sums $L^q \oplus \cdots \oplus L^q$ equipped with the norm of the right-hand side and $\mathsf{D}_q(D) \oplus \cdots \oplus \mathsf{D}_q(D)$ equipped with

$$\left\| \left(\sum_{j=1}^k |u_j|^2 \right)^{1/2} \right\|_q + \left\| \left(\sum_{j=1}^k |\lambda_j|^2 |Du_j|^2 \right)^{1/2} \right\|_q .$$

To obtain the R-lower bound (replacing δ), one linearizes using the Kahane-Kintchine inequality with the Rademacher functions

$$\left(\sum_{j=1}^k |u_j|^2 \right)^{1/2} \sim \left(\int_0^1 \left| \sum_{j=1}^k r_j(t) u_j \right|^q dt \right)^{1/q},$$

valid for any $q \in (1,\infty)$ (see, for example, [15] and follow the argument above). Details are left to the reader. □

Remark 4.6. The sets $\mathcal{B}(BD)$ and $\mathcal{R}(BD)$ may not be intervals. They are (possibly empty) intervals when restricted to each connected component of $\mathcal{I}(BD)$ because (R-)bisectoriality interpolates in L^p scales. See [13, Corollary 3.9] for a proof concerning R-sectoriality. In particular, if $\mathcal{I}(BD) = (1,\infty)$ these sets are (possibly empty) open intervals.

Theorem 4.7. *For $p \in \mathcal{I}(BD)$, the following assertions are equivalent:*

(i) *BD is R-bisectorial in L^p.*

(ii) *BD is bisectorial and has an H^∞-calculus in L^p.*

Moreover, the angles in (i) and (ii) are the same. Furthermore, if one of the items holds, then they hold as well for DB, and also for B^*D^* and D^*B^* in $L^{p'}$.

Proof. The implication (ii) $\Rightarrow$ (i) is a general fact proved in [14]. Assume conversely that (i) holds. Then, there is an interval (p_1, p_2) around p for which (i) holds with the same angle by Proposition 4.5. Note also that (2) and (3) of Proposition 2.1 apply with $\mathcal{X} = L^q$ for each $q \in (p_1, p_2)$. Hence, B^* has a lower bound on $\overline{\mathsf{R}_{q'}(D^*)}$. We may apply Corollary 8.17 of [11], which states that D^*B^* satisfies (ii) on $L^{q'}$. By duality, we conclude that BD satisfies (ii) in L^q.

The last part of the statement now follows from Corollary 2.6. □

Remark 4.8. As $p \in \mathcal{R}(BD)$ if and only if $p' \in \mathcal{R}(B^*D^*)$, Proposition 4.5 and Theorem 4.7 can be compared to Theorem 2.5 of [9] for the stability of R-bisectoriality and the equivalence with H^∞-calculus. The argument here is much easier and fairly general once we have Proposition 3.1. However, the argument in [9] is useful since it contains a quantitative estimate on how far one can move from p. We come back to this below. Recall that the motivation of [9, Theorem 2.5], thus reproved here, is to complete the theory developed in [11].

5. Relation to kernel/range decomposition

For a closed unbounded operator $\mathcal{A}$ on a reflexive Banach space $\mathcal{X}$, recall that $\mathcal{A}$ kernel/range decomposes if $\mathcal{X} = \mathsf{N}(\mathcal{A}) \oplus \overline{\mathsf{R}(\mathcal{A})}$ and that it is implied by bisectoriality. The converse is not true (the shift on $\ell^2(\mathbb{Z})$ is invertible, so the kernel/range decomposition is trivial, but it is not bisectorial as its spectrum is the unit circle). For the class of BD operators in the previous section, we shall show that a converse holds.

For a set $A \subseteq (1, \infty)$, let $A' = \{q' \,;\, q \in A\}$.

Consider D and B as in Section 4. Recall that $p \in \mathcal{R}(BD)$ if and only if $p' \in \mathcal{R}(B^*D^*)$. That is, $\mathcal{R}(B^*D^*)' = \mathcal{R}(BD)$. Recall also that $\mathcal{R}(BD) \subseteq \mathcal{I}(BD)$, hence $\mathcal{R}(BD) \subseteq \mathcal{I}(B^*D^*)'$ as well.

Assume $p_0 \in \mathcal{R}(BD)$ and let I_0 be the connected component of $\mathcal{I}(BD) \cap \mathcal{I}(B^*D^*)'$ that contains p_0. It is an open interval.

Let

$$\mathcal{B}_0(BD) = \{q \in I_0 \,;\, BD \text{ bisectorial in } L^q\}$$

$$\mathcal{R}_0(BD) = \{q \in I_0 \,;\, BD\ R\text{–bisectorial in } L^q\}$$

$$\mathcal{H}_0(BD) = \{q \in I_0 \,;\, BD \text{ bisectorial in } L^q \text{ with } H^\infty\text{–calculus}\}$$

$$\mathcal{S}_0(BD) = cc_{p_0}\{q \in I_0 \,;\, BD \text{ kernel/range decomposes in } L^q\}$$

The notation cc_{p_0} means the connected component that contains p_0.

Theorem 5.1. *Assume $\mathcal{R}_0(BD)$ is not empty. Then, the four sets above are equal open intervals.*

Proof. It is clear that $\mathcal{H}_0(BD) \subseteq \mathcal{R}_0(BD) \subseteq \mathcal{B}_0(BD)$. By Proposition 4.5 and the discussion in Remark 4.6, $\mathcal{R}_0(BD)$ and $\mathcal{B}_0(BD)$ are open subintervals of I_0. By Theorem 4.7, we also know that $\mathcal{H}_0(BD) = \mathcal{R}_0(BD)$.

As bisectoriality implies kernel/range decomposition, $\mathcal{B}_0(BD)$ is contained in the set $\{q \in I_0\,;\, BD$ kernel/range decomposes in $L^q\}$. As $\mathcal{B}_0(BD)$ contains p_0, we have $\mathcal{B}_0(BD) \subseteq \mathcal{S}_0(BD)$. Thus it remains to show that $\mathcal{S}_0(BD) \subseteq \mathcal{R}_0(BD)$, which is done in the next results. □

For $1 < p < \infty$, let p^*, p_* be the upper and lower Sobolev exponents: $p^* = \frac{np}{n-p}$ if $p < n$ and $p^* = \infty$ if $p \geq n$, while $p_* = \frac{np}{n+p}$.

Lemma 5.2. *Let $p \in \mathcal{R}_0(BD)$. Then $BD|_{\overline{\mathsf{R}_q(BD)}}$ is R-bisectorial (in $\overline{\mathsf{R}_q(BD)}$) for $q \in I_0 \cap (p_*, p^*)$.*

Proof. The (non-trivial) argument to extrapolate R-bisectoriality at p to R-bisectoriality at any $q \in I_0 \cap (p_*, p)$ is exactly what is proved in Sections 3 and 4 of [9], taken away the arguments related to kernel/range decomposition which are not assumed here. We next provide the argument for $q \in I_0 \cap (p, p^*)$. By symmetry of the assumptions, we obtain that $B^*D^*|_{\overline{\mathsf{R}_{q'}(B^*D^*)}}$ is R-bisectorial. By duality of R-bisectoriality in subspaces of reflexive Lebesgue spaces and Proposition 2.1, item (5), $DB|_{\overline{\mathsf{R}_q(D)}}$ is R-bisectorial. By Proposition 2.1, item (2), this implies that $BD|_{\overline{\mathsf{R}_q(BD)}}$ is R-bisectorial. □

Corollary 5.3. $\mathcal{S}_0(BD) \subseteq \mathcal{R}_0(BD)$.

Proof. The set $\{q \in I_0\,;\, BD$ kernel/range decomposes in $L^q\}$ is open (this was observed in [9], again as a consequence of Šneĭberg's result). Thus, as a connected component, $\mathcal{S}_0(BD)$ is an open interval. Write $\mathcal{R}_0(BD) = (r_-, r_+)$ and $\mathcal{S}_0(BD) = (s_-, s_+)$ and recall that $(r_-, r_+) \subseteq (s_-, s_+)$. Assume $s_- < r_-$. One can find p, q with $q \in I_0 \cap (p_*, p)$ and $s_- < q \leq r_- < p < r_+$. By the previous lemma, we have that $BD|_{\overline{\mathsf{R}_q(BD)}}$ is R-bisectorial in $\overline{\mathsf{R}_q(BD)}$. Also $BD|_{\mathsf{N}_q(BD)} = 0$ is R-bisectorial. As $q \in \mathcal{S}_0(BD) = (s_-, s_+)$, we have $L^q = \overline{\mathsf{R}_q(BD)} \oplus \mathsf{N}_q(BD)$. Hence, BD is R-bisectorial in L^q. This is a contradiction as $q \notin \mathcal{R}_0(BD)$. Thus $r_- \leq s_-$. The argument to obtain $s_+ \leq r_+$ is similar. □

Remark 5.4. It was observed and heavily used in [9] that for a given p, L^p boundedness of the resolvent of BD self-improves to off-diagonal estimates. Thus, the set of those $p \in I_0$ for which one has such estimates in addition to bisectoriality in L^p is equal to $\mathcal{B}_0(BD)$ as well.

6. Self-adjoint D and accretive B

The operators D and B are still as in Section 4. In addition, assume that D is self-adjoint on L^2 and that B is strictly accretive in $\overline{\mathsf{R}_2(D)}$, that is for some $\kappa > 0$,

$$\mathrm{Re}\langle u, Bu\rangle \geq \kappa \|u\|_2^2, \quad \forall u \in \overline{\mathsf{R}_2(D)}.$$

Then, B and B^* have lower bound κ on $\overline{\mathsf{R}_2(D)}$ and $\overline{\mathsf{R}_2(D^*)} = \overline{\mathsf{R}_2(D)}$. In this case, BD and $DB = (B^*D)^*$ (replacing B by B^*) are bisectorial operators in L^2. Moreover, using that B is multiplication and D a coercive first-order differential operator with constant coefficients, [8, Theorem 3.1] (see [4] for a direct proof) shows that BD and DB have H^∞-calculus in L^2. Thus, Theorem 5.1 applies and one has the

Theorem 6.1. *There exists an open interval $I(BD) = (q_-(BD), q_+(BD)) \subseteq (1, \infty)$, containing 2, with the following dichotomy: H^∞-calculus, R-bisectoriality, bisectoriality and kernel/range decomposition hold for BD in L^p if $p \in I(BD)$ and all fail if $p = q_\pm(BD)$ except, may be, when $q_\pm(BD)$ is already an endpoint of the set I_0 of Section 5 with $p_0 = 2$. The same property holds for DB with $I(DB) = I(BD)$. The same property holds for B^*D and DB^* in the dual interval $I(DB^*) = I(B^*D) = (I(BD))'$.*

In applications, one tries to find an interval of p for bisectoriality, which is the easiest property to check.

The example that motivated the study of perturbed Dirac operators is the following setup, introduced in [7] and exploited in [8] to reprove the Kato square root theorem obtained in [5] for second-order operators and in [6] for systems. Let $A \in L^\infty(\mathbb{R}^n; \mathcal{L}(\mathbb{C}^m \otimes \mathbb{C}^n))$ satisfy

$$\int_{\mathbb{R}^n} \nabla \bar{u}(x) \cdot A(x) \nabla u(x) dx \gtrsim \|\nabla u\|_2^2, \tag{1}$$

for all $u \in W^{1,2}(\mathbb{R}^n; \mathbb{C}^m)$. Then BD, with $B = \begin{pmatrix} I & 0 \\ 0 & A \end{pmatrix}$ and $D = \begin{pmatrix} 0 & -\mathrm{div} \\ \nabla & 0 \end{pmatrix}$, has a bounded H^∞-calculus in $L^p(\mathbb{R}^n; \mathbb{C}^m \oplus [\mathbb{C}^m \otimes \mathbb{C}^n])$ for all $p \in (q_-(BD), q_+(BD))$, with angle at most equal to the accretivity angle of A.

Let us finish with the interpretation of the kernel/range decomposition in this particular example. As $BD = \begin{pmatrix} 0 & -\mathrm{div} \\ A\nabla & 0 \end{pmatrix}$, we see that

$$\mathsf{N}_p(BD) = \{u = (0, g) \in L^p(\mathbb{R}^n; \mathbb{C}^m \oplus [\mathbb{C}^m \otimes \mathbb{C}^n])\, ;\, \mathrm{div} g = 0\}$$

and

$$\overline{\mathsf{R}_p(BD)} = \{u = (f, g) \in L^p(\mathbb{R}^n; \mathbb{C}^m \oplus [\mathbb{C}^m \otimes \mathbb{C}^n])\, ;\, g = A\nabla h, h \in \dot{W}^{1,p}(\mathbb{R}^n; \mathbb{C}^m)\},$$

where $\dot{W}^{1,p}(\mathbb{R}^n; \mathbb{C}^m)$ is the homogeneous Sobolev space. Thus,

$$L^p(\mathbb{R}^n; \mathbb{C}^m \oplus [\mathbb{C}^m \otimes \mathbb{C}^n]) = \mathsf{N}_p(BD) \oplus \overline{\mathsf{R}_p(BD)} \tag{2}$$

is equivalent to the *Hodge splitting adapted to A for vector fields*

$$L^p(\mathbb{R}^n;\mathbb{C}^m\otimes\mathbb{C}^n)=\mathsf{N}_p(\mathrm{div})\oplus A\nabla\dot{W}^{1,p}(\mathbb{R}^n;\mathbb{C}^m). \tag{3}$$

Writing details for DB instead we arrive at the equivalence between

$$L^p(\mathbb{R}^n;\mathbb{C}^m\oplus[\mathbb{C}^m\otimes\mathbb{C}^n])=\mathsf{N}_p(DB)\oplus\overline{\mathsf{R}_p(DB)} \tag{4}$$

and a second *Hodge splitting adapted to A for vector fields*

$$L^p(\mathbb{R}^n;\mathbb{C}^m\otimes\mathbb{C}^n)=\mathsf{N}_p(\mathrm{div}A)\oplus\nabla\dot{W}^{1,p}(\mathbb{R}^n;\mathbb{C}^m). \tag{5}$$

As $q_\pm(BD)=q_\pm(DB)$, we obtain that (3) and (5) hold for $p\in(q_-(BD),q_+(BD))$ and fail at the endpoints.

Let $L=-\mathrm{div}A\nabla$. It was shown in [2, Corollary 4.24] that (5) holds if and only if $p\in(q_+(L^*)',q_+(L))$, where the number $q_+(L)$ is defined as the supremum of those $p>2$ for which $t^{1/2}\nabla e^{-tL}$ is uniformly bounded on L^p for $t>0$ (Strictly speaking, this is done when $m=1$, and Section 7.2 in [2] gives an account of the extension to systems). As a consequence, we have shown that $q_+(BD)=q_+(DB)=q_+(L)$ and $q_-(BD)=q_-(DB)=q_+(L^*)'$.

In the previous example, the matrix B is block-diagonal. If B is a full matrix, then DB and BD happen to be in relation with a second-order system in $\mathbb{R}^{n+1}_+$ as first shown in [3]. Their study brought new information to the boundary value problems associated to such systems when $p=2$. Details when $p\neq 2$ will appear in the forthcoming PhD thesis of the second author.

Acknowledgment

This work is part of the forthcoming PhD thesis of the second author. The first author thanks the organizers of the IWOTA 2012 conference in Sydney for a stimulating environment. Both authors were partially supported by the ANR project "Harmonic Analysis at its Boundaries", ANR-12-BS01-0013-01.

References

[1] Sergey S. Ajiev. Extrapolation of the functional calculus of generalized Dirac operators and related embedding and Littlewood–Paley-type theorems. I. *J. Aust. Math. Soc.*, 83(3):297–326, 2007.

[2] Pascal Auscher. On necessary and sufficient conditions for L^p-estimates of Riesz transforms associated to elliptic operators on $\mathbb{R}^n$ and related estimates. *Mem. Amer. Math. Soc.*, 186(871):xviii+75, 2007.

[3] Pascal Auscher, Andreas Axelsson, and Alan McIntosh. Solvability of elliptic systems with square integrable boundary data. *Ark. Mat. 48* (2010), 253–287.

[4] Pascal Auscher, Andreas Axelsson, and Alan McIntosh. On a quadratic estimate related to the Kato conjecture and boundary value problems. *Contemp. Math.*, 505:105–129, 2010.

[5] Pascal Auscher, Steve Hofmann, Michael Lacey, Alan McIntosh, and Philippe Tchamitchian. The solution of the Kato square root problem for second order elliptic operators on $\mathbb{R}^n$. *Ann. of Math.* (2) 156, 2 (2002), 633–654.

[6] Pascal Auscher, Steve Hofmann, Alan McIntosh, and Philippe Tchamitchian. The Kato square root problem for higher order elliptic operators and systems on $\mathbb{R}^n$. *J. Evol. Equ.*, 1(4):361–385, 2001. Dedicated to the memory of Tosio Kato.

[7] Pascal Auscher, Alan McIntosh and Andrea Nahmod. The square root problem of Kato in one dimension, and first order elliptic systems. *Indiana Univ. Math. J.* 46, 3 (1997), 659–695.

[8] Andreas Axelsson, Stephen Keith, and Alan McIntosh. Quadratic estimates and functional calculi of perturbed Dirac operators. *Invent. Math.*, 163(3):455–497, 2006.

[9] Tuomas Hytönen and Alan McIntosh. Stability in p of the H^∞-calculus of first-order systems in L^p. *The AMSI-ANU Workshop on Spectral Theory and Harmonic Analysis*, 167–181, Proc. Centre Math. Appl. Austral. Nat. Univ., 44, Austral. Nat. Univ., Canberra, 2010.

[10] Tuomas Hytönen, Alan McIntosh, and Pierre Portal. Kato's square root problem in Banach spaces. *J. Funct. Anal.*, 254(3):675–726, 2008.

[11] Tuomas Hytönen, Alan McIntosh, and Pierre Portal. Holomorphic functional calculus of Hodge–Dirac operators in L^p. *J. Evol. Equ.*, 11 (2011), 71–105.

[12] Nigel Kalton and Marius Mitrea. Stability results on interpolation scales of quasi-Banach spaces and applications. *Trans. Amer. Math. Soc.*, 350(10):3903–3922, 1998.

[13] Nigel Kalton, Peer Kunstmann and Lutz Weis. Perturbation and interpolation theorems for the H^∞-calculus with applications to differential operators. *Math. Ann.* 336 (2006), no. 4, 747–801.

[14] Nigel Kalton and Lutz Weis. The H^∞-calculus and sums of closed operators. *Math. Ann.* 321 (2001), no. 2, 319–345.

[15] Peer Kunstmann and Lutz Weis. Maximal L_p-regularity for parabolic equations, Fourier multiplier theorems and H^∞-functional calculus. *Functional analytic methods for evolution equations*, 65–311, Lecture Notes in Math., 1855, Springer, Berlin, 2004.

[16] I. Ja. Šneĭberg. Spectral properties of linear operators in interpolation families of Banach spaces. *Mat. Issled.*, 9(2(32)):214–229, 254–255, 1974.

Pascal Auscher and Sebastian Stahlhut
Univ. Paris-Sud
laboratoire de Mathématiques
UMR 8628 du CNRS
F-91405 Orsay
e-mail: pascal.auscher@math.u-psud.fr
sebastian.stahlhut@math.u-psud.fr

Operator Theory:
Advances and Applications, Vol. 240, 45–68

(m,λ)-Berezin Transform and Approximation of Operators on Weighted Bergman Spaces over the Unit Ball

Wolfram Bauer, Crispin Herrera Yañez and Nikolai Vasilevski

Abstract. We establish various results on norm approximations of bounded linear operators acting on the weighted Bergman space $\mathcal{A}^2_\lambda(\mathbb{B}^n)$ over the unit ball by means of Toeplitz operators with bounded measurable symbols. The main tool here is the so-called (m,λ)-Berezin transform defined and studied in the paper. In a sense, this is a further development of the ideas and results of [6, 7, 9] to the case of operators acting on $\mathcal{A}^2_\lambda(\mathbb{B}^n)$.

Mathematics Subject Classification (2010). Primary 47B35; Secondary 30H20, 30E05.

Keywords. Toeplitz operator; unit ball; (m,λ)-Berezin transform; norm approximation.

1. Introduction

Let $\mathbb{B}^n$ be the open Euclidean unit ball in $\mathbb{C}^n$, and let

$$dv_\lambda(z) = c_\lambda(1-|z|^2)^\lambda dv(z), \tag{1.1}$$

be a family of standard weighted probability measures on $\mathbb{B}^n$, where the weight parameter fulfills $\lambda > -1$, the normalizing constant c_λ is given in (2.1) below, and dv is the standard volume form on $\mathbb{B}^n$. In this paper we consider the family of weighted Bergman spaces $\mathcal{A}^2_\lambda(\mathbb{B}^n)$ over $\mathbb{B}^n$ which consist of all complex analytic functions that are square integrable with respect to dv_λ. As is well known $\mathcal{A}^2_\lambda(\mathbb{B}^n)$ forms a closed subspace of $L_2(\mathbb{B}^n, dv_\lambda)$ and has the structure of a reproducing kernel Hilbert space. We denote by $\mathbf{B}_\lambda$ the (orthogonal) Bergman projection of $L_2(\mathbb{B}^n, dv_\lambda)$ onto $\mathcal{A}^2_\lambda(\mathbb{B}^n)$.

The first named author has been supported by an "Emmy-Noether scholarship" of DFG (*Deutsche Forschungsgemeinschaft*).
The third named author has been partially supported by CONACYT Project 102800, México.

Given an essentially bounded measurable function $a \in L_\infty(\mathbb{B}^n)$, we write T_a for the Toeplitz operator with symbol a, which acts on $\mathcal{A}^2_\lambda(\mathbb{B}^n)$ as $T_a f = \mathbf{B}_\lambda(af)$. That is, the Toeplitz operator is defined as the compression of a multiplication operator on $L_2(\mathbb{B}^n, dv_\lambda)$ onto the Bergman space. For simplicity we suppress the dependence on λ in the notation T_a.

Due to their simple structure Toeplitz operators form an important, tractable and intensively studied subclass in the algebra $\mathcal{L}(\mathcal{A}^2_\lambda(\mathbb{B}^n))$ of all bounded linear operators on $\mathcal{A}^2_\lambda(\mathbb{B}^n)$. Moreover, this subclass is dense in $\mathcal{L}(\mathcal{A}^2_\lambda(\mathbb{B}^n))$ with respect to the strong operator topology (shortly: SOT). In the case of the unweighted Bergman space (i.e., if $\lambda = 0$) and for a more general type of domains this property was established in [4]. However, the proof in [4] almost literally generalizes to the weighted case of $\lambda \in (-1, \infty)$.

A further challenging task is to characterize the closure of the above subclass of Toeplitz operators in the norm topology; or, in other words, to characterize the operators $S \in \mathcal{L}(\mathcal{A}^2_\lambda(\mathbb{B}^n))$ that can be approximated in norm by Toeplitz operators.

One of the tools, that proves to be useful and efficient here, is the so-called m-Berezin transform. The m-Berezin transform for any bounded linear operator acting on the unweighted Bergman space over the unit disk was defined in [9] and goes back to the work of Berezin (cf. [1]). For the unweighted Bergman space over the unit ball it was defined in [7], again for all bounded linear operators. The case of the weighted Bergman space has not been covered so far.

The (k, α)-Berezin transform for complex-valued regular measures (and in particular for bounded measurable symbols) on the weighted p-Bergman space over the unit ball was defined and studied in detail in [6]. This definition does not depend on the value of a weight parameter characterizing the space, and is not applicable to any bounded linear operator. Note in this context that an operator, being perfectly defined and bounded on a weighted Bergman space with the specific weight parameter, may easily have no sense for other values of the weight parameter. Thus any definition of the Berezin transform applicable for bounded linear operator acting on the specific weighted Bergman must forcibly include the value of a weight parameter in its definition.

Following the recipe in [7, proof of Proposition 2.1] we define in (3.3) the (m, λ)-Berezin transform for general bounded operators acting on the weighted Bergman space $\mathcal{A}^2_\lambda(\mathbb{B}^n)$. Note that our definition restricted to measures (more precisely: restricted to Toeplitz operators with measure symbols) does not coincide with the one in [6], which, being defined for the weighted Bergman space, does not depend on a weight parameter. Hence a further extension of the approach of [6] to general bounded linear operators forcibly involves some modifications.

Note that our (m, λ)-Berezin transform of a Toeplitz operator T_a acting on $\mathcal{A}^2_\lambda(\mathbb{B}^n)$ coincides with the (standard) $(0, \lambda + m)$-Berezin transform for T_a now considered on the weighted Bergman space $\mathcal{A}^2_{\lambda+m}(\mathbb{B}^n)$. Consequently the Berezin quantization procedure together with its correspondence principle suggest that the limit of the (m, λ)-Berezin transform as $m \to \infty$ may serve (and it does!) as a good approximation tool for certain classes of operators acting on the initial Bergman space.

The aim of this paper is to establish various results on norm approximations via the (m,λ)-Berezin transform. More precisely, we describe conditions under which a bounded linear operator S (Theorem 4.7 and Proposition 4.9) can be approximated in norm by Toeplitz operators whose symbols are bounded functions that are explicitly given as the (m,λ)-Berezin transforms of the initial operator S. In Appendix we remark that the approximation results of [6] remain valid for our definition of the (m,λ)-Berezin transform.

We would like to point out that these results generalize ideas and theorems in [6, 7, 9] and, in particular, form an essential tool in the explicit description of the norm closure of the set of Toeplitz operators with bounded radial symbols and the C^*-algebra generated by such Toeplitz operators, being considered as a subset of $\mathcal{L}(\mathcal{A}^2_\lambda(\mathbb{B}^n))$. Such applications and further aspects of the approximation procedure in the case of the so-called radial operators will be the subject of a forthcoming paper (cf. [2]).

2. Preliminaries

Let $\mathbb{B}^n := \{z \in \mathbb{C}^n : |z|^2 := |z_1|^2 + \cdots + |z_n|^2 < 1\}$ be the open unit ball in $\mathbb{C}^n$ equipped with the standard weighted measure (1.1), where $\lambda > -1$ is fixed. Here c_λ is given by

$$c_\lambda := \frac{\Gamma(n+\lambda+1)}{\pi^n \Gamma(\lambda+1)}, \tag{2.1}$$

so that $v_\lambda(\mathbb{B}^n) = 1$. We write $L_2(\mathbb{B}^n, dv_\lambda)$ for the Hilbert space of all functions that are square-integrable with respect to dv_λ. The corresponding norm and inner product are denoted by $\|\cdot\|_\lambda$ and $\langle\cdot,\cdot\rangle_\lambda$, respectively.

Let $\mathbb{Z}_+ := \{0, 1, \dots\}$ be the set of non-negative integers. With $\alpha \in \mathbb{Z}^n_+$ we use the standard notations $z^\alpha := z_1^{\alpha_1} \cdots z_n^{\alpha_n}$, $\alpha! := \alpha_1! \cdots \alpha_n!$ and $|\alpha| := \alpha_1 + \cdots + \alpha_n$. As is well known, for all $\alpha \in \mathbb{Z}^n_+$ we have

$$\|w^\alpha\|_\lambda = \sqrt{\frac{\alpha!\Gamma(n+\lambda+1)}{\Gamma(n+|\alpha|+\lambda+1}}, \tag{2.2}$$

and we write $[e_\alpha := w^\alpha \|w^\alpha\|_\lambda^{-1} : \alpha \in \mathbb{Z}^n_+]$ for the standard orthonormal basis of $\mathcal{A}^2_\lambda(\mathbb{B}^n)$.

The Bergman (orthogonal) projection $\mathbf{B}_\lambda$ from $L_2(\mathbb{B}^n, dv_\lambda)$ onto $\mathcal{A}^2_\lambda(\mathbb{B}^n)$ can be expressed as an integral operator in the explicit form

$$[\mathbf{B}_\lambda\varphi](z) = \int_{\mathbb{B}^n} \frac{\varphi(w)}{(1-\langle z,w\rangle)^{n+\lambda+1}} dv_\lambda(w) \qquad \text{with} \qquad \varphi \in L_2(\mathbb{B}^n, dv_\lambda),$$

where $\langle z,w\rangle := z_1\overline{w_1}+\cdots+z_n\overline{w_n}$ denotes the Euclidean inner product on $\mathbb{C}^n$. The reproducing kernel of the Bergman space $\mathcal{A}^2_\lambda(\mathbb{B}^n)$ is given by

$$K_z^\lambda(w) = \frac{1}{(1-\langle w,z\rangle)^{n+\lambda+1}} = \sum_{|\alpha|=0}^{\infty} \frac{\Gamma(n+|\alpha|+\lambda+1)}{\alpha!\Gamma(n+\lambda+1)}\overline{z}^\alpha w^\alpha. \tag{2.3}$$

We frequently use the normalized version of the Bergman kernel and write

$$k_z^\lambda(w) = \frac{K_z^\lambda(w)}{\|K_z^\lambda\|_\lambda} = \frac{(1-|z|^2)^{\frac{n+\lambda+1}{2}}}{(1-\langle w,z\rangle)^{n+\lambda+1}}.$$

By $\phi_z(w)$ we denote a biholomorphism of $\mathbb{B}^n$ that interchanges 0 and z. More precisely, we choose the explicit form of $\phi_z(w)$ given, for example, in [12, p.5] such that $\phi_0(w) = -w$. Recall [12, p.37] that the complex Jacobian $\det(\phi_z')$ of ϕ_z has the form

$$\det(\phi_z'(w)) = (-1)^n \frac{(1-|z|^2)^{\frac{n+1}{2}}}{(1-\langle w,z\rangle)^{n+1}} = (-1)^n k_z^0(w). \tag{2.4}$$

It is standard that the kernel K_z^λ transforms under the biholomorphisms ϕ_u as

$$K_z^\lambda(w) = \overline{k_u^\lambda(z)} K_{\phi_u(z)}^\lambda(\phi_u(w)) k_u^\lambda(w). \tag{2.5}$$

Given $z\in\mathbb{B}^n$, we introduce the unitary operator U_z on $\mathcal{A}^2_\lambda(\mathbb{B}^n)$ which acts as the weighted composition operator

$$\begin{aligned}(U_z f)(w) &:= \frac{(1-|z|^2)^{\frac{n+\lambda+1}{2}}}{(1-\langle w,z\rangle)^{n+\lambda+1}}(f\circ\phi_z)(w)\\ &= k_z^\lambda(w)\cdot f\circ\phi_z(w).\end{aligned} \tag{2.6}$$

It is easy to check that U_z is self-adjoint and so $U_z^2 = I$. Since ϕ_0 induces a reflection at the origin we have

$$(U_0 f)(w) = f(-w).$$

For a fixed $z\in\mathbb{B}^n$ we define an automorphism on the algebra $\mathcal{L}(\mathcal{A}^2_\lambda(\mathbb{B}^n))$ of all bounded operator on $\mathcal{A}^2_\lambda(\mathbb{B}^n)$ by

$$\mathcal{L}(\mathcal{A}^2_\lambda(\mathbb{B}^n)) \ni S \longmapsto S_z := U_z S U_z \in \mathcal{L}(\mathcal{A}^2_\lambda(\mathbb{B}^n)). \tag{2.7}$$

In particular, if $S = T_a$ is a Toeplitz operator then $(T_a)_z = T_{a\circ\phi_z}$.

Throughout the paper and as a convention we will denote by C a positive constant appearing in various estimates and whose value may change from place to place.

3. The (m,λ)-Berezin transform

Recall that the m-Berezin transform for the unweighted Bergman space over the unit disk and over the unit ball were defined in [9] and [7], respectively. In the case where $\lambda\neq 0$ the notion of the (k,α)-Berezin transform for measures on the weighted p-Bergman space over $\mathbb{B}^n$ was introduced in [6].

A generalization of the concept of the m-Berezin transform to an arbitrary bounded operator on the Bergman space $\mathcal{A}^2_\lambda(\mathbb{B}^n)$ requires a modification of the definition in [6]. We will follow the recipe in [7] and first introduce some notation. Put

$$C_{m,\alpha} := \binom{m}{|\alpha|} (-1)^{|\alpha|} \frac{|\alpha|!}{\alpha_1! \cdots \alpha_n!} = \frac{(-1)^{|\alpha|}}{\alpha!} \frac{m!}{(m-|\alpha|)!}, \tag{3.1}$$

so that

$$\sum_{|\alpha|=0}^{m} C_{m,\alpha}\, z^\alpha \overline{w}^\alpha = (1 - \langle z, w\rangle)^m . \tag{3.2}$$

Definition 3.1. We define the (m, λ)-*Berezin transform* of $S \in \mathcal{L}(\mathcal{A}^2_\lambda(\mathbb{B}^n))$ by

$$(B_{m,\lambda} S)(z) := \frac{c_{\lambda+m}}{c_\lambda} \sum_{|\alpha|=0}^{m} C_{m,\alpha} \left\langle S_z w^\alpha, w^\alpha \right\rangle_\lambda . \tag{3.3}$$

Note that a direct application of the Cauchy–Schwarz inequality gives the following pointwise estimate

$$\left| (B_{m,\lambda} S)(z) \right| \le \|S\| \frac{c_{\lambda+m}}{c_\lambda} \sum_{|\alpha|=0}^{m} \left| C_{m,\alpha} \right| \|w^\alpha\|^2_\lambda =: C(\lambda, m, n)\, \|S\|,$$

where the constant $C(\lambda, m, n) > 0$ is independent of $z \in \mathbb{B}^n$. That is, $B_{m,\lambda} S$ is a bounded function on $\mathbb{B}^n$ with

$$\|B_{m,\lambda} S\|_\infty \le C(\lambda, m, n)\, \|S\|. \tag{3.4}$$

As usual we define the (m, λ)-Berezin transform of a function $a \in L_\infty(\mathbb{B}^n)$ by

$$\begin{aligned} B_{m,\lambda}(a)(z) := (B_{m,\lambda} T_a)(z) &= \frac{c_{\lambda+m}}{c_\lambda} \sum_{|\alpha|=0}^{m} C_{m,\alpha} \left\langle (a \circ \phi_z) w^\alpha, w^\alpha \right\rangle_\lambda \\ &= \frac{c_{\lambda+m}}{c_\lambda} \int_{\mathbb{B}^n} (a \circ \phi_z)(w)\, c_\lambda (1 - |w|^2)^{\lambda+m} dv(w) \\ &= \int_{\mathbb{B}^n} (a \circ \phi_z)(w)\, dv_{\lambda+m}(w). \end{aligned} \tag{3.5}$$

As was mentioned earlier Definition 3.1 is different from the one in [6], where the (m, λ)-Berezin transform $\widetilde{B}_{m,\lambda}$ for finite, complex-valued, regular measures ν on $\mathbb{B}^n$ was introduced. In fact, in the special case of $\nu := a dv_\lambda$ with $a \in L_\infty(\mathbb{B}^n)$ the latter one gives

$$\widetilde{B}_{m,\lambda}(\nu)(z) = \int_{\mathbb{B}^n} (a \circ \phi_z)(w)\, dv_m(w),$$

which, differently from (3.5), is independent of the weight parameter λ. This seems to be inadequate as the initial data (measures and, more generally, operators) are defined on the specific weighted Bergman space $\mathcal{A}^2_\lambda(\mathbb{B}^n)$.

The next two propositions give alternative formulas for the (m, λ)-Berezin transform that, from time to time, are more suitable to work with. Note that the formula of the second proposition, in the particular case when $n = 1$ and $\lambda = 0$, coincides with the definition of the m-Berezin transform on the unit disk by Suárez [9].

We study then the properties of the (m, λ)-Berezin transform which will be used both in this and in a subsequent paper. The majority of our proofs use arguments similar to the ones of the unweighted case [7].

Proposition 3.2. *Let $S \in \mathcal{L}\left(\mathcal{A}^2_\lambda(\mathbb{B}^n)\right)$, $m \geq 0$ and $z \in \mathbb{B}^n$. Then*

$$(B_{m,\lambda}S)(z) = \frac{c_{\lambda+m}}{c_\lambda}\left(1-|z|^2\right)^{m+\lambda+n+1} \times \int_{\mathbb{B}^n}\int_{\mathbb{B}^n}(1-\langle u,w\rangle)^m K_z^{m+\lambda}(u)\overline{K_z^{m+\lambda}(w)S^*K_w^\lambda(u)}dv_\lambda(u)dv_\lambda(w).$$

Proof. We have

$$\begin{aligned}(B_{m,\lambda}S)(z) &= \frac{c_{\lambda+m}}{c_\lambda}\sum_{|\alpha|=0}^m C_{m,\alpha}\left\langle S_z w^\alpha, w^\alpha\right\rangle_\lambda \\ &= \frac{c_{\lambda+m}}{c_\lambda}\sum_{|\alpha|=0}^m C_{m,\alpha}\int_{\mathbb{B}^n} S\left(\phi_z^\alpha k_z^\lambda\right)(w)\overline{\phi_z^\alpha(w)k_z^\lambda(w)}dv_\lambda(w) \qquad (3.6)\\ &= \frac{c_{\lambda+m}}{c_\lambda}\sum_{|\alpha|=0}^m C_{m,\alpha}\int_{\mathbb{B}^n}\int_{\mathbb{B}^n}\phi_z^\alpha(u)k_z^\lambda(u)\overline{\phi_z^\alpha(w)k_z^\lambda(w)S^*K_w^\lambda(u)}dv_\lambda(u)dv_\lambda(w).\end{aligned}$$

In the last equality we use that

$$S\left(\phi_z^\alpha k_z^\lambda\right)(w) = \langle S\left(\phi_z^\alpha k_z^\lambda\right), K_w^\lambda\rangle_\lambda = \langle \phi_z^\alpha k_z^\lambda, S^*K_w^\lambda\rangle_\lambda.$$

Then, by (3.2) and (2.5), the expression (3.6) equals to

$$\begin{aligned}&\frac{c_{\lambda+m}}{c_\lambda}\int_{\mathbb{B}^n}\int_{\mathbb{B}^n}\left(1-\langle\phi_z(u),\phi_z(w)\rangle\right)^m k_z^\lambda(u)\overline{k_z^\lambda(w)S^*K_w^\lambda(u)}dv_\lambda(u)dv_\lambda(w)\\ &\quad= \frac{c_{\lambda+m}}{c_\lambda}\int_{\mathbb{B}^n}\int_{\mathbb{B}^n}\left(\frac{k_z^\lambda(u)\overline{k_z^\lambda(w)}}{K_w^\lambda(u)}\right)^{\frac{m}{\lambda+n+1}} k_z^\lambda(u)\overline{k_z^\lambda(w)S^*K_w^\lambda(u)}dv_\lambda(u)dv_\lambda(w)\\ &\quad= \frac{c_{\lambda+m}}{c_\lambda}\left(1-|z|^2\right)^{m+\lambda+n+1}\times\\ &\qquad\times\int_{\mathbb{B}^n}\int_{\mathbb{B}^n}\left(1-\langle u,w\rangle\right)^m K_z^{m+\lambda}(u)\overline{K_z^{m+\lambda}(w)S^*K_w^\lambda(u)}dv_\lambda(u)dv_\lambda(w),\end{aligned}$$

which finishes the proof. □

Proposition 3.3. *Let $S \in \mathcal{L}\left(\mathcal{A}^2_\lambda(\mathbb{B}^n)\right)$, $m \geq 0$ and $z \in \mathbb{B}^n$. Then*

$$(B_{m,\lambda}S)(z) = \frac{c_{\lambda+m}}{c_\lambda}\left(1-|z|^2\right)^{m+\lambda+n+1}\sum_{|\alpha|=0}^m C_{m,\alpha}\left\langle S(w^\alpha K_z^{m+\lambda}), w^\alpha K_z^{m+\lambda}\right\rangle_\lambda. \qquad (3.7)$$

Proof. We have

$$\begin{aligned}
&\int_{\mathbb{B}^n}\int_{\mathbb{B}^n}(1-\langle u,w\rangle)^m K_z^{m+\lambda}(u)\overline{K_z^{m+\lambda}(w)S^*K_w^{\lambda}(u)}dv_\lambda(u)dv_\lambda(w)\\
&=\sum_{|\alpha|=0}^{m} C_{m,\alpha}\int_{\mathbb{B}^n}\int_{\mathbb{B}^n} u^\alpha\overline{w^\alpha}K_z^{m+\lambda}(u)\overline{K_z^{m+\lambda}(w)S^*K_w^{\lambda}(u)}dv_\lambda(u)dv_\lambda(w)\\
&=\sum_{|\alpha|=0}^{m} C_{m,\alpha}\int_{\mathbb{B}^n} S(u^\alpha K_z^{m+\lambda})(w)\overline{w^\alpha K_z^{m+\lambda}(w)}dv_\lambda(w).
\end{aligned}$$

Thus the result follows from Proposition 3.2. □

Lemma 3.4. *Given $z,w\in\mathbb{B}^n$, the automorphism $\mathcal{U}:=\phi_{\phi_w(z)}\circ\phi_w\circ\phi_z$ of $\mathbb{B}^n$ extends to a unitary transformation of $\mathbb{C}^n$, and*

$$U_zU_w=[(-1)^n\det\mathcal{U}]^{\frac{n+\lambda+1}{n+1}}\cdot V_{\mathcal{U}}U_{\phi_w(z)},$$

where the operator $V_{\mathcal{U}}$ is given by

$$(V_{\mathcal{U}}f)(u):=f(\mathcal{U}u).$$

Proof. Since $\mathcal{U}$ in an automorphism of the unit ball having 0 as a fixed point it follows by the Cartan theorem that $\mathcal{U}$ acts by multiplication on a unitary matrix. This matrix will also be denoted by $\mathcal{U}$, i.e., $\mathcal{U}(u)=\mathcal{U}u$.

Differentiating the equality $\phi_{\phi_w(z)}\circ\mathcal{U}=\phi_w\circ\phi_z$ we have

$$\phi'_{\phi_w(z)}(\mathcal{U}(u))\,\mathcal{U}'(u)=\phi'_w(\phi_z(u))\,\phi'_z(u),$$

which implies

$$(-1)^n k^0_{\phi_w(z)}(\mathcal{U}u)\ \det\mathcal{U}=(-1)^n k^0_w(\phi_z(u))\cdot(-1)^n k^0_z(u).$$

As $k_z^\lambda=(k_z^0)^{\frac{n+\lambda+1}{n+1}}$ the application of identity (2.6) together with the last formula gives

$$\begin{aligned}
(U_zU_wf)(u)&=k_z^\lambda(u)\cdot k_w^\lambda(\phi_z(u))\cdot(f\circ\phi_w\circ\phi_z)(u)\\
&=(\det\mathcal{U})^{\frac{n+\lambda+1}{n+1}}\cdot(-1)^{\frac{n(n+\lambda+1)}{n+1}}k^\lambda_{\phi_w(z)}(\mathcal{U}u)\cdot(f\circ\phi_{\phi_w(z)}\circ\mathcal{U})(u)\\
&=[(-1)^n\det\mathcal{U}]^{\frac{n+\lambda+1}{n+1}}\cdot(V_{\mathcal{U}}U_{\phi_w(z)}f)(u).
\end{aligned}$$

Note that $[(-1)^n\det\mathcal{U}]^{\frac{n+\lambda+1}{n+1}}$ is a complex number of modulus one. □

Theorem 3.5. *Let $S\in\mathcal{L}(\mathcal{A}^2_\lambda(\mathbb{B}^n))$, $m\geq 0$ and $z\in\mathbb{B}^n$. Then $B_{m,\lambda}S_z=(B_{m,\lambda}S)\circ\phi_z$.*

Proof. By definition

$$
\begin{aligned}
(B_{m,\lambda}S_z)(0) &= \frac{c_{\lambda+m}}{c_\lambda}\sum_{|\alpha|=0}^{m} C_{m,\alpha}\langle U_0 S_z U_0 w^\alpha, w^\alpha\rangle_\lambda \\
&= \frac{c_{\lambda+m}}{c_\lambda}\sum_{|\alpha|=0}^{m} C_{m,\alpha}\langle S_z(-w)^\alpha, (-w)^\alpha\rangle_\lambda \\
&= \frac{c_{\lambda+m}}{c_\lambda}\sum_{|\alpha|=0}^{m} C_{m,\alpha}\langle S_z w^\alpha, w^\alpha\rangle_\lambda = B_{m,\lambda}S(z) = (B_{m,\lambda}S)\circ\phi_z(0).
\end{aligned}
$$

For any $\eta \in \mathbb{B}^n$, by Proposition 3.2 and Lemma 3.4 we have

$$
\begin{aligned}
&(B_{m,\lambda}S_z)\circ\phi_\eta(0) = B_{m,\lambda}((S_z)_\eta)(0) \\
&= \frac{c_{\lambda+m}}{c_\lambda}\int_{\mathbb{B}^n}\int_{\mathbb{B}^n}(1-\langle u,w\rangle)^m\,\overline{((S_z)_\eta)^* K_w^\lambda(u)}\,dv_\lambda(u)dv_\lambda(w) \\
&= \frac{c_{\lambda+m}}{c_\lambda}\int_{\mathbb{B}^n}\int_{\mathbb{B}^n}(1-\langle u,w\rangle)^m\,\overline{U_\eta U_z S^* U_z U_\eta K_w^\lambda(u)}\,dv_\lambda(u)dv_\lambda(w) \\
&= \frac{c_{\lambda+m}}{c_\lambda}\int_{\mathbb{B}^n}\int_{\mathbb{B}^n}(1-\langle u,w\rangle)^m\,\overline{V_{\mathcal{U}} U_{\phi_z(\eta)} S^* U_{\phi_z(\eta)} V_{\mathcal{U}}^* K_w^\lambda(u)}\,dv_\lambda(u)dv_\lambda(w) \\
&= B_{m,\lambda}S_{\phi_z(\eta)}(0),
\end{aligned}
$$

where $V_{\mathcal{U}}$ is the unitary operator of Lemma 3.4. This implies the statement. □

The next two lemmas are preparatory for the proof of Proposition 3.8, which plays a crucial role and states the commutativity of the (m,λ)-Berezin transforms for different values of the parameter m.

Lemma 3.6. *Let $S \in \mathcal{L}\left(\mathcal{A}_\lambda^2(\mathbb{B}^n)\right)$ and $m, j \geq 0$. If $|S^* K_z^\lambda(w)| \leq C$ for any $w \in \mathbb{B}^n$, then*

$$B_{m,\lambda}B_{j,\lambda}S = B_{j,\lambda}B_{m,\lambda}S.$$

Proof. Due to Theorem 3.5, we only need to check that $(B_{m,\lambda}B_{j,\lambda}S)(0) = (B_{j,\lambda}B_{m,\lambda}S)(0)$. Property (3.5), Proposition 3.2, and Fubini's theorem imply that

$$
\begin{aligned}
&B_{m,\lambda}(B_{j,\lambda}S)(0) = B_{m,\lambda}(T_{B_{j,\lambda}S})(0) = c_{m+\lambda}\int_{\mathbb{B}^n} B_{j,\lambda}S(z)(1-|z|^2)^{m+\lambda}dv(z) \\
&= \int_{\mathbb{B}^n}\frac{c_{m+\lambda}c_{j+\lambda}}{c_\lambda}(1-|z|^2)^{m+j+2\lambda+n+1} \\
&\quad\times\int_{\mathbb{B}^n}\int_{\mathbb{B}^n}(1-\langle u,w\rangle)^j K_z^{j+\lambda}(u)\overline{K_z^{j+\lambda}(w) S^* K_w^\lambda(u)}\,dv_\lambda(u)dv_\lambda(w)dv(z) \\
&= \int_{\mathbb{B}^n}\int_{\mathbb{B}^n}\frac{c_{m+\lambda}c_{j+\lambda}}{c_\lambda}F_{m,j}(u,w)\overline{S^* K_w^\lambda(u)}\,dv_\lambda(u)dv_\lambda(w),
\end{aligned}
$$

where the function $F_{m,j}(u,w)$ in the integrand is defined by

$$F_{m,j}(u,w) := (1-\langle u,w\rangle)^j \int_{\mathbb{B}^n} (1-|z|^2)^{m+j+2\lambda+n+1} K_z^{j+\lambda}(u)\overline{K_z^{j+\lambda}(w)}dv(z). \quad (3.8)$$

Observe that (3.8) can be represented as a finite sum

$$F_{m,j}(u,w) = \sum_{i=1}^{l} H_i(u)\overline{G_i(w)}$$

for certain holomorphic functions H_i and G_i. By [3, Lemma 10], it is sufficient to show that $F_{m,j}(w,w) = F_{j,m}(w,w)$, where $w \in \mathbb{B}^n$, which can be easily verified by a direct calculation

$$\begin{aligned}
F_{m,j}(w,w) &= (1-|w|^2)^j \int_{\mathbb{B}^n} (1-|z|^2)^{m+j+2\lambda+n+1} |K_z^{j+\lambda}(w)|^2 dv(z) \\
&= (1-|w|^2)^j \int_{\mathbb{B}^n} (1-|\phi_w(z)|^2)^{m+j+2\lambda+n+1} |K_w^{j+\lambda}(\phi_w(z))|^2 |k_w^0(z)|^2 dv(z) \\
&= (1-|w|^2)^m \int_{\mathbb{B}^n} (1-|z|^2)^{m+j+2\lambda+n+1} |K_z^{m+\lambda}(w)|^2 dv(z) \\
&= F_{j,m}(w,w).
\end{aligned}$$

In the second equality we have changed variables using (2.4). □

By $S_1 = S_1(\mathcal{A}_\lambda^2(\mathbb{B}^n))$ denote the set of all trace class operators acting on $\mathcal{A}_\lambda^2(\mathbb{B}^n)$. Given $A \in S_1$, we write $\mathrm{tr}[A]$ for its trace, and recall that the trace norm of A is given by

$$\|A\|_{S_1} := \mathrm{tr}\left[\sqrt{A^*A}\right].$$

Given $f, g \in \mathcal{A}_\lambda^2(\mathbb{B}^n)$, the rank-one-operator $f \otimes g$, acting on $\mathcal{A}_\lambda^2(\mathbb{B}^n)$ by the formula $(f \otimes g)h = \langle h, g\rangle_\lambda f$ obviously belongs to S_1. Furthermore,

$$\|f \otimes g\|_{S_1} = \|f\|_\lambda \cdot \|g\|_\lambda$$

and $\mathrm{tr}\,[f \otimes g] = \langle f, g\rangle_\lambda$. Recall as well that if $A \in S_1$ has rank m, then one has the inequality

$$\|A\|_{S_1} \le \sqrt{m}\,(\mathrm{tr}\,[A^*A])^{\frac{1}{2}}.$$

Lemma 3.7. *For any $S \in \mathcal{L}\left(\mathcal{A}_\lambda^2(\mathbb{B}^n)\right)$, there exist sequences $\{S_\alpha\}$, satisfying the property*

$$|S_\alpha^* K_z^\lambda(w)| \le C(\alpha), \quad (3.9)$$

such that $B_{m,\lambda}S_\alpha$ point-wisely converges to $B_{m,\lambda}S$.

Proof. Let $H^\infty = H^\infty(\mathbb{B}^n)$ denote the algebra of bounded holomorphic functions on $\mathbb{B}^n$. Both, the density of H^∞ in $\mathcal{A}_\lambda^2(\mathbb{B}^n)$ and the density of finite rank operators in the ideal $\mathcal{K}$ of compact operators on $\mathcal{L}(\mathcal{A}_\lambda^2(\mathbb{B}^n))$, imply that the set

$$\mathcal{F} := \left\{\sum_{j=1}^{l} f_j \otimes g_j : f_j, g_j \in H^\infty\right\}$$

is norm-dense in $\mathcal{K}$. At the same time $\mathcal{K}$ is dense in $\mathcal{L}(\mathcal{A}^2_\lambda(\mathbb{B}^n))$ with respect to the strong operator topology. Thus, for each $S \in \mathcal{L}\left(\mathcal{A}^2_\lambda(\mathbb{B}^n)\right)$ there exists a sequence $\{S_\alpha\}$ of finite rank operators

$$S_\alpha = \sum_{j=1}^{l(\alpha)} f_{\alpha,j} \otimes g_{\alpha,j}$$

converging strongly to S. The representation (3.7) shows that $B_{m,\lambda}S_\alpha$ converges to $B_{m,\lambda}S$ point-wise. To finish the proof we estimate

$$\begin{aligned}|S_\alpha^* K_z^\lambda(w)| &= \Big|\sum_{j=1}^{l(\alpha)} (g_{\alpha,j} \otimes f_{\alpha,j}) K_z^\lambda(w)\Big| = \Big|\sum_{j=1}^{l(\alpha)} \langle K_z^\lambda(w), f_{\alpha,j}(w)\rangle_\lambda g_{\alpha,j}(w)\Big| \\ &\leq \sum_{j=1}^{l(\alpha)} |f_{\alpha,j}(z)||g_{\alpha,j}(w)| \leq \sum_{j=1}^{l(\alpha)} \|f_{\alpha,j}\|_\infty \|g_{\alpha,j}\|_\infty = C(\alpha)\end{aligned}$$

uniformly for $z \in \mathbb{B}^n$. □

Proposition 3.8. *Let $S \in \mathcal{L}\left(\mathcal{A}^2_\lambda(\mathbb{B}^n)\right)$ and $m, j \geq 0$, then $B_{m,\lambda}B_{j,\lambda}S = B_{j,\lambda}B_{m,\lambda}S$.*

Proof. Let $S \in \mathcal{L}\left(\mathcal{A}^2_\lambda(\mathbb{B}^n)\right)$. According to Lemma 3.7 there exists a sequence $\{S_\alpha\}$ of operators that satisfy (3.9) and the point-wise convergence $B_{m,\lambda}S_\alpha \to B_{m,\lambda}S$ holds. Lemma 3.6 implies that

$$\big(B_{m,\lambda}B_{j,\lambda}S_\alpha\big)(z) = \big(B_{j,\lambda}B_{m,\lambda}S_\alpha\big)(z). \tag{3.10}$$

By representation (3.5),

$$\big(B_{m,\lambda}B_{j,\lambda}S_\alpha\big)(z) = \int_{\mathbb{B}^n} (B_{j,\lambda}S_\alpha) \circ \phi_z(u) dv_{m+\lambda}(u).$$

As the sequence $\{S_\alpha\}$ converges to S strongly we have by construction

$$\|(B_{j,\lambda}S_\alpha) \circ \phi_z\|_\infty = \|B_{j,\lambda}S_\alpha\|_\infty \leq \|B_{j,\lambda}\| \cdot \|S_\alpha\| \leq C(j,\lambda) \cdot \|S\|.$$

Furthermore $(B_{j,\lambda}S_\alpha) \circ \phi_z(u)$ converges to $(B_{j,\lambda}S) \circ \phi_z(u)$. As a consequence the functions $(B_{m,\lambda}B_{j,\lambda}S_\alpha)(z)$ and $(B_{j,\lambda}B_{m,\lambda}S_\alpha)(z)$ converge to $(B_{m,\lambda}B_{j,\lambda}S)(z)$ and $(B_{j,\lambda}B_{m,\lambda}S)(z)$, respectively. Passing to the limit in (3.10) finishes the proof. □

Corollary 3.9. *For all $\lambda > -1$ and $m \in \mathbb{Z}_+$ the (m,λ)-Berezin transform is one-to-one on bounded operators on $\mathcal{A}^2_\lambda(\mathbb{B}^n)$.*

Proof. Since $B_{m,\lambda}$ restricted to functions coincides with the usual Berezin transform on $\mathcal{A}^2_{\lambda+m}(\mathbb{B}^n)$ (cf. (3.5)) it is one-to-one on functions (on Toeplitz operators). Now assume that $S \in \mathcal{L}(\mathcal{A}^2_\lambda(\mathbb{B}^n))$ such that $B_{m,\lambda}S \equiv 0$. From Proposition 3.8 we obtain that

$$0 = B_{0,\lambda}B_{m,\lambda}S = B_{m,\lambda}B_{0,\lambda}S$$

and therefore $B_{0,\lambda}S \equiv 0$. Since $B_{0,\lambda}$ is known to be one-to-one on bounded operators over $\mathcal{A}^2_\lambda(\mathbb{B}^n)$ we conclude that $S = 0$, which finishes the proof. □

Recall that the pseudo-hyperbolic metric on the unit ball is defined as

$$\rho(z, w) := |\phi_z(w)| = |\phi_w(z)|.$$

As is well known $\rho(\cdot, \cdot)$ is invariant under the automorphisms ϕ_u of $\mathbb{B}^n$. The next result shows the Lipschitz continuity of $B_{0,\lambda} S$ with respect to this metric.

Theorem 3.10. *Let $S \in \mathcal{L}(\mathcal{A}^2_\lambda(\mathbb{B}^n))$. Then there exists a constant $C(n, \lambda) > 0$ such that*

$$\big|(B_{0,\lambda}S)(z) - (B_{0,\lambda}S)(w)\big| \leq C(n, \lambda)\|S\|\, \rho(z, w).$$

Proof. By definition and the above-mentioned properties of trace class operators we have

$$\begin{aligned}
\big|(B_{0,\lambda}S)(z) - (B_{0,\lambda}S)(w)\big| &= |\langle S_z 1, 1\rangle_\lambda - \langle S_w 1, 1\rangle_\lambda| \\
&= |\mathrm{tr}\ [S_z(1 \otimes 1)] - \mathrm{tr}\ [S_w(1 \otimes 1)]| \\
&= |\mathrm{tr}\ [S_z(1 \otimes 1) - SU_w(1 \otimes 1)U_w]| \\
&= |\mathrm{tr}\ [S_z(1 \otimes 1) - SU_z(U_zU_w1 \otimes U_zU_w1)U_z]| = (*).
\end{aligned}$$

By Lemma 3.4, we estimate

$$\begin{aligned}
(*) &< \|S_z\| \|1 \otimes 1 - U_{\phi_w(z)}1 \otimes U_{\phi_w(z)}1\|_{S_1} \\
&\leq \sqrt{2}\|S_z\| \big(2 - 2|\langle 1, k^\lambda_{\phi_w(z)}\rangle_\lambda|^2\big)^{1/2} \\
&= 2\|S\| \big[1 - (1 - |\phi_w(z)|^2)^{n+\lambda+1}\big]^{1/2} \\
&\leq C(n, \lambda)\|S\| |\phi_w(z)|,
\end{aligned}$$

which according to the definition of the pseudo-hyperbolic metric shows the result. □

Now representation (3.5) yields

Corollary 3.11. *Let $S \in \mathcal{L}(\mathcal{A}^2_\lambda(\mathbb{B}^n))$, and $a := B_{0,\lambda}S \in L_\infty(\mathbb{B}^n)$. Then*

$$\lim_{m \to \infty} \|B_{m,\lambda}(a) - a\|_\infty = 0.$$

Proof. Let $\varepsilon > 0$ and choose $\delta > 0$ with $|a(z) - a(w)| < \varepsilon$ whenever $z, w \in \mathbb{B}^n$ with $\rho(z, w) < \delta$. If $w \in \mathbb{B}^n$ and $m \in \mathbb{N}$, we have according to (3.5)

$$\begin{aligned}
&|B_{m,\lambda}(a)(w) - a(w)| \\
&\leq c_{\lambda+m} \int_{\mathbb{B}^n} \big|a \circ \phi_w(z) - a \circ \phi_w(0)\big|(1 - |z|^2)^{\lambda+m} dv(z) \\
&\leq c_{\lambda+m} \left\{\int_{0 \leq |z| < \delta} + \int_{1 > |z| \geq \delta}\right\} \big|a \circ \phi_w(z) - a \circ \phi_w(0)\big|(1 - |z|^2)^{\lambda+m} dv(z).
\end{aligned}$$

Since $\rho(\cdot, \cdot)$ is invariant under the automorphisms ϕ_w and $\rho(z, 0) < |z|$ (see, for example, [12, page 28]), we have $\rho(\phi_w(z), \phi_z(0)) = \rho(z, 0) < \delta$ in the first integral, and therefore by the Lipschitz continuity of a

$$c_{\lambda+m}\int_{0\le|z|<\delta}\big|a\circ\phi_w(z)-a\circ\phi_w(0)\big|(1-|z|^2)^{\lambda+m}dv(z)<\varepsilon. \tag{3.11}$$

Now, we estimate the second integral above.

$$\begin{aligned}c_{\lambda+m}&\int_{1>|z|\ge\delta}\big|a\circ\phi_w(z)-a\circ\phi_w(0)\big|(1-|z|^2)^{\lambda+m}dv(z)\\&\le 2c_{\lambda+m}\|a\|_\infty\int_{1>|z|\ge\delta}(1-|z|^2)^{\lambda+m}dv(z)\\&\le 2c_{\lambda+m}\|a\|_\infty(1-\delta)^{\lambda+m}\mathrm{vol}(\mathbb{B}^n).\end{aligned} \tag{3.12}$$

Since the normalizing constant $c_{\lambda+m}$ has at most polynomial growth as $m\to\infty$ (see the definition (2.1) and [5, Formula 8.328.2]) it is clear that the right-hand side converges to zero as $m\to\infty$. The assertion follows by combining the estimates (3.11) and (3.12). □

4. Approximation by Toeplitz operators

We start this section with a technical statement which is taken from [8, Proposition 1.4.10] and also stated as Lemma 3.1 in [7].

Lemma 4.1. *Suppose $a<1$ and $a+b<n+1$. Then*

$$\sup_{z\in\mathbb{B}^n}\int_{\mathbb{B}^n}\frac{dv(w)}{(1-|w|^2)^a\,|1-\langle w,z\rangle|^b}<\infty.$$

Let $1<q<\infty$ and p be the conjugate exponent of q. Note that the inequality

$$q=1+\frac{1}{p-1}<\frac{n+2(1+\lambda)}{n+1+\lambda}=1+\frac{1+\lambda}{n+1+\lambda}=:R \tag{4.1}$$

is equivalent to

$$p>2+\frac{n}{1+\lambda}.$$

In what follows we use the standard weighted L^p-norm $\|\cdot\|_{p,\lambda}$ defined by

$$\|f\|_{p,\lambda}=\left(\int_{\mathbb{B}^n}|f(z)|^p dv_\lambda(z)\right)^{\frac{1}{p}}.$$

Lemma 4.2. *Let $S\in\mathcal{L}\left(\mathcal{A}^2_\lambda(\mathbb{B}^n)\right)$, $p>2+\frac{n}{1+\lambda}$, and put $h(z)=(1-|z|^2)^{-a}$ with*

$$a=\frac{(1+\lambda)(n+1+\lambda)}{n+2(1+\lambda)}=\frac{1+\lambda}{R}.$$

Then there exists $C(n,p,\lambda)>0$ such that

$$\int_{\mathbb{B}^n}|(SK_z^\lambda)(w)|h(w)dv_\lambda(w)\le C(n,p,\lambda)\|S_z1\|_{p,\lambda}h(z), \tag{4.2}$$

for all $z \in \mathbb{B}^n$, *and*

$$\int_{\mathbb{B}^n} |(SK_z^\lambda)(w)| h(z) dv_\lambda(z) \leq C(n,p,\lambda) \|S_w^* 1\|_{p,\lambda} h(w), \tag{4.3}$$

for all $w \in \mathbb{B}^n$.

Proof. Given $z \in \mathbb{B}^n$, the equality

$$U_z 1 = (1-|z|^2)^{\frac{n+\lambda+1}{2}} K_z^\lambda$$

implies that

$$\begin{aligned} SK_z^\lambda &= \frac{1}{(1-|z|^2)^{\frac{n+\lambda+1}{2}}} SU_z 1 \\ &= \frac{1}{(1-|z|^2)^{\frac{n+\lambda+1}{2}}} U_z S_z 1 = (S_z 1 \circ \phi_z) K_z^\lambda. \end{aligned}$$

We change the variable $u = \phi_z(w)$ and apply the Hölder inequality

$$\begin{aligned} &\int_{\mathbb{B}^n} \frac{|(SK_z^\lambda)(w)|}{(1-|w|^2)^a} dv_\lambda(w) = c_\lambda \int_{\mathbb{B}^n} \frac{|S_z 1 \circ \phi_z(w)||K_z^\lambda(w)|(1-|w|^2)^\lambda}{(1-|w|^2)^a} dv(w) \\ &= \frac{1}{(1-|z|^2)^a} \int_{\mathbb{B}^n} \frac{|S_z 1(u)|}{|1-\langle u,z\rangle|^{n+\lambda+1-2a}(1-|u|^2)^a} dv_\lambda(u) \\ &\leq \frac{\|S_z 1\|_{p,\lambda}}{(1-|z|^2)^a} \left(c_\lambda \int_{\mathbb{B}^n} \frac{dv(u)}{(1-|u|^2)^{aq-\lambda}|1-\langle u,z\rangle|^{(n+\lambda+1-2a)q}} \right)^{1/q}. \end{aligned}$$

According to (4.1) we have $aq-\lambda < 1$ and $aq - \lambda + (n+\lambda+1-2a)q < n+1$. Hence inequality (4.2) follows from Lemma 4.1.

The second inequality (4.3) follows from (4.2) after replacing S by S^*, interchanging w and z, and making use of

$$\left(S^* K_w^\lambda\right)(z) = \left\langle S^* K_w^\lambda, K_z^\lambda \right\rangle_\lambda = \left\langle K_w^\lambda, SK_z^\lambda \right\rangle_\lambda = \overline{SK_z^\lambda}(w), \tag{4.4}$$

which holds for all $z, w \in \mathbb{B}^n$. □

Lemma 4.3. *Let* $S \in \mathcal{L}(\mathcal{A}_\lambda^2(\mathbb{B}^n))$ *and* $p > 2 + \frac{n}{1+\lambda}$. *Then*

$$\|S\| \leq C(n,p,\lambda) \Big(\sup_{z\in\mathbb{B}^n} \|S_z 1\|_{p,\lambda} \Big)^{1/2} \Big(\sup_{z\in\mathbb{B}^n} \|S_z^* 1\|_{p,\lambda} \Big)^{1/2},$$

where $C(n,p,\lambda)$ *is the constant of Lemma* 4.2.

Proof. By (4.4) we have that

$$(Sf)(w) = \langle Sf, K_w^\lambda \rangle_\lambda = \int_{\mathbb{B}^n} f(z) \overline{(S^* K_w^\lambda)}(z) dv_\lambda(z) = \int_{\mathbb{B}^n} f(z)(SK_z^\lambda)(w) dv_\lambda(z),$$

for $f \in \mathcal{A}_\lambda^2(\mathbb{B}^n)$ and $w \in \mathbb{B}^n$. Now Lemma 4.2 and the Schur theorem (see, for example, [11, Corollary 3.2.3]) imply the result. □

Lemma 4.4. *Let $\{S_m\}$ be a bounded sequence in $\mathcal{L}(\mathcal{A}^2_\lambda(\mathbb{B}^n))$ with $\lim_{m\to\infty}\|B_{0,\lambda}S_m\|_\infty = 0$. Then*

$$\sup_{z\in\mathbb{B}^n} \left|\left\langle (S_m)_z 1, f\right\rangle_\lambda\right| \to 0$$

as $m\to\infty$ for any $f\in\mathcal{A}^2_\lambda(\mathbb{B}^n)$, and

$$\sup_{z\in\mathbb{B}^n} |(S_m)_z 1(\cdot)| \to 0 \tag{4.5}$$

uniformly on compact subsets of $\mathbb{B}^n$ as $m\to\infty$.

Proof. To prove the first statement it is sufficient to check that for each multi-index k

$$\sup_{z\in\mathbb{B}^n} \left|\left\langle (S_m)_z 1, w^k\right\rangle_\lambda\right| \to 0 \qquad \text{as} \qquad m\to\infty.$$

Using (2.3) we calculate

$$\begin{aligned} B_{0,\lambda}S_m(\phi_z(u)) &= B_{0,\lambda}(S_m)_z(u) = (1-|u|^2)^{n+\lambda+1}\left\langle (S_m)_z K_u^\lambda, K_u^\lambda\right\rangle_\lambda \\ &= (1-|u|^2)^{n+\lambda+1}\sum_{|\alpha|=0}^{\infty}\sum_{|\beta|=0}^{\infty}\frac{\Gamma(n+|\alpha|+\lambda+1)}{\alpha!\Gamma(n+\lambda+1)}\frac{\Gamma(n+|\beta|+\lambda+1)}{\beta!\Gamma(n+\lambda+1)} \\ &\quad\times \left\langle (S_m)_z w^\alpha, w^\beta\right\rangle_\lambda \overline{u}^\alpha u^\beta. \end{aligned}$$

Given a multi-index k and $r\in(0,1)$, we obtain by using (2.1)

$$\begin{aligned} &\int_{|u|<r}\frac{B_{0,\lambda}S_m(\phi_z(u))\overline{u}^k}{(1-|u|^2)^{n+\lambda+1}}dv_\lambda(u) \\ &= \sum_{|\alpha|=0}^{\infty}\sum_{|\beta|=0}^{\infty}\frac{\Gamma(n+|\alpha|+\lambda+1)}{\alpha!\Gamma(n+\lambda+1)}\frac{\Gamma(n+|\beta|+\lambda+1)}{\beta!\Gamma(n+\lambda+1)}\left\langle (S_m)_z w^\alpha, w^\beta\right\rangle_\lambda \\ &\quad\times\int_{|u|<r}\overline{u}^{\alpha+k}u^\beta dv_\lambda(u) \\ &= \sum_{|\alpha|=0}^{\infty}\frac{\Gamma(n+|\alpha|+\lambda+1)}{\alpha!\Gamma(n+\lambda+1)}\frac{\Gamma(n+|\alpha|+|k|+\lambda+1)}{(\alpha+k)!\Gamma(n+\lambda+1)}\left\langle (S_m)_z w^\alpha, w^{\alpha+k}\right\rangle_\lambda \\ &\quad\times\int_{|u|<r}|u^{\alpha+k}|^2 dv_\lambda(u) \\ &= \sum_{|\alpha|=0}^{\infty}\frac{\Gamma(n+|\alpha|+\lambda+1)}{\alpha!\Gamma(n+\lambda+1)}\frac{\Gamma(n+|\alpha|+|k|+\lambda+1)}{(\alpha+k)!\pi^n\Gamma(\lambda+1)}\left\langle (S_m)_z w^\alpha, w^{\alpha+k}\right\rangle_\lambda \\ &\quad\times\int_{|u|<r}|u^{\alpha+k}|^2(1-|u|^2)^\lambda dv(u). \end{aligned}$$

Passing in the last integral to the polar coordinates $u = s\xi$, where $s\in\mathbb{R}_+$ and $\xi\in S^{2n-1}$, and making use of the next formulas (here dS is the surface measure on $S^{2n-1} = \partial\mathbb{B}^n$)

$$\int_{\mathbb{B}^n} f(u)dv(u) = \int_0^1 s^{2n-1}dr \int_{S^{2n-1}} f(s\xi)dS(\xi),$$
$$\int_{S^{2n-1}} |\xi^m|^2 dS(\xi) = \frac{2\pi^n m!}{(n-1+|m|)!}$$

the last expression is equal to

$$\sum_{|\alpha|=0}^{\infty} \frac{\Gamma(n+|\alpha|+\lambda+1)}{\alpha!\Gamma(n+\lambda+1)} \frac{\Gamma(n+|\alpha|+|k|+\lambda+1)}{\Gamma(\lambda+1)\Gamma(n+|\alpha|+|k|)}$$
$$\times \left\langle (S_m)_z w^\alpha, w^{\alpha+k} \right\rangle_\lambda 2\int_0^r s^{2n+2|\alpha|+2|k|-1}(1-s^2)^\lambda ds$$
$$= \sum_{|\alpha|=0}^{\infty} \frac{\Gamma(n+|\alpha|+\lambda+1)}{\alpha!\Gamma(n+\lambda+1)} \frac{\Gamma(n+|\alpha|+|k|+\lambda+1)}{\Gamma(\lambda+1)\Gamma(n+|\alpha|+|k|)}$$
$$\times \left\langle (S_m)_z w^\alpha, w^{\alpha+k} \right\rangle_\lambda \int_0^{r^2} s^{n+|\alpha|+|k|-1}(1-s)^\lambda ds$$
$$= \sum_{|\alpha|=0}^{\infty} \frac{\Gamma(n+|\alpha|+\lambda+1)}{\alpha!\Gamma(n+\lambda+1)} \left\langle (S_m)_z w^\alpha, w^{\alpha+k} \right\rangle_\lambda I_{r^2}(n+|\alpha|+|k|, \lambda+1)$$
$$= \left\langle (S_m)_z 1, w^k \right\rangle_\lambda I_{r^2}(n+|k|, \lambda+1)$$
$$+ \sum_{|\alpha|=1}^{\infty} \frac{\Gamma(n+|\alpha|+\lambda+1)}{\alpha!\Gamma(n+\lambda+1)} \left\langle (S_m)_z w^\alpha, w^{\alpha+k} \right\rangle_\lambda I_{r^2}(n+|\alpha|+|k|, \lambda+1).$$

Here the function $I_x(a,b)$ is defined in the standard way (see, for example, [5, Formula 8.392],

$$I_x(a,b) = \frac{\Gamma(a+b)}{\Gamma(a)\Gamma(b)} \int_0^x t^{a-1}(1-t)^{b-1}dt.$$

Then we have

$$|\left\langle (S_m)_z 1, w^k \right\rangle_\lambda| \le \frac{1}{I_{r^2}(n+|k|, \lambda+1)} \left| \int_{|u|<r} \frac{B_{0,\lambda} S_m(\phi_z(u))\overline{u}^k}{(1-|u|^2)^{n+\lambda+1}} dv_\lambda(u) \right|$$
$$+ \left| \sum_{|\alpha|=1}^{\infty} \frac{\Gamma(n+|\alpha|+\lambda+1)}{\alpha!\Gamma(n+\lambda+1)} \left\langle (S_m)_z w^\alpha, w^{\alpha+k} \right\rangle_\lambda \frac{I_{r^2}(n+|\alpha|+|k|, \lambda+1)}{I_{r^2}(n+|k|, \lambda+1)} \right|$$
$$\le \frac{1}{I_{r^2}(n+|k|, \lambda+1)} \|B_{0,\lambda} S_m\|_\infty c_\lambda \int_{|u|<r} \frac{|u^k|}{(1-|u|^2)^{n+1}} dv(u)$$
$$+ \sum_{|\alpha|=1}^{\infty} \frac{\Gamma(n+|\alpha|+\lambda+1)}{\alpha!\Gamma(n+\lambda+1)} \|(S_m)_z\| \|w^\alpha\|_\lambda \|w^{\alpha+k}\|_\lambda \frac{I_{r^2}(n+|\alpha|+|k|, \lambda+1)}{I_{r^2}(n+|k|, \lambda+1)}$$

$$\leq \|B_{0,\lambda}S_m\|_\infty \frac{c_\lambda}{I_{r^2}(n+|k|,\lambda+1)} \int_{|u|<r} \frac{|u^k|}{(1-|u|^2)^{n+1}} dv(u)$$
$$+ C \sum_{|\alpha|=1}^{\infty} \frac{I_{r^2}(n+|\alpha|+|k|,\lambda+1)}{I_{r^2}(n+|k|,\lambda+1)} = I + \Sigma,$$

where $C > 0$ is a constant independent of m and z. In the last estimate we used the boundedness of the sequence $\{S_m\}$ and the inequality

$$\frac{\Gamma(n+|\alpha|+\lambda+1)}{\alpha!\Gamma(n+\lambda+1)} \|w^\alpha\|_\lambda \|w^{\alpha+k}\|_\lambda \leq 1,$$

which easily follows from (2.2). The first summand I above tends to zero as $m \to \infty$ due to the assumptions of the lemma. We estimate now the series in the second summand Σ. By [5, Formula 8.328.2]

$$\lim_{|\alpha|\to\infty} \frac{\Gamma(n+|\alpha|+|k|+\lambda+1)}{\Gamma(n+|\alpha|+|k|)} \frac{1}{(n+|\alpha|+|k|)^{\lambda+1}} = 1,$$

and thus there exists $C > 0$ such that for all $\alpha \in \mathbb{Z}_+^n$

$$\frac{\Gamma(n+|\alpha|+|k|+\lambda+1)}{\Gamma(n+|\alpha|+|k|)} \frac{1}{(n+|\alpha|+|k|)^{\lambda+1}} < C.$$

Then

$$\Sigma_1 := \sum_{|\alpha|=1}^{\infty} \frac{I_{r^2}(n+|\alpha|+|k|,\lambda+1)}{I_{r^2}(n+|k|,\lambda+1)}$$
$$= \frac{\Gamma(n+|k|)\Gamma(\lambda+1)}{\Gamma(n+|k|+\lambda+1)} \left(\int_0^{r^2} t^{n+|k|-1}(1-t)^\lambda dt \right)^{-1}$$
$$\times \sum_{|\alpha|=1}^{\infty} \frac{\Gamma(n+|\alpha|+|k|+\lambda+1)}{\Gamma(n+|\alpha|+|k|)\Gamma(\lambda+1)} \int_0^{r^2} t^{n+|\alpha|+|k|-1}(1-t)^\lambda dt$$
$$\leq C \frac{\Gamma(n+|k|)}{\Gamma(n+|k|+\lambda+1)} \left(\int_0^{r^2} t^{n+|k|-1}(1-t)^\lambda dt \right)^{-1}$$
$$\times \sum_{|\alpha|=1}^{\infty} (n+|\alpha|+|k|)^{\lambda+1} \int_0^{r^2} t^{n+|\alpha|+|k|-1}(1-t)^\lambda dt.$$

Estimating the multiple $(1-t)^\lambda$ in both integrals

$$(1-r^2)^\lambda \leq (1-t)^\lambda \leq 1, \quad \text{for} \quad \lambda \geq 0,$$
$$1 \leq (1-t)^\lambda \leq (1-r^2)^\lambda, \quad \text{for} \quad \lambda \in (-1,0),$$

we come to the following inequality

$$\Sigma_1 \leq C\,\frac{\Gamma(n+|k|+1)}{\Gamma(n+|k|+\lambda+1)}(1-r^2)^{-|\lambda|}\sum_{|\alpha|=1}^{\infty}(n+|\alpha|+|k|)^{\lambda}r^{2|\alpha|}$$
$$= C\,\frac{\Gamma(n+|k|+1)}{\Gamma(n+|k|+\lambda+1)}(1-r^2)^{-|\lambda|}\sum_{m=1}^{\infty}\binom{m+n-1}{n}(n+m+|k|)^{\lambda}\,r^{2m}.$$

The power series in r in the last line has radius of convergence equal to 1 and vanishes at 0. Thus the value of Σ becomes small if one takes r sufficiently closed to 0.

Both above estimates, on I and on Σ, are independent of $z \in \mathbb{B}^n$, which proves the first statement of the lemma. In order to prove the second statement we use the series representation (2.3) and again (2.2)

$$\begin{aligned}
\left|(S_m)_z 1(u)\right| &= \left|\left\langle (S_m)_z 1, K_u^{\lambda}\right\rangle_{\lambda}\right| \\
&\leq \sum_{|\alpha|=0}^{\infty}\frac{\Gamma(n+|\alpha|+\lambda+1)}{\alpha!\Gamma(n+\lambda+1)}\left|\left\langle (S_m)_z 1, w^{\alpha}\right\rangle_{\lambda}\right|\cdot|u^{\alpha}| \\
&\leq \sum_{|\alpha|=0}^{l-1}\frac{\Gamma(n+|\alpha|+\lambda+1)}{\alpha!\Gamma(n+\lambda+1)}\left|\left\langle (S_m)_z 1, w^{\alpha}\right\rangle_{\lambda}\right| \\
&\quad+\sum_{|\alpha|=l}^{\infty}\frac{\Gamma(n+|\alpha|+\lambda+1)}{\alpha!\Gamma(n+\lambda+1)}\|S_m\|\|w^{\alpha}\|_{\lambda}|u^{\alpha}| \\
&\leq \sum_{|\alpha|=0}^{l-1}\frac{\Gamma(n+|\alpha|+\lambda+1)}{\alpha!\Gamma(n+\lambda+1)}\left|\left\langle (S_m)_z 1, w^{\alpha}\right\rangle_{\lambda}\right| \\
&\quad+C\sum_{|\alpha|=l}^{\infty}\left(\frac{\Gamma(n+|\alpha|+\lambda+1)}{\alpha!\Gamma(n+\lambda+1)}\right)^{\frac{1}{2}}|u^{\alpha}| \\
&= \widetilde{\Sigma}_1+\widetilde{\Sigma}_2.
\end{aligned}$$

To estimate $\widetilde{\Sigma}_2$ we use the Cauchy–Schwarz inequality,

$$\begin{aligned}
\widetilde{\Sigma}_2 &\leq C\sum_{j=l}^{\infty}\left(\frac{\Gamma(n+j+\lambda+1)}{j!\Gamma(n+\lambda+1)}\right)^{\frac{1}{2}}\sum_{|\alpha|=j}\left[\frac{j!}{\alpha!}\right]^{\frac{1}{2}}|u^{\alpha}| \\
&\leq C\sum_{j=l}^{\infty}\left(\frac{\Gamma(n+j+\lambda+1)}{j!\Gamma(n+\lambda+1)}\right)^{\frac{1}{2}}\left(\sum_{|\alpha|=j}\frac{j!}{\alpha!}|u^{\alpha}|^2\right)^{\frac{1}{2}}\left(\sum_{|\alpha|=j}1\right)^{\frac{1}{2}} \\
&= C\sum_{j=l}^{\infty}\left(\frac{\Gamma(n+j+\lambda+1)}{j!\Gamma(n+\lambda+1)}\right)^{\frac{1}{2}}\left(\frac{(n+j-1)!}{j!(n-1)!}\right)^{\frac{1}{2}}\left(\sum_{|\alpha|=j}\frac{j!}{\alpha!}|u^{\alpha}|^2\right)^{\frac{1}{2}}.
\end{aligned}$$

If we assume that $|u| \leq r < 1$ and use the multi-nomial theorem for the expression in the last brackets

$$\sum_{|\alpha|=j} \frac{j!}{\alpha!} |u^\alpha|^2 = |u|^{2j},$$

we finally have

$$\widetilde{\Sigma}_2 \leq C \sum_{j=l}^{\infty} \left(\frac{\Gamma(n+j+\lambda+1)}{j!\Gamma(n+\lambda+1)} \right)^{\frac{1}{2}} \left(\frac{(n+j-1)!}{j!(n-1)!} \right)^{\frac{1}{2}} r^j.$$

By choosing l sufficiently large we can make $\widetilde{\Sigma}_2$ as small as needed. Note that by the first statement of the lemma the expression $\widetilde{\Sigma}_1$, with l being already fixed, uniformly tends to zero as $m \to \infty$. This ends the proof. □

Lemma 4.5. *Let $\{S_m\}$ be a sequence in $\mathcal{L}(\mathcal{A}^2_\lambda(\mathbb{B}^n))$ such that $\|B_{0,\lambda} S_m\|_\infty \to 0$ as $m \to \infty$. Assume that for some $p > 2 + \frac{n}{1+\lambda}$ the following inequalities hold*

$$\sup_{z \in \mathbb{B}^n} \|(S_m)_z 1\|_{p,\lambda} \leq C \qquad \text{and} \qquad \sup_{z \in \mathbb{B}^n} \|(S^*_m)_z 1\|_{p,\lambda} \leq C, \tag{4.6}$$

where $C > 0$ is independent of m. Then $S_m \to 0$ as $m \to \infty$ in the $\mathcal{L}(\mathcal{A}^2_\lambda(\mathbb{B}^n))$-norm.

Proof. By Lemma 4.3 and (4.5) we have

$$\|S_m\| \leq C(n,p,\lambda) \left(\sup_{z \in \mathbb{B}^n} \|(S_m)_z 1\|_{p,\lambda} \right)^{1/2} \left(\sup_{z \in \mathbb{B}^n} \|(S^*_m)_z 1\|_{p,\lambda} \right)^{1/2} \leq C(n,p,\lambda).$$

Then, for $2 + \frac{n}{1+\lambda} < s < p$, Hölder's inequality gives

$$\begin{aligned} &\sup_{z \in \mathbb{B}^n} \|(S_m)_z 1\|^s_{s,\lambda} \\ &\quad \leq \sup_{z \in \mathbb{B}^n} \int_{|w|>r} |(S_m)_z 1(w)|^s dv_\lambda(w) + \sup_{z \in \mathbb{B}^n} \int_{|w|\leq r} |(S_m)_z 1(w)|^s dv_\lambda(w) \\ &\quad \leq \sup_{z \in \mathbb{B}^n} \|(S_m)_z 1\|^s_{p,\lambda} \left(\int_{|w|>r} dv_\lambda(w) \right)^{1-\frac{s}{p}} + \sup_{z \in \mathbb{B}^n} \int_{|w|\leq r} |(S_m)_z 1(w)|^s dv_\lambda(w), \end{aligned}$$

where, by (4.5), the second term tends to 0 as $m \to \infty$. By the first inequality in (4.6), the first term above can be made arbitrarily small by taking r sufficiently close to 1. Finally Lemma 4.3 and the second inequality in (4.6) yield

$$\|S_m\| \leq C(n,s,\lambda) \left(\sup_{z \in \mathbb{B}^n} \|(S_m)_z 1\|_{s,\lambda} \right)^{1/2} \left(\sup_{z \in \mathbb{B}^n} \|(S^*_m)_z 1\|_{s,\lambda} \right)^{1/2} \to 0,$$

as $m \to \infty$ proving the statement of the lemma. □

Corollary 4.6. *Let $S \in \mathcal{L}(\mathcal{A}^2_\lambda(\mathbb{B}^n))$ such that for some $p > 2 + \frac{n}{1+\lambda}$,*

$$\sup_{z \in \mathbb{B}^n} \|S_z 1 - (T_{B_{m,\lambda} S})_z 1\|_{p,\lambda} \leq C \qquad \text{and} \qquad \sup_{z \in \mathbb{B}^n} \|S^*_z 1 - (T_{B_{m,\lambda} S^*})_z 1\|_{p,\lambda} \leq C, \tag{4.7}$$

where $C > 0$ is independent of m. Then $T_{B_{m,\lambda}S} \to S$ as $m \to \infty$ in $\mathcal{L}(\mathcal{A}^2_\lambda(\mathbb{B}^n))$-norm.

Proof. We set $S_m = S - T_{B_{m,\lambda}S}$. By Proposition 3.8 we have

$$B_{0,\lambda}S_m = B_{0,\lambda}S - B_{0,\lambda}(T_{B_{m,\lambda}S}) = B_{0,\lambda}S - B_{m,\lambda}B_{0,\lambda}S,$$

which, by Corollary 3.11, uniformly tends to 0 as $m \to \infty$, i.e., $\|B_{0,\lambda}S_m\|_\infty \to 0$. An application of Lemma 4.5 finishes the proof. □

Theorem 4.7. *Let $S \in \mathcal{L}(\mathcal{A}^2_\lambda(\mathbb{B}^n))$. If there is $p > 2 + \frac{n}{1+\lambda}$ such that*

$$\sup_{z\in\mathbb{B}^n} \|T_{(B_{m,\lambda}S)\circ\phi_z}1\|_{p,\lambda} \leq C \qquad \text{and} \qquad \sup_{z\in\mathbb{B}^n} \|T^*_{(B_{m,\lambda}S)\circ\phi_z}1\|_{p,\lambda} \leq C, \tag{4.8}$$

where $C > 0$ is independent of m. Then $T_{B_{m,\lambda}S} \to S$ as $m \to \infty$ in the $\mathcal{L}(\mathcal{A}^2_\lambda(\mathbb{B}^n))$-norm.

Proof. We prove first that

$$\sup_{z\in\mathbb{B}^n} \|S_z 1\|_{p,\lambda} < \infty. \tag{4.9}$$

The equality $T_{(B_{m,\lambda}S)\circ\phi_z} = (T_{B_{m,\lambda}S})_z$, together with Lemma 4.3, implies

$$\begin{aligned}\|T_{B_{m,\lambda}S}\| &\leq C(n,p,\lambda)\Big(\sup_{z\in\mathbb{B}^n} \|T_{(B_{m,\lambda}S)\circ\phi_z}1\|_{p,\lambda}\Big)^{1/2}\Big(\sup_{z\in\mathbb{B}^n} \|T^*_{(B_{m,\lambda}S)\circ\phi_z}1\|_{p,\lambda}\Big)^{1/2},\\ &< C, \end{aligned} \tag{4.10}$$

where C is independent of m. Let $S_m = S - T_{B_{m,\lambda}S}$, then by the arguments in the proof of Corollary 4.6 we have

$$\lim_{m\to\infty} \|B_{0,\lambda}S_m\|_\infty = 0.$$

According to (4.10) the sequence $\{S_m\}$ is bounded; thus taking a polynomial f with $\|f\|_{q,\lambda} = 1$ we obtain from Lemma 4.4 that

$$\sup_{z\in\mathbb{B}^n} |\langle (S_m)_z 1, f\rangle_\lambda| \to 0, \qquad \text{as} \qquad m \to \infty.$$

Then for any $z_0 \in \mathbb{B}^n$ and any $\varepsilon > 0$, there is (a sufficiently large) m such that

$$|\langle S_{z_0}1, f\rangle_\lambda| \leq \sup_{z\in\mathbb{B}^n} |\langle (S_m)_z 1, f\rangle_\lambda| + |\langle (T_{B_{m,\lambda}S})_{z_0}1, f\rangle_\lambda| \leq \varepsilon + C,$$

with C being independent of m and z_0. This proves (4.9). Further, the equality

$$T^*_{(B_{m,\lambda}S)\circ\phi_z} = T_{\overline{B_{m,\lambda}S_z}} = T_{B_{m,\lambda}S^*_z} = T_{(B_{m,\lambda}S^*)\circ\phi_z},$$

together with (4.8) and (4.9), implies (4.7), and Corollary 4.6 finishes the proof. □

Another approach to approximation theorems involves the *invariant Laplacian* and its application to the (m,λ)-Berezin transform. Recall that the invariant Laplacian $\widetilde{\Delta}$ on $\mathbb{B}^n$, defined for $u \in C^2(\mathbb{B}^n)$ and $z \in \mathbb{B}^n$, is given by

$$(\widetilde{\Delta}u)(z) := \Delta(u \circ \phi_z)(0), \qquad \text{where} \qquad \Delta := 4\sum_{j=1}^n \frac{\partial^2}{\partial z_j \partial \overline{z}_j}.$$

Here $\frac{\partial}{\partial z_j}$ and $\frac{\partial}{\partial \bar{z}_j}$ denote the Cauchy–Riemann operators with respect to the complex coordinate z_j, $j = 1, \dots, n$, and Δ is the standard Laplacian on $\mathbb{C}^n \cong \mathbb{R}^{2n}$. Let $S \in \mathcal{L}(\mathcal{A}^2_\lambda(\mathbb{B}^n))$ and $m \in \mathbb{Z}_+$, then we wish to calculate the function

$$\widetilde{\Delta} B_{m,\lambda} S \in C^\infty(\mathbb{B}^n).$$

Note that in the case $\lambda = 0$ and $n = 1$ this calculation was done in Proposition 2.4. of [9]. According to Theorem 3.5 we have

$$\widetilde{\Delta}\big[B_{m,\lambda}S\big](z) = \Delta\big(B_{m,\lambda}S \circ \phi_z\big)(0) = \Delta\big(B_{m,\lambda}S_z\big)(0)$$

and therefore we can assume that $z = 0$. We intend to use the form of $B_{m,\lambda}S$ in Proposition 3.2. We apply Δ to the z-dependent part of $B_{m,\lambda}S$ in the integral representation given there. Hence we have to evaluate the derivative

$$\begin{aligned} &\Delta\left[\frac{(1-|z|^2)^{m+\lambda+n+1}}{(1-\langle u,z\rangle)^{n+m+\lambda+1}(1-\langle z,w\rangle)^{n+m+\lambda+1}}\right](0) \\ &\quad = -4(m+n+\lambda+1) + 4(m+n+\lambda+1)^2\langle u,w\rangle. \end{aligned}$$

Inserting this relation into the expression of $B_{m,\lambda}S$ given in Proposition 3.2 shows

$$\begin{aligned} \Delta(B_{m,\lambda}S)(0) &= -4(m+n+\lambda+1)(B_{m,\lambda}S)(0) \\ &+ 4(m+n+\lambda+1)^2\frac{c_{\lambda+m}}{c_\lambda}\int_{\mathbb{B}^n}\int_{\mathbb{B}^n}\big(1-\langle u,w\rangle\big)^m\langle u,w\rangle\overline{S^*K_w^\lambda(u)}\,dv_\lambda(u)dv_\lambda(w). \end{aligned} \tag{4.11}$$

On the other hand the same proposition implies that

$$\begin{aligned} &\frac{c_\lambda}{c_{\lambda+m}}\left(B_{m,\lambda}S\right)(0) - \frac{c_\lambda}{c_{\lambda+m+1}}\left(B_{m+1,\lambda}S\right)(0) \\ &\quad = \int_{\mathbb{B}^n}\int_{\mathbb{B}^n}\big(1-\langle u,w\rangle\big)^m\langle u,w\rangle\overline{S^*K_w^\lambda(u)}\,dv_\lambda(u)dv_\lambda(w). \end{aligned} \tag{4.12}$$

Combining the equations (4.11) and (4.12) now gives

$$\begin{aligned} \Delta(B_{m,\lambda}S)(0) &= 4(m+n+\lambda+1)(m+n+\lambda)\big(B_{m,\lambda}S\big)(0) \\ &\quad - 4\frac{c_{\lambda+m}}{c_{\lambda+m+1}}(m+n+\lambda+1)^2\big(B_{m+1,\lambda}S\big)(0). \end{aligned}$$

According to (2.1) we have

$$\frac{c_{\lambda+m}}{c_{\lambda+m+1}}(m+n+\lambda+1)^2 = (n+m+\lambda+1)(\lambda+m+1)$$

and we have shown the following relation:

Proposition 4.8. *Let $S \in \mathcal{L}(\mathcal{A}^2_\lambda(\mathbb{B}^n))$. For all $m \in \mathbb{Z}_+$ and $\lambda > -1$ it holds*

$$\widetilde{\Delta}\big[B_{m,\lambda}S\big] = 4(m+n+\lambda+1)\Big[(m+n+\lambda)\big(B_{m,\lambda}S\big) - (m+\lambda+1)\big(B_{m+1,\lambda}S\big)\Big]. \tag{4.13}$$

Moreover, for all k, m we have

$$\widetilde{\Delta}B_{m,\lambda}\big(B_{k,\lambda}S\big) = B_{m,\lambda}\big(\widetilde{\Delta}B_{k,\lambda}S\big). \tag{4.14}$$

Proof. It suffices to prove (4.14). According to Proposition 3.8 and using (4.13) we have

$$\begin{aligned}
&\widetilde{\Delta} B_{m,\lambda}(B_{k,\lambda} S) \\
&\quad = \widetilde{\Delta} B_{k,\lambda}(B_{m,\lambda} S) \\
&\quad = 4(k+n+\lambda+1)\big[(k+n+\lambda) B_{k,\lambda} B_{m,\lambda} S - (k+\lambda+1) B_{k+1,\lambda} B_{m,\lambda} S\big] \\
&\quad = 4(k+n+\lambda+1)\big[(k+n+\lambda) B_{m,\lambda} B_{k,\lambda} S - (k+\lambda+1) B_{m,\lambda} B_{k+1,\lambda} S\big] \\
&\quad = B_{m,\lambda}(\widetilde{\Delta} B_{k,\lambda} S),
\end{aligned}$$

which shows the assertion. □

For the remaining part of the section we specialize to the case of dimension $n = 1$. Proposition 4.8 then implies

$$B_{m,\lambda} S - B_{m+1,\lambda} S = \frac{\widetilde{\Delta}[B_{m,\lambda} S]}{4(m+\lambda+2)(m+\lambda+1)} \tag{4.15}$$

and we can prove an analogue of Lemma 4.1 in [10]. We write $\mathbb{D} := \mathbb{B}^1 \subset \mathbb{C}$ for the open unit disc.

Proposition 4.9. *Let $S \in \mathcal{L}(\mathcal{A}^2_\lambda(\mathbb{D}))$ where $\lambda > -1$. Assume that $\|T_{\widetilde{\Delta}(B_{m,\lambda} S)}\| \leq C$ where C does not depend on m. Then we have the norm convergence*

$$\lim_{m \to \infty} T_{B_{m,\lambda} S} = S, \tag{4.16}$$

Proof. According to (4.15) we can write

$$\begin{aligned}
T_{B_{m+1,\lambda} S} &= T_{B_{0,\lambda} S} - \sum_{k=0}^{m} \left\{ T_{B_{k,\lambda} S} - T_{B_{k+1,\lambda} S} \right\} \\
&= T_{B_{0,\lambda} S} - \sum_{k=0}^{m} \frac{T_{\widetilde{\Delta}(B_{k,\lambda} S)}}{4(k+\lambda+2)(k+\lambda+1)}.
\end{aligned}$$

From the boundedness assumption on $\|T_{\widetilde{\Delta}(B_{k,\lambda} S)}\|$ we conclude that the right-hand side of the equation converges in norm to some operator $R \in \mathcal{L}(\mathcal{A}^2_\lambda(\mathbb{D}))$. The continuity property of the usual Berezin transform $B_{0,\lambda}$, cf. (3.4) implies that

$$\lim_{m \to \infty} B_{0,\lambda} T_{B_{m,\lambda} S} = B_{0,\lambda} R.$$

On the other hand note that Proposition 3.8 and Corollary 3.11 imply the pointwise convergence

$$B_{0,\lambda}(T_{B_{m,\lambda} S}) = B_{0,\lambda} B_{m,\lambda}(S) = B_{m,\lambda} B_{0,\lambda}(S) \longrightarrow B_{0,\lambda} S$$

and it follows that $B_{0,\lambda} S = B_{0,\lambda} R$. Finally the injectivity of $B_{0,\lambda}$ shows that $S = R$. □

Appendix: Toeplitz operators with measure symbols

The (k, λ)-Berezin transform for complex-valued regular measures on the weighted p-Bergman space over the unit ball was defined and studied in detail in [6], where, in particular, the approximation results were obtained. As was mentioned, this (k, λ)-Berezin transform cannot be applied to any bounded liner operator acting on $\mathcal{A}^2_\lambda(\mathbb{B}^n)$, which forced us to introduce the modified definition (3.3). In this appendix, following arguments of [6], we remark that the approximation results of [6] remain valid for our definition of the (m, λ)-Berezin transform.

Recall that a positive finite regular measure ν on $\mathbb{B}^n$ is called *Carleson measure* for $\mathcal{A}^2_\lambda(\mathbb{B}^n)$ if there is a constant $C > 0$ independent of $h \in \mathcal{A}^2_\lambda(\mathbb{B}^n)$ such that

$$\int_{\mathbb{B}^n} |h(z)|^2 d\nu(z) \leq C \int_{\mathbb{B}^n} |h(z)|^2 dv_\lambda(z).$$

This means that the inclusion of $\mathcal{A}^2_\lambda(\mathbb{B}^n)$ into $L_2(\mathbb{B}^n, d\nu)$ is well defined and continuous. Consider the operator T_ν defined for all $h \in H^\infty(\mathbb{B}^n)$ and $z \in \mathbb{B}^n$ by

$$T_\nu h(z) := \int_{\mathbb{B}^n} \frac{h(w)}{(1 - \langle z, w\rangle)^{n+\lambda+1}} d\nu(w). \tag{A.1}$$

Then T_ν is densely defined on $\mathcal{A}^2_\lambda(\mathbb{B}^n)$ and it admits a bounded extension to an element in $\mathcal{L}(\mathcal{A}^2_\lambda(\mathbb{B}^n))$ if and only if ν is a Carlson measure. In the case where μ is a complex measure such that $\nu = |\mu|$ is Carleson we can define T_μ in an analogous way and it can be checked that one obtains a bounded operator on $\mathcal{A}^2_\lambda(\mathbb{B}^n)$, as well. We call T_μ the Toeplitz operator with measure symbol μ.

Now we calculate the (m, λ)-Berezin transform for Toeplitz operators of the form (A.1). First note that by a direct application of Fubini's theorem we have

$$\langle T_\nu f, g\rangle_\lambda = \int_{\mathbb{B}^n} f(w)\overline{g(w)} d\nu(w) =: \langle f, g\rangle_\nu.$$

In particular, it holds

$$\begin{aligned}
\langle (T_\nu)_z w^k, w^k\rangle_\lambda &= \langle T_\nu U_z w^k, U_z w^k\rangle_\lambda \\
&= \langle U_z w^k, U_z w^k\rangle_\nu \\
&= \int_{\mathbb{B}^n} \frac{(1 - |z|^2)^{n+\lambda+1}}{|1 - \langle z, w\rangle|^{2(n+\lambda+1)}} |\phi_z^k(w)|^2 d\nu(w).
\end{aligned}$$

By plugging the last relation into the definition of the (m, λ)-Berezin transform and using (3.2) one obtains the integral transform

$$\begin{aligned}
B_{m,\lambda}(\nu)(z) :&= (B_{m,\lambda} T_\nu)(z) \\
&= \frac{c_{\lambda+m}}{c_\lambda} \sum_{|k|=0}^{m} C_{m,k} \int_{\mathbb{B}^n} \left|\phi_z^k(w)\right|^2 \frac{(1 - |z|^2)^{n+\lambda+1}}{|1 - \langle z, w\rangle|^{2(n+\lambda+1)}} d\nu(w) \\
&= \frac{c_{\lambda+m}}{c_\lambda} \int_{\mathbb{B}^n} (1 - |\phi_z(w)|^2)^m \frac{(1 - |z|^2)^{n+\lambda+1}}{|1 - \langle z, w\rangle|^{2(n+\lambda+1)}} d\nu(w).
\end{aligned}$$

We insert the well-known relation

$$1 - |\phi_z(w)|^2 = \frac{(1-|z|^2)(1-|w|^2)}{|1-\langle z,w\rangle|^2} \tag{A.2}$$

and finally obtain

$$B_{m,\lambda}(\nu)(z) = \frac{c_{\lambda+m}}{c_\lambda} \int_{\mathbb{B}^n} \frac{(1-|z|^2)^{n+\lambda+m+1}}{|1-\langle z,w\rangle|^{2(n+\lambda+m+1)}} (1-|w|^2)^m d\nu(w). \tag{A.3}$$

In [6], Section 4 the (m, λ)-Berezin transform $\widetilde{B}_m(\nu)$ of ν was defined as

$$\widetilde{B}_m(\nu)(z) := \frac{c_m}{c_\lambda} \int_{\mathbb{B}_n} \frac{(1-|\phi_z(w)|^2)^{n+1+m}}{(1-|w|^2)^{n+1+\lambda}} d\nu(w).$$

A comparison with (A.3) shows that the restriction of $B_{m,\lambda}$ to Toeplitz operators T_ν with measure symbols ν is related to $\widetilde{B}_m(\nu)(z)$ by a simple shift in the parameter m:

$$B_{m,\lambda}(\nu) = \widetilde{B}_{m+\lambda}(\nu).$$

Hence an approximation result completely analogous to Theorem 4.7 in [6] holds for $B_{m,\lambda}(\nu)$. We only restate the result in the case $p = q = 2$ according to the notation in [6].

Theorem A.1. *Let ν be a complex measure such that $|\nu|$ is Carleson. Fix $f, g \in \mathcal{A}^2_\lambda(\mathbb{B}^n)$ and put $h := f\overline{g} \in L^1(\mathbb{B}^n, dv_\lambda)$. Then*

$$\left| \int_{\mathbb{B}^n} B_{m,\lambda}(\nu) h dv_\lambda - \int_{\mathbb{B}^n} h d\nu \right| \leq \widetilde{C}(m) \|B_{0,\lambda}(|\nu|)\|_\infty \|f\|_\lambda \|g\|_\lambda, \tag{A.4}$$

where $\lim_{m\to\infty} \widetilde{C}(m) = 0$. In particular, for all $\lambda > -1$ we have the following norm convergence

$$\lim_{m\to\infty} T_{B_{m,\lambda}(\nu)} = T_\nu.$$

References

[1] F.A. Berezin, *Covariant and contravariant symbols of operators* (Russian). Izv. Akad. Nauk. SSSR Ser. Mat. 36 (1972), 1134–1167.

[2] W. Bauer, C. Herrera Yañez, N. Vasilevski, *Eigenvalue characterization of radial operators on weighted Bergman spaces over the unit ball*, Integr. Equ. Oper. Theory, 78(2) (2014), 271–300.

[3] B.R. Choe, Y.J. Lee, *Pluriharmonic symbols of commuting Toeplitz operators*, Illinois J. Math. 37 (1993), 424–436.

[4] M. Engliš, *Density of algebras generated by Toeplitz operators on Bergman spaces*, Ark. Mat. 30 no. 2 (1992), 227–243.

[5] I.S. Gradshteyn, I.M. Ryzhik, *Tables of integrals, series, and products*, Academic Press, 1980.

[6] M. Mitkovski, D. Suárez, and B.D. Wick, *The essential norm of operators on $A^p_\alpha(\mathbb{B}_n)$*, Integr. Equ. Oper. Theory 75, no. 2 (2013), 197–233.

[7] K. Nam, D. Zheng, and C. Zhong, *m-Berezin transform and compact operators*, Rev. Mat. Iberoamericana, 22(3) (2006), 867–892.

[8] W. Rudin, *Function theory in the unit ball of* $\mathbb{C}^n$, Fundamental principles of Mathematical Science 241, Springer-Verlag, New York-Berlin, 1980.

[9] D. Suárez, *Approximation and symbolic calculus for Toeplitz algebras on the Bergman space*, Rev. Mat. Iberoamericana 20(2) (2004), 563–610.

[10] D. Suárez , *The eigenvalues of limits of radial Toeplitz operators*, Bull. London Math. Soc. v. 40 (2008), 631–641.

[11] K. Zhu, *Operator theory in function spaces*, Marcel Dekker, Inc., 1990.

[12] K. Zhu, *Spaces of holomorphic functions in the unit ball*, Springer-Verlag, 2005.

Wolfram Bauer
Mathematisches Institut
Georg-August-Universität
Bunsenstr. 3–5
D-37073 Göttingen, Germany
e-mail: wbauer@uni-math.gwdg.de

Crispin Herrera Yañez and Nikolai Vasilevski
Departamento de Matemáticas
CINVESTAV del I.P.N.
Av. IPN 2508, Col. San Pedro Zacatenco
México D.F. 07360, México
e-mail: cherrera@math.cinvestav.mx
nvasilev@math.cinvestav.mx

Operator Theory:
Advances and Applications, Vol. 240, 69–85

Normal and Cohyponormal Weighted Composition Operators on H^2

Carl C. Cowen, Sungeun Jung and Eungil Ko

Abstract. In this paper we study normal and cohyponormal weighted composition operators on the Hardy space H^2. We show that if $W_{f,\varphi}$ is cohyponormal, then f is outer and φ is univalent. Moreover, we prove that when the composition map φ has the Denjoy–Wolff point in the open unit disk, $W_{f,\varphi}$ is cohyponormal if and only if it is normal; in this case, f and φ can be expressed as linear fractional maps. As a corollary, we find the polar decomposition of the cohyponormal operator $W_{f,\varphi}$. Finally, we examine the commutant of a cohyponormal weighted composition operator.

Mathematics Subject Classification (2010). 47B20, 47B38, 47B33.

Keywords. Weighted composition operator, composition operator, normal operator, hyponormal operator, cohyponormal operator, polar decomposition, commutant, inner-outer factorization.

1. Introduction

Let $\mathcal{L}(\mathcal{H})$ be the algebra of all bounded linear operators on a separable complex Hilbert space $\mathcal{H}$. If $T \in \mathcal{L}(\mathcal{H})$, we write $r(T)$, $\sigma(T)$, and $\sigma_p(T)$ for the spectral radius, the spectrum, and the point spectrum of T, respectively.

Let $\mathbb{D}$ denote the open unit disk in the complex plane $\mathbb{C}$. The Hardy space $H^2(\mathbb{D}) = H^2$ consists of all the analytic functions on $\mathbb{D}$ having power series representation with square summable complex coefficients. The space $H^\infty(\mathbb{D}) = H^\infty$ consists of all the functions that are analytic and bounded on $\mathbb{D}$. If φ is an analytic mapping from $\mathbb{D}$ into itself, the *composition operator* C_φ on H^2 is defined by $C_\varphi h = h \circ \varphi$ for all $h \in H^2$. The composition operator C_φ is bounded on H^2 by Littlewood subordination theorem (see [10] for more details). It is well known that

This research was supported by Basic Science Research Program through the National Research Foundation of Korea (NRF) funded by the Ministry of Education, Science and Technology (2012R1A2A2A02008590).

the composition operator C_φ defined on H^2 is normal if and only if $\varphi(z) = \gamma z$ where $|\gamma| \leq 1$ (see [9] or [10]).

For an analytic function f on $\mathbb{D}$ and an analytic selfmap φ of $\mathbb{D}$, the operator $W_{f,\varphi} : H^2 \to H^2$ given by

$$W_{f,\varphi}h = f \cdot (h \circ \varphi),\ \ h \in H^2$$

is called a *weighted composition operator*. If f is bounded on $\mathbb{D}$, then $W_{f,\varphi}$ is clearly bounded on H^2. For $f \in H^\infty$, the *multiplication operator* on H^2 is given by $M_f h = fh$ for all $h \in H^2$. Remark that $W_{f,\varphi}$ can be written by $W_{f,\varphi} = M_f C_\varphi$ if $f \in H^\infty$. We refer to [1], [7], and [8] for more details about weighted composition operators. In particular, C.C. Cowen and E. Ko characterized self-adjoint weighted composition operators on H^2 (see [8]), and P. Bourdon and S.K. Narayan provided several properties of normal weighted composition operators on H^2 (see [1]).

An operator $T \in \mathcal{L}(H)$ is said to be *hyponormal* if $T^*T \geq TT^*$. We say that $T \in \mathcal{L}(H)$ is *cohyponormal* if T^* is hyponormal, i.e.,

$$TT^* \geq T^*T, \text{ or equivalently, } \|T^*x\| \geq \|Tx\| \text{ for all } x \in \mathcal{H}.$$

We refer the reader to [4] and [11] for hyponormal operators. In this paper we study normal and cohyponormal weighted composition operators on the Hardy space H^2. We show that if $W_{f,\varphi}$ is cohyponormal, then f is outer and φ is univalent. Moreover, we prove that when the composition map φ has the Denjoy–Wolff point in the open unit disk, $W_{f,\varphi}$ is cohyponormal if and only if it is normal; in this case, f and φ can be expressed as linear fractional maps. As a corollary, we find the polar decomposition of the cohyponormal operator $W_{f,\varphi}$. Finally, we examine the commutant of a cohyponormal weighted composition operator.

2. Preliminaries

In this section, we recall some definitions needed for our program. Let $\partial\mathbb{D}$ denote the unit circle in the complex plane $\mathbb{C}$. For each $\beta \in \mathbb{D}$, the function $K_\beta(z) = \frac{1}{1-\overline{\beta}z} \in H^2$, called the *reproducing kernel* for H^2 at β, has the property that $\langle f, K_\beta \rangle = f(\beta)$ for every $f \in H^2$ and $\beta \in \mathbb{D}$. It is well known that the linear span of the reproducing kernels $\{K_\beta : \beta \in \mathbb{D}\}$ is dense in H^2. In [6], C.C. Cowen gave an adjoint formula of a composition operator whose symbol is a linear fractional selfmap of $\mathbb{D}$. If $\varphi(z) = \frac{az+b}{cz+d}$ is a linear fractional selfmap of $\mathbb{D}$, then $C_\varphi^* = M_g C_\sigma M_h^*$ where $g(z) = \frac{1}{-\overline{b}z+\overline{d}}$, $\sigma(z) = \frac{\overline{a}z-\overline{c}}{-\overline{b}z+\overline{d}}$, and $h(z) = cz + d$. It follows from [6] that σ is a selfmap of $\mathbb{D}$ and $g \in H^\infty$. Notice that $W_{f,\varphi}^* K_\beta = \overline{f(\beta)} K_{\varphi(\beta)}$ when $W_{f,\varphi}$ is bounded on H^2 and $\beta \in \mathbb{D}$ (see [8]); indeed, for any $\beta \in \mathbb{D}$ and $f \in H^2$

$$\langle f, W_{f,\varphi}^* K_\beta \rangle = \langle f \cdot (h \circ \varphi), K_\beta \rangle = f(\beta)h(\varphi(\beta)) = \langle h, \overline{f(\beta)} K_{\varphi(\beta)} \rangle.$$

In particular, $C_\varphi^* K_\beta = K_{\varphi(\beta)}$ because $C_\varphi = W_{1,\varphi}$.

For any selfmap φ of $\mathbb{D}$ and each positive integer n, we write $\varphi_1 := \varphi$ and $\varphi_{n+1} := \varphi \circ \varphi_n$, which is called the *iterate* of φ for n. We also denote by φ_0 the identical function on $\mathbb{D}$. When φ is any analytic selfmap of $\mathbb{D}$, we call $w \in \overline{\mathbb{D}}$ a *fixed point* of φ if $\lim_{r\to 1^-} \varphi(rw) = w$. We say that a function f has a *nontangential limit* at $\zeta \in \partial\mathbb{D}$ if $\lim_{z\to\zeta} f(z)$ exists in the nontangential region

$$\Gamma(\zeta,\alpha) = \{z \in \mathbb{D} : |z-\zeta| < \alpha(1-|z|)\}$$

for each $\alpha > 1$. An analytic selfmap φ of $\mathbb{D}$ is said to have a *finite angular derivative* at $\zeta \in \partial\mathbb{D}$ if there exists $\eta \in \partial\mathbb{D}$ such that $\frac{\varphi(z)-\eta}{z-\zeta}$ has a finite nontangential limit as $z \to \zeta$. This limit is written as $\varphi'(\zeta)$ if it exists (see [10, page 50] for more details).

It is well known that for an analytic function $\varphi : \mathbb{D} \to \mathbb{D}$, if φ is neither the identity map nor an elliptic automorphism of $\mathbb{D}$, then there is a point w of $\overline{\mathbb{D}}$ so that the iterates of φ converges uniformly to w on compact subsets of $\mathbb{D}$. Moreover, w is the unique fixed point of φ in $\overline{\mathbb{D}}$ for which $|\varphi'(w)| \leq 1$. We say that the unique fixed point w is the *Denjoy–Wolff point* of φ.

Let φ be an automorphism of $\mathbb{D}$. Then φ is of the form

$$\varphi(z) = \frac{az+\overline{b}}{bz+\overline{a}}$$

for all $z \in \mathbb{D}$, where a and b in $\mathbb{C}$ with $|a|^2 - |b|^2 = 1$. When $b \neq 0$, it is easy to calculate that

$$\frac{iIm(a) \pm \sqrt{|b|^2 - (\mathrm{Im}(a))^2}}{b}$$

are the fixed points of φ. If $|\,\mathrm{Im}(a)| = |b|$, then φ is called *parabolic*, and we say that φ is *hyperbolic* if $|\,\mathrm{Im}(a)| < |b|$. Remark that φ is parabolic if and only if it has one fixed point lying on $\partial\mathbb{D}$, while φ is hyperbolic if and only if it has two fixed points lying on $\partial\mathbb{D}$. If $|\,\mathrm{Im}(a)| > |b|$, then φ is said to be *elliptic*. We note that φ is elliptic if and only if one of its fixed points is inside $\mathbb{D}$ and another is outside $\mathbb{D}$. In this sense, this type also includes the case when $b = 0$, i.e., when 0 and ∞ are the fixed points of φ.

3. Main results

Throughout this section, we examine several properties of the weight function f and the composition symbol φ of a cohyponormal weighted composition operator $W_{f,\varphi} \in \mathcal{L}(H^2)$.

Lemma 3.1. *If $f \in H^\infty$, then* $\ker(M_f^*) = (\psi H^2)^\perp$ *where $f = \psi F$, ψ is an inner function, and F is an outer function.*

Proof. If $g \in (\psi H^2)^\perp$, then for any $h \in H^2$ we get that

$$\langle M_f^* g, h\rangle = \langle g, M_f h\rangle = \langle g, \psi F h\rangle = 0.$$

Hence $M_f^* g = 0$ and so $(\psi H^2)^\perp \subseteq \ker(M_f^*)$.

On the other hand, if there exists $g \in \ker(M_f^*) \setminus (\psi H^2)^\perp$, then there exists $u \in H^2$ such that $\langle g, \psi u\rangle \neq 0$ but $M_f^* g = 0$. Since F is outer, we have $\frac{1}{F} \in H^\infty$ and so it follows that

$$0 = \left\langle M_f^* g, \frac{u}{F}\right\rangle = \langle g, \psi u\rangle \neq 0,$$

which is a contradiction. Thus $(\psi H^2)^\perp \supseteq \ker(M_f^*)$. □

Theorem 3.2. *Let φ be a nonconstant analytic selfmap of $\mathbb{D}$ and let $f \in H^\infty$ be not identically zero. If $W_{f,\varphi}$ is cohyponormal, then f is outer; hence never vanishes on $\mathbb{D}$ and φ is univalent.*

Proof. Suppose that $W_{f,\varphi}$ is cohyponormal. If $W_{f,\varphi}h = 0$ for some $h \in H^2$, then $f(z)h(\varphi(z)) \equiv 0$. Since f is not identically zero, $h(\varphi(z)) \equiv 0$ on $\mathbb{D}$. Since φ is nonconstant, $h(z) \equiv 0$ on $\mathbb{D}$. Hence $\ker(W_{f,\varphi}) = \{0\}$. Since $W_{f,\varphi}$ is cohyponormal, $\ker(W_{f,\varphi}^*) \subseteq \ker(W_{f,\varphi}) = \{0\}$. Since $\ker(M_f^*) \subseteq \ker(W_{f,\varphi}^*)$, we have $\ker(M_f^*) = \{0\}$. Since $f \in H^2$ is not identically zero, f can be factorized as $f = \psi F$ where ψ is inner and F is outer. Then $M_f = M_\psi M_F = M_F M_\psi$ and $M_f^* = M_F^* M_\psi^*$. Since ψ is inner, M_ψ is an isometry of H^2 onto ψH^2. If ψ is not constant, then $H^2 \neq \psi H^2$. Hence Lemma 3.1 implies that $\ker(M_f^*) = (\psi H^2)^\perp \neq \{0\}$, a contradiction. Thus ψ must be constant, and so f is an outer function. In particular, f never vanishes on $\mathbb{D}$.

Assume that φ is not univalent on $\mathbb{D}$. Then we can choose distinct points $\alpha, \beta \in \mathbb{D}$ such that $\varphi(\alpha) = \varphi(\beta)$. Since f never vanishes on $\mathbb{D}$, we obtain that $f(\alpha) \neq 0$ and $f(\beta) \neq 0$. Set $g = \frac{1}{\overline{f(\alpha)}}K_\alpha - \frac{1}{\overline{f(\beta)}}K_\beta$. If g is the zero function in H^2, then we have $\frac{\overline{f(\alpha)}}{1-\overline{\alpha}z} = \frac{\overline{f(\beta)}}{1-\overline{\beta}z}$ for all $z \in \mathbb{D}$, but it gives the contradiction that $\alpha = \beta$. Hence g is not the zero function in H^2. Since $W_{f,\varphi}^* g = K_{\varphi(\alpha)} - K_{\varphi(\beta)} = 0$ and $W_{f,\varphi}^*$ is hyponormal, it holds that $W_{f,\varphi} g = 0$. This means that $f(z)g(\varphi(z)) = 0$ for all $z \in \mathbb{D}$. Moreover, f has no zeros in $\mathbb{D}$ by the above argument. Thus $g(\varphi(z)) = 0$ for all $z \in \mathbb{D}$. Since $\varphi(\mathbb{D})$ is a nonempty open subset of $\mathbb{D}$, we get that $g \equiv 0$ on $\mathbb{D}$, which is a contradiction. Therefore, φ should be univalent. □

From the proof of Theorem 3.2 we observe that a cohyponormal weighted composition operator $W_{f,\varphi}$ is Fredholm if and only if it has closed range. For an operator $T \in \mathcal{L}(H)$, a vector $x \in \mathcal{H}$ is said to be *cyclic* if the linear span of the orbit $O(x,T) := \{T^n x\}_{n=0}^\infty$ is norm dense in $\mathcal{H}$. If there is a cyclic vector x for T, then we say that T is a *cyclic* operator. If the orbit $O(x,T)$ is norm dense in $\mathcal{H}$ for some $x \in \mathcal{H}$, i.e., $\overline{O(x,T)} = \mathcal{H}$, then T is called *hypercyclic* operator and x is called a *hypercyclic vector*. It is obvious that every hypercyclic operator is cyclic.

Corollary 3.3. *Let φ be an analytic selfmap, not an elliptic automorphism, of $\mathbb{D}$ with a fixed point in $\mathbb{D}$ and let $f \in H^\infty \setminus \{0\}$. If $W_{f,\varphi}$ is cohyponormal, then $W_{f,\varphi}^*$ is cyclic.*

Proof. We note that there exists a point $z_0 \in \mathbb{D}$ so that the sequence $\{\varphi_n(z_0)\}_{n=0}^\infty$ consists of pairwise distinct points in $\mathbb{D}$ which converges to the interior fixed point of φ by [17, Lemma 1] and [16, Section 5.2, Proposition 1]. Let $g \in H^2$ be such that $g \perp \bigvee_{n=0}^{\infty}\{(W_{f,\varphi}^*)^n K_{z_0}\}$. Since it holds that for any positive integer n

$$W_{f,\varphi}^n = W_{f\cdot(f\circ\varphi)\cdot(f\circ\varphi_2)\cdots(f\circ\varphi_{n-1}),\varphi_n},$$

we obtain that

$$\begin{aligned} 0 &= \langle g, (W_{f,\varphi}^*)^n K_{z_0}\rangle = \langle W_{f,\varphi}^n g, K_{z_0}\rangle \\ &= f(z_0)f(\varphi(z_0))f(\varphi_2(z_0))\cdots f(\varphi_{n-1}(z_0))g(\varphi_n(z_0)) \end{aligned}$$

for any positive integer n. Since f has no zeros in $\mathbb{D}$ by Theorem 3.2, it follows that $g(\varphi_n(z_0)) = 0$ for any positive integer n. Thus $g \equiv 0$ on $\mathbb{D}$ and so we have $\bigvee_{n=0}^{\infty}\{(W_{f,\varphi}^*)^n K_{z_0}\} = H^2$. □

Next we give a lower bound and an upper bound for the spectral radius of a cohyponormal weighted composition operator.

Lemma 3.4. *Let φ be an analytic selfmap of $\mathbb{D}$ and let $f \in H^\infty$. Then*

$$\frac{|f(0)f(\varphi(0))\cdots f(\varphi_{n-1}(0))|}{\sqrt{1-|\varphi_n(0)|^2}} \le \|W_{f,\varphi}^n\| \le \frac{2\|f\|_\infty^n}{\sqrt{1-|\varphi_n(0)|^2}}$$

for each positive integer n.

Proof. It follows from [5, Theorem 2.1] that

$$\begin{aligned} \|W_{f,\varphi}^n\| &= \|W_{f\cdot(f\circ\varphi)\cdots(f\circ\varphi_{n-1}),\varphi_n}\| \le \|M_{f\cdot(f\circ\varphi)\cdots(f\circ\varphi_{n-1})}\|\|C_{\varphi_n}\| \\ &\le \|f\|_\infty^n \frac{1+|\varphi_n(0)|}{\sqrt{1-|\varphi_n(0)|^2}} \le \frac{2\|f\|_\infty^n}{\sqrt{1-|\varphi_n(0)|^2}}. \end{aligned}$$

On the other hand, it holds that

$$\begin{aligned} \|W_{f,\varphi}^n\| &= \|W_{f,\varphi}^{*n}\| \ge \|W_{f,\varphi}^{*n}K_0\| = \|W_{f\cdot(f\circ\varphi)\cdots(f\circ\varphi_{n-1}),\varphi_n}^* K_0\| \\ &= |f(0)f(\varphi(0))\cdots f(\varphi_{n-1}(0))|\, \|K_{\varphi_n(0)}\| \\ &= \frac{|f(0)f(\varphi(0))\cdots f(\varphi_{n-1}(0))|}{\sqrt{1-|\varphi_n(0)|^2}}. \end{aligned}$$

Hence we complete the proof. □

Theorem 3.5. *Let φ be an analytic selfmap of $\mathbb{D}$ with the Denjoy–Wolff point w, and let $f \in H^\infty$. If $W_{f,\varphi}$ is cohyponormal, then the following statements hold:*

(i) *If $w \in \mathbb{D}$, then*

$$|f(w)| \le r(W_{f,\varphi}) \le \|f\|_\infty.$$

(ii) *If $w \in \partial\mathbb{D}$ and f is continuous at w, then*

$$\varphi'(w)^{-\frac{1}{2}}|f(w)| \le r(W_{f,\varphi}) \le \varphi'(w)^{-\frac{1}{2}}\|f\|_\infty.$$

Proof. We may assume that f is not identically zero on $\mathbb{D}$. By Lemma 3.4 we have

$$\frac{|f(0)f(\varphi(0))\cdots f(\varphi_{n-1}(0))|^{\frac{1}{n}}}{(1-|\varphi_n(0)|^2)^{\frac{1}{2n}}} \leq \|W^n_{f,\varphi}\|^{\frac{1}{n}} \leq \frac{2^{\frac{1}{n}}\|f\|_\infty}{(1-|\varphi_n(0)|^2)^{\frac{1}{2n}}}$$

for each positive integer n. Put $x_n = \log|f(\varphi_n(0))|$ for each nonnegative integer n; here, we note that $f(\varphi_n(0)) \neq 0$ and $f(w) \neq 0$ from Theorem 3.2. In addition, $\lim_{n\to\infty} x_n = \log|f(w)|$, which implies that

$$\lim_{n\to\infty} |f(0)f(\varphi(0))\cdots f(\varphi_{n-1}(0))|^{\frac{1}{n}} = |f(w)|. \tag{1}$$

On the other hand, since it holds for all n that

$$\frac{1}{(1-|\varphi_n(0)|^2)^{\frac{1}{2n}}} \leq \|C^n_\varphi\|^{\frac{1}{n}} \leq \frac{2^{\frac{1}{n}}}{(1-|\varphi_n(0)|^2)^{\frac{1}{2n}}}$$

from [5, Theorem 2.1] or Lemma 3.4, we ensure that

$$r(C_\varphi) = \lim_{n\to\infty} \frac{1}{(1-|\varphi_n(0)|^2)^{\frac{1}{2n}}}. \tag{2}$$

Hence we obtain from (1) and (2) that $r(C_\varphi)|f(w)| \leq r(W_{f,\varphi}) \leq r(C_\varphi)\|f\|_\infty$, and so the proof follows from [10, Theorem 3.9]. □

Recall that for any $\alpha \in \mathbb{D}$ and any positive integer j, the jth derivative evaluation kernel at α, $K^{[j]}_\alpha$, is the function in H^2 so that

$$\langle f, K^{[j]}_\alpha\rangle = f^{(j)}(\alpha)$$

for $f \in H^2$. In particular, $K^{[0]}_\alpha := K_\alpha$.

Lemma 3.6. *Let φ be an analytic selfmap of $\mathbb{D}$ with a fixed point $w \in \mathbb{D}$ and let $f \in H^\infty$. If $W_{f,\varphi}$ is a cohyponormal operator on H^2, then*

$$\{f(w), f(w)\varphi'(w), f(w)(\varphi'(w))^2, f(w)(\varphi'(w))^3, \dots\} \subseteq \sigma_p(W_{f,\varphi}).$$

Proof. Let g be any function in H^2 and let n be any positive integer. Then

$$\begin{aligned}\langle g, W^*_{f,\varphi}K^{[n]}_w\rangle = \langle W_{f,\varphi}g, K^{[n]}_w\rangle &= \langle f\cdot(g\circ\varphi), K^{[n]}_w\rangle = (f\cdot(g\circ\varphi))^{(n)}(w)\\ &= \sum_{j=0}^{n-1}\gamma_j(w)g^{(j)}(w) + f(w)g^{(n)}(w)(\varphi'(w))^n\\ &= \langle g, \sum_{j=0}^{n-1}\overline{\gamma_j(w)}K^{[j]}_w + \overline{f(w)(\varphi'(w))^n}K^{[n]}_w\rangle\end{aligned}$$

where each γ_j is a function consisting of sums of products of derivatives of f and φ at w. Since g is arbitrary in H^2, we have

$$W^*_{f,\varphi}K^{[n]}_w = \sum_{j=0}^{n-1}\overline{\gamma_j(w)}K^{[j]}_w + \overline{f(w)(\varphi'(w))^n}K^{[n]}_w \tag{3}$$

for any positive integer n. Moreover, we get that

$$W_{f,\varphi}^* K_w = \overline{f(w)} K_{\varphi(w)} = \overline{f(w)} K_w. \tag{4}$$

For any nonnegative integer n, set $M_n := \text{span}\{K_w, K_w^{[1]}, K_w^{[2]}, \ldots, K_w^{[n]}\}$. Then $W_{f,\varphi}^* M_n \subseteq M_n$ from (3) and (4). Since M_n is finite dimensional, it is closed and so an invariant subspace for $W_{f,\varphi}^*$. Hence the operator $W_{f,\varphi}^*$ can be written as

$$W_{f,\varphi}^* = \begin{pmatrix} W_{f,\varphi}^*|_{M_n} & B \\ 0 & C \end{pmatrix} \quad \text{on } H^2 = M_n \oplus M_n^{\perp}$$

where $W_{f,\varphi}^*|_{M_n}$ stands for the restriction of $W_{f,\varphi}^*$ to the invariant subspace M_n. With respect to the basis $\{K_w, K_w^{[1]}, K_w^{[2]}, \ldots, K_w^{[n]}\}$ of M_n, the operator $W_{f,\varphi}^*|_{M_n}$ has the following matrix representation:

$$W_{f,\varphi}^*|_{M_n} = \begin{pmatrix} \overline{f(w)} & * & \cdots & * \\ & \overline{f(w)\varphi'(w)} & & \vdots \\ & & \ddots & \vdots \\ & 0 & & \overline{f(w)(\varphi'(w))^n} \end{pmatrix}.$$

Since $W_{f,\varphi}^*|_{M_n}$ is a finite-dimensional upper triangular matrix, we obtain that

$$\sigma_p(W_{f,\varphi}^*) \supseteq \sigma_p(W_{f,\varphi}^*|_{M_n}) = \{\overline{f(w)}, \overline{f(w)\varphi'(w)}, \ldots, \overline{f(w)(\varphi'(w))^n}\}.$$

Such an inclusion holds for every nonnegative integer n, and hence we have

$$\{\overline{f(w)}, \overline{f(w)\varphi'(w)}, \overline{f(w)(\varphi'(w))^2}, \overline{f(w)(\varphi'(w))^3}, \ldots\} \subseteq \sigma_p(W_{f,\varphi}^*).$$

Since $W_{f,\varphi}$ is cohyponormal, we complete the proof. □

Theorem 3.7. *Let φ be an analytic selfmap of $\mathbb{D}$ with the Denjoy–Wolff point $w \in \mathbb{D}$ and let $f \in H^\infty \setminus \{0\}$. Then the following statements are equivalent:*

(i) *$W_{f,\varphi}$ is normal.*
(ii) *$W_{f,\varphi}$ is cohyponormal.*
(iii) *The functions f and φ have the forms of*

$$f(z) = f(w)\frac{1-\overline{w}\varphi(z)}{1-\overline{w}z} \quad \textit{and} \quad \varphi(z) = a_0 + \frac{d_1 z}{1-\overline{d_0}z} \tag{5}$$

where

$$a_0 = \varphi(0) = \frac{w(1-\varphi'(w))}{1-|w|^2\varphi'(w)}, \quad d_0 = \frac{w(1-\overline{\varphi'(w)})}{1-|w|^2\overline{\varphi'(w)}},$$

and

$$d_1 = \varphi'(0) = \frac{\varphi'(w)(1-|w|^2)^2}{(1-|w|^2\varphi'(w))^2}.$$

Proof. Since every normal operator is cohyponormal, it suffices to show that (ii) $\Rightarrow$ (iii) and (iii) $\Rightarrow$ (i). Suppose that $W_{f,\varphi}$ is cohyponormal. Since φ has the Denjoy–Wolff point $w \in \mathbb{D}$, we have $|\varphi'(w)| < 1$. In addition, $f(w) \neq 0$ by Theorem 3.2. Thus $\overline{f(w)}$ and $\overline{f(w)\varphi'(w)}$ are distinct eigenvalues of $W_{f,\varphi}^*$ from the proof of Lemma 3.6. Recall that if T is a hyponormal operator on a Hilbert space and u and v are eigenvectors corresponding to distinct eigenvalues of T, then $u \perp v$ (see [4, Proposition 4.4]). Since $W_{f,\varphi}$ is cohyponormal, we get that

$$\ker(W_{f,\varphi}^* - \overline{f(w)}) \perp \ker(W_{f,\varphi}^* - \overline{f(w)\varphi'(w)}).$$

Since $\varphi(w) = w$, we obtain that $W_{f,\varphi}^* K_w = \overline{f(w)} K_w$. Due to the cohyponormality of $W_{f,\varphi}$, we have $W_{f,\varphi} K_w = f(w) K_w$, which ensures that

$$f(z) = f(w)\frac{1-\overline{w}\varphi(z)}{1-\overline{w}z}. \tag{6}$$

From the proof of Lemma 3.6, there exists an eigenvector $g = \beta_0 K_w + \beta_1 K_w^{[1]}$ of $W_{f,\varphi}^*|_{M_1}$ corresponding to the eigenvalue $\overline{f(w)\varphi'(w)}$. Then g is orthogonal to K_w, and so it holds that

$$0 = \langle g, K_w\rangle = \beta_0 K_w(w) + \beta_1 K_w^{[1]}(w).$$

Since $K_w(z) = \frac{1}{1-\overline{w}z}$ and $K_w^{[1]}(z) = \frac{z}{(1-\overline{w}z)^2}$, we obtain that $\beta_0 = -\frac{w}{1-|w|^2}\beta_1$. Therefore, we have $g = -\beta_1(\frac{w}{1-|w|^2}K_w - K_w^{[1]})$. Here, it is easy to see that the function $\frac{w}{1-|w|^2}K_w - K_w^{[1]}$ is not the zero map on $\mathbb{D}$. Indeed, if $\frac{w}{1-|w|^2}K_w = K_w^{[1]}$, then $w(1-\overline{w}z)^2 = (1-|w|^2)(1-\overline{w}z)z$ for all $z \in \mathbb{D}$. Then we have $w = 0$ and $-2|w|^2 = 1-|w|^2$ by comparing the constant terms and the coefficients of z, which is a contradiction. Thus $\frac{w}{1-|w|^2}K_w - K_w^{[1]}$ is an eigenvector for $W_{f,\varphi}^*$ corresponding to the eigenvalue $\overline{f(w)\varphi'(w)}$. Since $W_{f,\varphi}$ is cohyponormal, $\frac{w}{1-|w|^2}K_w - K_w^{[1]}$ is an eigenvector for $W_{f,\varphi}$ corresponding to the eigenvalue $f(w)\varphi'(w)$. Hence we get that

$$W_{f,\varphi}\left(\frac{w}{1-|w|^2}K_w - K_w^{[1]}\right) = f(w)\varphi'(w)\left(\frac{w}{1-|w|^2}K_w - K_w^{[1]}\right). \tag{7}$$

Since $K_w^{[1]}(z) = \frac{z}{(1-\overline{w}z)^2}$, it follows from (7) that

$$\begin{aligned} f(z)&\left\{\frac{w}{(1-|w|^2)(1-\overline{w}\varphi(z))} - \frac{\varphi(z)}{(1-\overline{w}\varphi(z))^2}\right\} \\ &= f(w)\varphi'(w)\left\{\left(\frac{w}{1-|w|^2}\right)\left(\frac{1}{1-\overline{w}z}\right) - \frac{z}{(1-\overline{w}z)^2}\right\}. \end{aligned}$$

Since $f(w) \neq 0$, we obtain that

$$\varphi(z) = \frac{(|w|^2-\varphi'(w))z - w(1-\varphi'(w))}{\overline{w}(1-\varphi'(w))z - (1-|w|^2\varphi'(w))} = \frac{-\frac{|w|^2-\varphi'(w)}{1-|w|^2\varphi'(w)}z + \frac{w(1-\varphi'(w))}{1-|w|^2\varphi'(w)}}{1-\frac{\overline{w}(1-\varphi'(w))}{1-|w|^2\varphi'(w)}z}. \tag{8}$$

We note that

$$\frac{w(1-\varphi'(w))}{1-|w|^2\varphi'(w)} = \varphi(0) \in \mathbb{D}.$$

Moreover, since

$$\varphi'(z) = \frac{\varphi'(w)(1-|w|^2)^2}{\{(1-|w|^2\varphi'(w)) - \bar{w}(1-\varphi'(w))z\}^2},$$

we get that

$$\varphi'(0) = \frac{\varphi'(w)(1-|w|^2)^2}{(1-|w|^2\varphi'(w))^2}.$$

Set $a_0 = \varphi(0)$, $d_1 = \varphi'(0)$, and $d_0 = \frac{w(1-\overline{\varphi'(w)})}{1-|w|^2\overline{\varphi'(w)}}$. Then

$$a_0\overline{d_0} - d_1 = \frac{|w|^2(1-\varphi'(w))^2}{(1-|w|^2\varphi'(w))^2} - \frac{\varphi'(w)(1-|w|^2)^2}{(1-|w|^2\varphi'(w))^2} = \frac{|w|^2-\varphi'(w)}{1-|w|^2\varphi'(w)}.$$

Thus $\varphi(z) = \frac{a_0-(a_0\overline{d_0}-d_1)z}{1-\overline{d_0}z} = a_0 + \frac{d_1 z}{1-\overline{d_0}z}$, which completes the proof of (ii) $\Rightarrow$ (iii).

Assume that f and φ have the forms specified in (5). We want to show that $W_{f,\varphi}$ is unitarily equivalent to $C_{\varphi'(w)z}$ that is a normal operator since $|\varphi'(w)| < 1$. From (8), we observe that

$$\varphi(z) = \frac{(|w|^2-\varphi'(w))z + w(\varphi'(w)-1)}{\overline{w}(1-\varphi'(w))z + (|w|^2\varphi'(w)-1)}.$$

Hence, we have the following expression of φ:

$$\varphi = \eta_w \circ (\varphi'(w)\eta_w)$$

where $\eta_w(z) = \frac{w-z}{1-\overline{w}z}$. Furthermore, it is obvious that

$$f = f(w)\frac{K_w}{K_w \circ \varphi}.$$

Since the hyponormality is invariant under scalar multiples, we may assume that $f(w) = 1$. This yields that

$$f(z) = \frac{K_w(z)}{K_w(\varphi(z))} = \frac{1-|w|^2}{1-|w|^2\varphi'(w) - \overline{w}(1-\varphi'(w))z}.$$

From [6, Theorem 2] we have

$$C^*_{\eta_w} = M_{K_w} C_{\eta_w} M^*_h \tag{9}$$

where $h(z) = 1 - \overline{w}z$. In addition, it holds that

$$M^*_h M^*_{\psi_w} = M^*_{h\psi_w} = \frac{1}{\|K_w\|} I \tag{10}$$

where $\psi_w(z) = \frac{K_w}{\|K_w\|}$. It follows from (9) and (10) that

$$W^*_{\psi_w,\eta_w} = C^*_{\eta_w} M^*_{\psi_w} = M_{K_w} C_{\eta_w} M^*_h M^*_{\psi_w} = \frac{1}{\|K_w\|} M_{K_w} C_{\eta_w}.$$

Thus we get that

$$\begin{aligned} W_{\psi_w,\eta_w} C_{\varphi'(w)z} W^*_{\psi_w,\eta_w} &= \frac{1}{\|K_w\|} M_{\psi_w} C_{\eta_w} C_{\varphi'(w)z} M_{K_w} C_{\eta_w} \\ &= \frac{K_w}{\|K_w\|^2} K_w \circ (\varphi'(w)\eta_w) C_{\eta_w \circ (\varphi'(w)\eta_w)} \\ &= \frac{K_w}{\|K_w\|^2} K_w \circ (\varphi'(w)\eta_w) C_\varphi. \end{aligned} \tag{11}$$

Since

$$\frac{K_w}{\|K_w\|^2} K_w \circ (\varphi'(w)\eta_w) = \Big(\frac{1-|w|^2}{1-\overline{w}z}\Big)\Big(\frac{1-\overline{w}z}{1-|w|^2\varphi'(w) - \overline{w}(1-\varphi'(w))z}\Big) = f(z),$$

we obtain from (11) that

$$W_{f,\varphi} = W_{\psi_w,\eta_w} C_{\varphi'(w)z} W^*_{\psi_w,\eta_w}.$$

Since W_{ψ_w,η_w} is unitary from [1, Theorem 6] and $C_{\varphi'(w)z}$ is normal, we conclude that $W_{f,\varphi}$ is also normal. Hence (iii) $\Rightarrow$ (i). ☐

Next we give an example which satisfies Theorem 3.7.

Example 3.8. Let φ be an analytic selfmap of $\mathbb{D}$ with the Denjoy–Wolff point $\frac{1}{2}$. From the proof of Theorem 3.7 we obtain that

$$\varphi(z) = \frac{\left(\frac{1}{4} - \varphi'\left(\frac{1}{2}\right)\right) z + \frac{1}{2}\left(\varphi'\left(\frac{1}{2}\right) - 1\right)}{\frac{1}{2}\left(1 - \varphi'\left(\frac{1}{2}\right)\right) z + \left(\frac{1}{4}\varphi'\left(\frac{1}{2}\right) - 1\right)}.$$

and

$$f(z) = \frac{3/4}{\left(1 - \frac{1}{4}\varphi'\left(\frac{1}{2}\right)\right) - \frac{1}{2}\left(1 - \varphi'\left(\frac{1}{2}\right)\right) z}.$$

Then $W_{f,\varphi}$ is unitarily equivalent to $C_{\varphi'(\frac{1}{2})z}$. Hence $W_{f,\varphi}$ is cohyponormal. In particular, if $\varphi'(\frac{1}{2})$ is not real, then $W_{f,\varphi}$ is not self-adjoint from [8, Theorem 5].

As an application of Theorem 3.7, we show that the symbol φ has an exterior fixed point when $b \neq 0$.

Corollary 3.9. *Let φ be an analytic selfmap of $\mathbb{D}$ with the Denjoy–Wolff point $w \in \mathbb{D} \setminus \{0\}$ and let $f \in H^\infty \setminus \{0\}$. If $W_{f,\varphi}$ is a cohyponormal operator on H^2, then φ has another fixed point $\frac{1}{\overline{w}}$.*

Proof. It follows from Theorem 3.7 that

$$\varphi(z) = a_0 + \frac{d_1 z}{1 - \overline{d_0} z}$$

where $a_0 = \varphi(0) = \frac{w(1-\varphi'(w))}{1-|w|^2\varphi'(w)}$, $d_0 = \frac{w(1-\overline{\varphi'(w)})}{1-|w|^2\overline{\varphi'(w)}}$, and $d_1 = \varphi'(0) = \frac{\varphi'(w)(1-|w|^2)^2}{(1-|w|^2\varphi'(w))^2}$. In order to find the fixed points of φ, we must solve the equation $\varphi(z) = z$, that is, $\overline{d_0}z^2 + (d_1 - a_0\overline{d_0} - 1)z + a_0 = 0$. Since $w \neq 0$ and $|\varphi'(w)| < 1$, we have $d_0 \neq 0$. Hence, if γ is another zero of the quadratic polynomial $\overline{d_0}z^2 + (d_1 - a_0\overline{d_0} - 1)z + a_0$, then $w\gamma = \frac{a_0}{\overline{d_0}} = \frac{w}{\overline{w}}$, i.e., $\gamma = \frac{1}{\overline{w}}$. ☐

Corollary 3.10. *Let φ be an analytic selfmap of $\mathbb{D}$ with the Denjoy–Wolff point $w \in \mathbb{D}$ and let $f \in H^\infty \setminus \{0\}$. If $W_{f,\varphi}$ is a cohyponormal operator on H^2, then it is compact and $\sigma(W_{f,\varphi}) = \{0, f(w)[\varphi'(w)]^n : n = 0, 1, 2, \dots\}$.*

Proof. As in the proof of Theorem 3.7, we can show that $W_{f,\varphi}$ is unitarily equivalent to $C_{\varphi'(w)z}$. Since $w \in \mathbb{D}$, we have $|\varphi'(w)| < 1$. Hence $C_{\varphi'(w)z}$ is compact by [12], and so $W_{f,\varphi}$ is also compact. The result for the spectrum follows from [12]. □

Corollary 3.11. *Let φ be an analytic selfmap of $\mathbb{D}$ with the Denjoy–Wolff point $w \in \mathbb{D}$ and $\varphi'(w) \neq 0$, and let $f \in H^\infty \setminus \{0\}$. If $W_{f,\varphi}$ is a cohyponormal operator on H^2, then*

$$W_{f,\varphi} = \sum_{n=0}^{\infty} f(w)[\varphi'(w)]^n P_n$$

where P_n is the orthogonal projection of H^2 onto $\ker(W_{f,\varphi} - f(w)[\varphi'(w)]^n I)$.

Proof. By Theorem 3.7 and Corollary 3.10, $W_{f,\varphi}$ is compact and normal. Since $f(w) \neq 0$ and $0 < |\varphi'(w)| < 1$, the set $\{f(w)(\varphi'(w))^n : n = 0, 1, 2, \dots\}$ consists of distinct nonzero eigenvalues of $W_{f,\varphi}$ by Corollary 3.10. Thus the proof follows from [3]. □

Corollary 3.12. *Let φ be an analytic selfmap of $\mathbb{D}$ with the Denjoy–Wolff point $w \in \mathbb{D}$ and $\varphi'(w) \neq 0$, and let $f \in H^\infty \setminus \{0\}$. If $W_{f,\varphi}$ is a cohyponormal operator on H^2, the following assertions hold:*

(i) *each P_n has finite rank,*
(ii) $\left(\vee \{P_n(H^2) : n \geq 0\}\right)^\perp = \{0\}$, *and*
(iii) $\|W_{f,\varphi}\| = |f(w)|$

where P_n is the orthogonal projection of H^2 onto $\ker(W_{f,\varphi} - f(w)[\varphi'(w)]^n I)$.

Proof. The proof follows from [3, Corollary II.7.8] and Theorem 3.2 (for (ii), use the fact that $\ker(W_{f,\varphi}) = \{0\}$). □

If $T = U|T|$ is the polar decomposition of $T \in \mathcal{L}(\mathcal{H})$, then $\widetilde{T} := |T|^{\frac{1}{2}} U |T|^{\frac{1}{2}}$ called the *Aluthge transform* of T.

Corollary 3.13. *Let φ be an analytic selfmap of $\mathbb{D}$ with the Denjoy–Wolff point $w \in \mathbb{D}$ and $\varphi'(w) \neq 0$, and let $f \in H^\infty \setminus \{0\}$. If $W_{f,\varphi}$ is a cohyponormal operator on H^2, then $\widetilde{W_{f,\varphi}} = W_{f,\varphi}$.*

Proof. Let $f(w) = |f(w)|e^{i\theta}$ and $\varphi'(w) = |\varphi'(w)|e^{i\tilde{\theta}}$. If $W_{f,\varphi} = U|W_{f,\varphi}|$ is the polar decomposition of the weighted composition operator $W_{f,\varphi}$, then it is trivial that $U = \sum_{n=0}^{\infty} e^{i\theta} e^{in\tilde{\theta}} P_n$ and $|W_{f,\varphi}|^{\frac{1}{2}} = \sum_{n=0}^{\infty} |f(w)|^{\frac{1}{2}} |\varphi'(w)|^{\frac{n}{2}} P_n$ where P_n denotes the orthogonal projection of H^2 onto $\ker(W_{f,\varphi} - f(w)[\varphi'(w)]^n I)$ for each n. Hence we get that

$$\widetilde{W_{f,\varphi}} = |W_{f,\varphi}|^{\frac{1}{2}} U |W_{f,\varphi}|^{\frac{1}{2}} = \sum_{n=0}^{\infty} f(w)(\varphi'(w))^n P_n = W_{f,\varphi}.$$

So we complete our proof. □

Corollary 3.14. *Let φ be an analytic selfmap of $\mathbb{D}$ with the Denjoy–Wolff point $w \in \mathbb{D}$ and $\varphi'(w) \neq 0$, and let $f \in H^\infty \setminus \{0\}$. If $W_{f,\varphi}$ is a cohyponormal operator on H^2, then it is diagonalizable.*

Proof. Since $W_{f,\varphi}$ is compact and normal by Theorem 3.7 and Corollary 3.10, the proof follows from [3, Corollary II.7.9]. □

Corollary 3.15. *Let $\varphi(z) = \frac{az+b}{cz+1}$ be a linear fractional transformation mapping $\mathbb{D}$ into itself such that $\varphi(1) = 1$ and $\varphi(\lambda) = \lambda$ for some $\lambda \in \mathbb{D}$, and let $f \in H^\infty \setminus \{0\}$. Then $W_{f,\varphi}$ is cohyponormal if and only if φ is the identity map and f is constant on $\mathbb{D}$.*

Proof. Since $1 = \varphi(1) = \frac{a+b}{c+1}$ and $\lambda = \varphi(\lambda) = \frac{a\lambda+b}{c\lambda+1}$, we get that $a + b - c = 1$ and $\lambda a + b - \lambda^2 c = \lambda$, and so

$$a = 1 + (\lambda + 1)c \text{ and } b = -\lambda c. \tag{12}$$

Thus it holds that

$$\begin{cases} \varphi(0) = -\lambda c, \\ \varphi'(0) = (c+1)(\lambda c + 1), \text{ and} \\ \varphi'(\lambda) = \frac{(c+1)(\lambda c+1)}{(\lambda c+1)^2} = \frac{c+1}{\lambda c+1}. \end{cases} \tag{13}$$

Here, $\lambda c + 1 \neq 0$ since $\varphi(0) \neq 1$. If $W_{f,\varphi}$ is cohyponormal, then φ has the form in (5) of Theorem 3.7. Assume $\lambda \neq 0$. Then we obtain from (13) that

$$-\lambda c = \varphi(0) = \frac{\lambda(1 - \varphi'(\lambda))}{1 - |\lambda|^2 \varphi'(\lambda)} = \frac{\lambda(\lambda - 1)c}{\lambda(1 - \overline{\lambda})c + (1 - |\lambda|^2)}$$

and thus $\lambda^2(\overline{\lambda} - 1)c(c+1) = 0$. Since $\lambda \neq 0$ and $\lambda \neq 1$, we have $c = 0$ or $c = -1$. However, since there exist at most two fixed points of φ in $\mathbb{C}$ and $\lambda \neq 0$, it follows that $\varphi(0) \neq 0$ and so $c \neq 0$. Thus $c = -1$ and then $\varphi'(1) = \varphi'(\lambda) = 0$, but this contradicts to the Denjoy–Wolff theorem. Hence $\lambda = 0$. This implies that $\varphi(z) = \frac{(c+1)z}{cz+1}$ by (12). We note that $c \neq -1$ since φ is a linear fractional map. On the other hand, with the same notations we used in (5), $a_0 = \varphi(0) = 0$, $d_0 = 0$, and $d_1 = \varphi'(0) = c + 1$ from (13). Then it follows that

$$\frac{(c+1)z}{cz+1} = \varphi(z) = a_0 + \frac{d_1 z}{1 - \overline{d_0} z} = (c+1)z \quad \text{for all} \quad z \in \mathbb{D},$$

which yields that $c = 0$. Thus φ is the identity map on $\mathbb{D}$. Moreover, the weight function f is constant on $\mathbb{D}$ by (5).

Conversely, if $\varphi(z) \equiv z$ and $f(z) \equiv \gamma$ on $\mathbb{D}$ for some constant $\gamma \in \mathbb{C}$, then $W_{f,\varphi} = \gamma I$ is trivially cohyponormal. □

Remark. Although replacing φ with an analytic selfmap of $\mathbb{D}$ in Corollary 3.15, we have the same result. Let $W_{f,\varphi}$ be a cohyponormal operator on H^2, where φ is an analytic selfmap of $\mathbb{D}$ with the Denjoy–Wolff point in the open unit disk and $\varphi(1) = 1$, and let $f \in H^\infty \setminus \{0\}$. By Theorem 3.7, φ should be a linear fractional transformation. Hence $W_{f,\varphi}$ is a scalar multiple of the identity operator on H^2 from Corollary 3.15.

Example 3.16. Set $\varphi(z) = \frac{sz}{1-(1-s)z}$ for some $0 < s < 1$. Then φ is an analytic selfmap of $\mathbb{D}$ with $\varphi(0) = 0$ and $\varphi(1) = 1$. Therefore it follows from Corollary 3.15 that $W_{f,\varphi}$ is not cohyponormal for any $f \in H^\infty \setminus \{0\}$.

Provided that φ is an elliptic automorphism of $\mathbb{D}$, we also get that cohyponormality and normality of a weighted composition operator $W_{f,\varphi}$ are equivalent.

Proposition 3.17. *Let φ be an elliptic automorphism of $\mathbb{D}$ and let $f \in H^\infty$. Then $W_{f,\varphi}$ is cohyponormal on H^2 if and only if it is normal on H^2.*

Proof. Suppose that $W_{f,\varphi}$ is cohyponormal on H^2. From [13, Lemma 3.0.6], $W_{f,\varphi}$ is similar to a weighted composition operator with composition map $\widetilde{\varphi}(z) = \varphi'(w)z$ where w is the fixed point of φ in $\mathbb{D}$ and $|\varphi'(w)| = 1$. In addition, if a hyponormal operator is similar to a normal operator, then it is normal by [15]. Hence it suffices to assume that $\varphi(z) = \gamma z$ for some $|\gamma| = 1$. Since $W^*_{f,\varphi}K_0 = \overline{f(0)}K_0$ and $W_{f,\varphi}$ is cohyponormal, we have $W_{f,\varphi}K_0 = f(0)K_0$. This implies that $f(z) \equiv f(0)$ on $\mathbb{D}$, and so $W_{f,\varphi} = f(0)C_{\gamma z}$ is normal. The converse statement is trivial. □

Now we consider normal and cohyponormal weighted composition operators whose symbols are linear fractional transformations. If $\varphi(z) = \frac{az+b}{cz+d}$ is a linear fractional selfmap of $\mathbb{D}$ and $f(z) = \frac{k}{cz+d}$, then we obtain from [1, Proposition 12] that $W_{f,\varphi}$ is normal if and only if

$$\frac{\overline{f(\alpha)}f(z)}{1-\overline{\varphi(\alpha)}\varphi(z)} = \frac{\overline{kg(\alpha)f(\sigma(\alpha))}}{1-\overline{\varphi(\sigma(\alpha))}z}$$

for any $\alpha, z \in \mathbb{D}$, where $g(z) = \frac{1}{-\bar{b}z+\bar{d}}$, $\sigma(z) = \frac{\bar{a}z-\bar{c}}{-\bar{b}z+\bar{d}}$, and $h(z) = cz + d$. In particular, if $\varphi(z) = a_0 + \frac{d_1 z}{1-\overline{d_0}z}$ is a selfmap of $\mathbb{D}$ and $f(z) = \frac{1}{1-\overline{d_0}z}$ where $d_0 \in \mathbb{D}$, then it is easy to see that $W_{f,\varphi}$ is normal if and only if

$$|a_0| = |d_0| \text{ and } d_0 + a_0\overline{d_1} - |a_0|^2 d_0 = a_0 + d_0 d_1 - a_0|d_0|^2. \tag{14}$$

Hence, if $\varphi = a_0 + \frac{d_1 z}{1-\overline{d_0}z}$ has a fixed point $w \in \mathbb{D}$ and $f(z) = \frac{1}{1-\overline{d_0}z}$ where

$$\begin{cases} a_0 = \varphi(0) = \frac{w(1-\varphi'(w))}{1-|w|^2\varphi'(w)} \\ d_0 = \frac{w(1-\overline{\varphi'(w)})}{1-|w|^2\overline{\varphi'(w)}} \in \mathbb{D} \\ d_1 = \varphi'(0) = \frac{\varphi'(w)(1-|w|^2)^2}{(1-|w|^2\varphi'(w))^2}, \end{cases}$$

then the equalities in (14) are true, which implies that $W_{f,\varphi}$ is normal and so cohyponormal.

Remark. Let $\varphi(z) = a_0 + \frac{d_1 z}{1-\overline{d_0}z}$ be a selfmap of $\mathbb{D}$ with a fixed point $w \in \mathbb{D}$, where $a_0 = \varphi(0) = \frac{w(1-\varphi'(w))}{1-|w|^2\varphi'(w)}$, $d_0 = \frac{w(1-\overline{\varphi'(w)})}{1-|w|^2\overline{\varphi'(w)}} \in \mathbb{D}$, and $d_1 = \varphi'(0) = \frac{\varphi'(w)(1-|w|^2)^2}{(1-|w|^2\varphi'(w))^2}$, and let $f \in H^\infty \setminus \{0\}$.

(i) Suppose that $W_{f,\varphi} = \gamma W_{g,\psi}$ for some $\gamma \in \mathbb{C} \setminus \{0\}$ where $\psi(z) = a_0 + \frac{a_1 z}{1-\overline{a_0}z}$ for some real a_1 and $g(z) = \frac{c}{1-\overline{a_0}z}$ for some real c. Since $\frac{1}{\gamma}W_{f,\varphi} = W_{g,\psi}$ is self-adjoint from [8, Theorem 5], we obtain that

$$\left(\frac{1}{\gamma}W_{f,\varphi}K_\alpha\right)(0) = \left(\frac{1}{\overline{\gamma}}W_{f,\varphi}^*K_\alpha\right)(0) \tag{15}$$

for any $\alpha \in \mathbb{D}$, which ensures that $\varphi'(w)$ is real and $f(z) = \frac{\gamma c}{1-\overline{a_0}z}$.

(ii) Assume that $\varphi'(w)$ is real and $f(z) = \frac{\gamma c}{1-\overline{a_0}z}$ for some $\gamma \in \mathbb{C}$. Since $\varphi'(w)$ is real, $a_0 = d_0$ and d_1 is real. Hence it follows from [8, Theorem 5] that $W_{h,\varphi}$ is self-adjoint where $h(z) = \frac{c}{1-\overline{a_0}z}$. Since $\gamma W_{h,\varphi} = W_{\gamma h,\varphi} = W_{f,\varphi}$, we conclude that $W_{f,\varphi}$ is normal and hence cohyponormal.

Next we examine the commutant of a cohyponormal weighted composition operator. We denote the commutant of an operator $T \in \mathcal{L}(H)$ by $\{T\}'$.

Theorem 3.18. *Let φ be an analytic selfmap of $\mathbb{D}$ with the Denjoy–Wolff point $w \in \mathbb{D}$ and let $f \in H^\infty \setminus \{0\}$. Suppose that $W_{f,\varphi}$ is cohyponormal. If ψ is an analytic selfmap of $\mathbb{D}$ and $g \in H^\infty$ such that $g(w) \neq 0$ and $W_{g,\psi} \in \{W_{f,\varphi}\}'$, then w is a fixed point of ψ.*

Proof. Since $\varphi(w) = w \in \mathbb{D}$, we have

$$\begin{cases} W_{g,\psi}^* W_{f,\varphi}^* K_w = \overline{f(w)} W_{g,\psi}^* K_w = \overline{f(w)g(w)} K_{\psi(w)} \\ W_{f,\varphi}^* W_{g,\psi}^* K_w = \overline{g(w)} W_{f,\varphi}^* K_{\psi(w)} = \overline{g(w)f(\psi(w))} K_{\varphi(\psi(w))}. \end{cases}$$

Since $W_{g,\psi} \in \{W_{f,\varphi}\}'$, we have $\overline{f(w)g(w)}K_{\psi(w)} = \overline{g(w)f(\psi(w))}K_{\varphi(\psi(w))}$. From the hypothesis that $g(w) \neq 0$, we obtain that
$\overline{f(w)}K_{\psi(w)} = \overline{f(\psi(w))}K_{\varphi(\psi(w))}$, i.e.,

$$\frac{\overline{f(w)}}{1-\overline{\psi(w)}z} = \frac{\overline{f(\psi(w))}}{1-\overline{\varphi(\psi(w))}z}$$

for all $z \in \mathbb{D}$. Hence $f(w) = f(\psi(w))$ and $f(w)\varphi(\psi(w)) = f(\psi(w))\psi(w)$. These identities yield that $f(w)\varphi(\psi(w)) = f(w)\psi(w)$. Since $W_{f,\varphi}$ is cohyponormal, f never vanishes on $\mathbb{D}$ from Theorem 3.2. In particular, $f(w) \neq 0$. This implies that $\varphi(\psi(w)) = \psi(w) \in \mathbb{D}$. By the Denjoy–Wolff theorem, $\varphi_n \to w$ and $\varphi_n \to \psi(w)$ uniformly on compact subsets of $\mathbb{D}$. Thus we get that $\psi(w) = w$. □

Corollary 3.19. *Let φ be an analytic selfmap of $\mathbb{D}$ with the Denjoy–Wolff point $w \in \mathbb{D}$ and let $f \in H^\infty$. Suppose that $W_{f,\varphi}$ is cohyponormal. If there exists an analytic selfmap ψ of $\mathbb{D}$, not an elliptic automorphism of $\mathbb{D}$, such that $C_\psi \in \{W_{f,\varphi}\}'$, then the following statements hold:*

(i) *f is constant.*
(ii) *If f is not identically zero on $\mathbb{D}$ and $w \neq 0$, then φ is the identity map on $\mathbb{D}$.*

Proof. (i) We may assume that f is not identically zero on $\mathbb{D}$. It follows from Theorem 3.18 that $\psi(w) = w$. From [2, Theorem 12] we obtain that $(W_{f,\varphi}h)(w) = h(w)W_{f,\varphi}(1)$ for any $h \in H^2$. This implies that $f(w)h(\varphi(w)) = h(w)f(z)$ for any $h \in H^2$ and $z \in \mathbb{D}$. Taking $h \equiv 1$, we have that $f(z) \equiv f(w)$ on $\mathbb{D}$.

(ii) Assume that $w \neq 0$. By (i) and Theorem 3.7, it holds that

$$f(w) = f = f(w)\frac{K_w}{C_\varphi K_w}.$$

Since $f(w) \neq 0$, we get that $C_\varphi K_w = K_w$, which yields that $\varphi(z) \equiv z$ on $\mathbb{D}$. □

Corollary 3.20. *Let φ be an analytic selfmap of $\mathbb{D}$ with the Denjoy–Wolff point $w \in \mathbb{D}$ and let $f \in H^\infty \setminus \{0\}$. Suppose that $W_{f,\varphi}$ is cohyponormal. If ψ is an analytic selfmap of $\mathbb{D}$ and $g \in H^\infty$ with $g(w) \neq 0$ such that $W_{g,\psi} \in \{W_{f,\varphi}\}'$, then the space $\{h \in H^2 : h(w) = 0\}$ is a nontrivial invariant subspace for $W_{g,\psi}$.*

Proof. It follows from Theorem 3.18 that

$$(W_{g,\psi}h)(w) = g(w)h(\psi(w)) = g(w)h(w) \tag{16}$$

for all $h \in H^2$. Hence $\{h \in H^2 : h(w) = 0\}$ is invariant for $W_{g,\psi}$. Moreover, $\{h \in H^2 : h(w) = 0\} = \{K_w\}^\perp$ is clearly nontrivial. □

Corollary 3.21. *Let φ be an analytic selfmap of $\mathbb{D}$ with the Denjoy–Wolff point $w \in \mathbb{D}$, and let $f \in H^\infty \setminus \{0\}$. Suppose that $W_{f,\varphi}$ is cohyponormal. If ψ is an analytic selfmap of $\mathbb{D}$ and $g \in H^\infty$ with $g(w) \neq 0$ such that $W_{g,\psi} \in \{W_{f,\varphi}\}'$, then*

$$\sigma_p(W_{g,\psi}) \subset \sigma_p(W_{g,\psi}|_{\mathcal{M}}) \cup \{g(w)\}$$

and

$$\sigma(W_{g,\psi}) = \sigma(W_{g,\psi}|_{\mathcal{M}}) \cup \{g(w)\}$$

where $\mathcal{M} = \{h \in H^2 : h(w) = 0\}$.

Proof. Corollary 3.20 implies that $\mathcal{M}$ is an invariant subspace for $W_{g,\psi}$. If $\lambda \in \sigma_p(W_{g,\psi})$, then $W_{g,\psi}h = \lambda h$ for some $h \in H^2 \setminus \{0\}$. It follows from Theorem 3.18 that

$$\lambda h(w) = (W_{g,\psi}h)(w) = g(w)h(w).$$

Hence either $h(w) = 0$ or $\lambda = g(w)$, and thus $\sigma_p(W_{g,\psi}) \subset \sigma_p(W_{g,\psi}|_{\mathcal{M}}) \cup \{g(w)\}$. In addition, we can write

$$W_{g,\psi} = \begin{pmatrix} W_{g,\psi}|_{\mathcal{M}} & A \\ 0 & B \end{pmatrix} \text{ on } H^2 = \mathcal{M} \oplus \mathcal{M}^\perp.$$

Since B is a linear operator on the one-dimensional space $\mathcal{M}^\perp = \operatorname{span}\{K_w\}$, choose $\lambda \in \mathbb{C}$ such that $\sigma(B) = \sigma_p(B) = \{\lambda\}$. Obviously, $\sigma(W_{g,\psi}|_{\mathcal{M}}) \cap \sigma(B)$ has no interior point. Thus the identity $\sigma(W_{g,\psi}) = \sigma(W_{g,\psi}|_{\mathcal{M}}) \cup \sigma(B)$ holds by [14, Corollary 8]. Since $\sigma(B^*) = \sigma_p(B^*) = \{\overline{\lambda}\}$, we get that

$$(W_{g,\psi}^* - \overline{\lambda})K_w = \begin{pmatrix} (W_{g,\psi}|_{\mathcal{M}})^* - \overline{\lambda} & 0 \\ A^* & B^* - \overline{\lambda} \end{pmatrix} \begin{pmatrix} 0 \\ K_w \end{pmatrix} = \begin{pmatrix} 0 \\ (B^* - \overline{\lambda})K_w \end{pmatrix} = 0,$$

which ensures that

$$0 = \overline{g(w)}K_{\psi(w)} - \overline{\lambda}K_w = (\overline{g(w)} - \overline{\lambda})K_w$$

by Theorem 3.18. This implies that $\lambda = g(w)$ and so $\sigma(W_{g,\psi}) = \sigma(W_{g,\psi}|_{\mathcal{M}}) \cup \{g(w)\}$. □

References

[1] P. Bourdon and S.K. Narayan, *Normal weighted composition operators on the Hardy space* $H^2(\mathbb{D})$, J. Math. Anal. Appl. **367**(2010), 278–286.

[2] B. Cload, *Commutants of composition operators*, Ph.D. thesis, Univ. Toronto, 1997.

[3] J.B. Conway, *A course in functional analysis*, Springer-Verlag, New York, second edition, 1990.

[4] J.B. Conway, *The theory of subnormal operators*, Mathematical Surveys and Monographs, **36**, Americal Mathematical Society, Providence, Rhode Island, 1991.

[5] C.C. Cowen, *Composition operators on* H^2, J. Operator Theory **9**(1983), 77–106.

[6] C.C. Cowen, *Linear fractional composition operator on* H^2, Int. Eq. Op. Th. **11**(1988), 151–160.

[7] C.C. Cowen, G. Gunatillke, and E. Ko, *Hermitian weighted composition operators and Bergman extremal functions*, Complex Anal. Oper. Theory, to appear.

[8] C.C. Cowen and E. Ko, *Hermitian weighted composition operator on* H^2, Trans. Amer. Math. Soc. **362**(2010), 5771–5801.

[9] C.C. Cowen and T. Kriete, *Subnormality and composition operators on* H^2, J. Funct. Analysis. **81**(1988), 298–319.

[10] C.C. Cowen and B.D. MacCluer, *Composition operators on spaces of analytic functions*, CRC Press, 1995.

[11] M. Martin and M. Putinar, *Lectures on hyponormal operators*, Operator Theory: Advances and Applications, **39**, Birkhäuser Verlag, Basel, 1989.

[12] G. Gunatillake, *Weighted composition operators*, Ph.D. thesis, Purdue Univ., 1992.

[13] G. Gunatillake, *Invertible weighted composition operators*, J. Funct. Anal. **261**(2011) 831–860.

[14] J.K. Han, H.Y. Lee, and W.Y. Lee, *Invertible completions of* 2×2 *upper triangular operator matrices*, Proc. Amer. Math. Soc. **128**(1999), 119–123.

[15] M.O. Otieno, *On quasi-similarity and w-hyponormal operators*, Opuscula Math. **27**(2007), 73–81.

[16] J.H. Shapiro, *Composition operators and classical function theory*, Springer-Verlag, New York, 1993.

[17] T. Worner, *Commutants of certain composition operators*, Acta Sci. Math. (Szeged) **68**(2002), 413–432.

Carl C. Cowen
Department of Mathematical Sciences
IUPUI
Indianapolis, Indiana 46202, USA
e-mail: ccowen@iupui.edu

Sungeun Jung
Institute of Mathematical Sciences
Ewha Womans University
120-750 Seoul, Korea
e-mail: ssung105@ewhain.net

Eungil Ko
Department of Mathematics
Ewha Womans University
120-750 Seoul, Korea
e-mail: eiko@ewha.ac.kr

Operator Theory:
Advances and Applications, Vol. 240, 87–110

A Subnormal Toeplitz Completion Problem

Raúl E. Curto, In Sung Hwang and Woo Young Lee

Abstract. We give a brief survey of subnormality and hyponormality of Toeplitz operators on the vector-valued Hardy space of the unit circle. We also solve the following subnormal Toeplitz completion problem: Complete the unspecified *rational* Toeplitz operators (i.e., the unknown entries are rational Toeplitz operators) of the partial block Toeplitz matrix

$$G := \begin{bmatrix} T_{\overline{\omega_1}} & ? \\ ? & T_{\overline{\omega_2}} \end{bmatrix} \quad (\omega_1 \text{ and } \omega_2 \text{ are finite Blaschke products})$$

to make G subnormal.

Mathematics Subject Classification (2010). Primary 47B20, 47B35.

Keywords. (Block) Toeplitz operators, bounded type functions, matrix-valued rational functions, Halmos' Problem 5, Abrahamse's Theorem, hyponormal, subnormal, completion.

1. Hyponormality and subnormality of Toeplitz operators: A brief survey

1.1. Which operators are subnormal?

Let $\mathcal{H}$ be a complex Hilbert space and let $\mathcal{B}(\mathcal{H})$ be the algebra of bounded linear operators acting on $\mathcal{H}$. An operator $T \in \mathcal{B}(\mathcal{H})$ is said to be *hyponormal* if its self-commutator $[T^*, T] := T^*T - TT^*$ is positive (semi-definite), and *subnormal* if there exists a normal operator N on some Hilbert space $\mathcal{K} \supseteq \mathcal{H}$ such that $\mathcal{H}$ is invariant under N and $N|_{\mathcal{H}} = T$. The notion of subnormality was introduced by P.R. Halmos in 1950 and the study of subnormal operators has been highly successful and fruitful: we refer to [Con] for details. Indeed, the theory of subnormal

The work of the first named author was partially supported by NSF Grant DMS-0801168. The work of the second author was supported by Basic Science Research Program through the National Research Foundation of Korea (NRF) funded by the Ministry of Education, Science and Technology (2011-0022577). The work of the third author was supported by Basic Science Research Program through the National Research Foundation of Korea (NRF) grant funded by the Korea government(MEST) (2012-0000939).

operators has made significant contributions to a number of problems in functional analysis, operator theory, mathematical physics, and other fields. Oddly however, the question "Which operators are subnormal ?" is difficult to answer. In general, it is quite intricate to examine whether a normal extension exists for an operator. Of course, there are a couple of constructive methods for determining subnormality; one of them is the Bram–Halmos criterion of subnormality ([Br]), which states that an operator $T \in \mathcal{B}(\mathcal{H})$ is subnormal if and only if $\sum_{i,j}(T^i x_j, T^j x_i) \geq 0$ for all finite collections $x_0, x_1, \ldots, x_k \in \mathcal{H}$. It is easy to see that this is equivalent to the following positivity test:

$$\begin{bmatrix} I & T^* & \ldots & T^{*k} \\ T & T^*T & \ldots & T^{*k}T \\ \vdots & \vdots & \ddots & \vdots \\ T^k & T^*T^k & \ldots & T^{*k}T^k \end{bmatrix} \geq 0 \qquad (1)$$

for all $k \geq 1$. Condition (1) provides a measure of the gap between hyponormality and subnormality. In fact, the positivity condition (1) for $k = 1$ is equivalent to the hyponormality of T, while subnormality requires the validity of (1) for all k. If we denote by $[A, B] := AB - BA$ the commutator of two operators A and B, and if we define T to be *k-hyponormal* whenever the $k \times k$ operator matrix

$$M_k(T) := ([T^{*j}, T^i])_{i,j=1}^k$$

is positive, or equivalently, the $(k+1) \times (k+1)$ operator matrix in (1) is positive (via the operator version of the Cholesky algorithm), then the Bram–Halmos criterion can be rephrased as saying that T is subnormal if and only if T is k-hyponormal for every $k \geq 1$ ([CMX]). But it still may not be possible to test the positivity condition (1) for *every* positive integer k, in general. Hence the following question is interesting and challenging:

Are there feasible tests for the subnormality of an operator ? (2)

Recall ([At], [CMX], [CoS]) that $T \in \mathcal{B}(\mathcal{H})$ is said to be *weakly k-hyponormal* if

$$LS(T, T^2, \ldots, T^k) := \left\{ \sum_{j=1}^k \alpha_j T^j : \alpha = (\alpha_1, \ldots, \alpha_k) \in \mathbb{C}^k \right\}$$

consists entirely of hyponormal operators. If $k = 2$ then T is called *quadratically hyponormal*, and if $k = 3$ then T is said to be *cubically hyponormal.* Similarly, $T \in \mathcal{B}(\mathcal{H})$ is said to be *polynomially hyponormal* if $p(T)$ is hyponormal for every polynomial $p \in \mathbb{C}[z]$. It is known that k-hyponormal $\Rightarrow$ weakly k-hyponormal, but the converse is not true in general. k-hyponormality and weak k-hyponormality have been considered by many authors with an aim at understanding the gap between hyponormality and subnormality ([Cu1], [Cu2], [CuF1], [CuF2], [CuF3], [CLL], [CL1], [CL2], [CL3], [CMX], [DPY], [McCP]). The study of this gap has been only partially successful. For example, such a gap is not yet well described for Toeplitz operators on the Hardy space of the unit circle. For weighted shifts, positive results appear in [Cu1] and [CuF3], although no concrete example of a weighted shift which is polynomially hyponormal but not subnormal has yet been found (the existence of such weighted shifts was established in [CP1] and [CP2]). The Bram–Halmos criterion on subnormality indicates that 2-hyponormality is generally far from subnormality. There are special classes of operators, however, for which these two notions are equivalent. For example, in [CL1, Theorem 3.2], it was shown that 2-hyponormality and subnormality coincide for Toeplitz operators T_φ with trigonometric polynomial symbols $\varphi \in L^\infty$. On the other hand, 2-hyponormality and subnormality enjoy some common properties. One of them is the following fact ([CL2]):

$$\text{If } T \in \mathcal{B}(\mathcal{H}) \text{ is 2-hyponormal then } \ker[T^*, T] \text{ is invariant for } T. \tag{3}$$

In fact, since the invariance of $\ker[T^*, T]$ for T is one of the most important properties for subnormal operators, we may, in view of (3), expect that 2-hyponormality and subnormality coincide for special classes of operators. Indeed, in Section 2, we shall see this phenomenon for a Toeplitz completion problem.

1.2. (Block) Toeplitz operators and bounded type functions

Toeplitz and Hankel operators arise in a variety of problems in several fields of mathematics and physics, and nowadays the theory of Toeplitz and Hankel operators has become a very wide area. Let $\mathbb{T} \equiv \partial\mathbb{D}$ be the unit circle in the complex plane $\mathbb{C}$. Let $L^2 \equiv L^2(\mathbb{T})$ be the set of all square-integrable measurable functions on $\mathbb{T}$ and let $H^2 \equiv H^2(\mathbb{T})$ be the corresponding Hardy space. Let $H^\infty \equiv H^\infty(\mathbb{T}) := L^\infty \cap H^2$, that is, H^∞ is the set of bounded analytic functions on the unit disk $\mathbb{D}$. Given $\varphi \in L^\infty$, the Toeplitz operator T_φ and the Hankel operator H_φ are defined by

$$T_\varphi g := P(\varphi g) \quad \text{and} \quad H_\varphi g := JP^\perp(\varphi g) \qquad (g \in H^2),$$

where P and $P^\perp$ denote the orthogonal projections that map from L^2 onto H^2 and $(H^2)^\perp$, respectively, and where J denotes the unitary operator on L^2 defined by $J(f)(z) = \overline{z} f(\overline{z})$.

We recall that a function $\varphi \in L^\infty$ is said to be of *bounded type* (or in the Nevanlinna class) if there are analytic functions $\psi_1, \psi_2 \in H^\infty$ such that

$$\varphi(z) = \frac{\psi_1(z)}{\psi_2(z)} \quad \text{for almost all } z \in \mathbb{T}.$$

It is well known [Ab, Lemma 3] that if $\varphi \in L^\infty$ then

$$\varphi \text{ is of bounded type} \iff \ker H_\varphi \neq \{0\}.$$

Assume that both φ and $\overline{\varphi}$ are of bounded type. Since $T_{\overline{z}} H_\psi = H_\psi T_z$ for all $\psi \in L^\infty$, it follows from Beurling's Theorem that $\ker H_{\overline{\varphi_-}} = \theta_0 H^2$ and $\ker H_{\overline{\varphi_+}} = \theta_+ H^2$ for some inner functions θ_0, θ_+. We thus have $b := \overline{\varphi_-}\theta_0 \in H^2$, and hence we can write

$$\varphi_- = \theta_0 \overline{b}, \text{ and similarly } \varphi_+ = \theta_+ \overline{a} \text{ for some } a \in H^2. \tag{4}$$

In the factorization (4), we will always assume that θ_0 and b are coprime and θ_+ and a are coprime. In (4), $\theta_0\overline{b}$ and $\theta_+\overline{a}$ are called *coprime factorizations* of φ_- and φ_+, respectively. By Kronecker's Lemma [Ni, p. 183], if $f \in H^\infty$ then $\overline{f}$ is a rational function if and only if rank $H_{\overline{f}} < \infty$, which implies that

$$\overline{f} \text{ is rational } \iff f = \theta\overline{b} \text{ with a finite Blaschke product } \theta. \tag{5}$$

For a Hilbert space $\mathcal{X}$, let $L^2_{\mathcal{X}} \equiv L^2_{\mathcal{X}}(\mathbb{T})$ be the Hilbert space of $\mathcal{X}$-valued norm square-integrable measurable functions on $\mathbb{T}$ and let $H^2_{\mathcal{X}} \equiv H^2_{\mathcal{X}}(\mathbb{T})$ be the corresponding Hardy space. We observe that $L^2_{\mathbb{C}^n} = L^2 \otimes \mathbb{C}^n$ and $H^2_{\mathbb{C}^n} = H^2 \otimes \mathbb{C}^n$. If Φ is a matrix-valued function in $L^\infty_{M_n} \equiv L^\infty_{M_n}(\mathbb{T})$ $(= L^\infty \otimes M_n)$ then $T_\Phi : H^2_{\mathbb{C}^n} \to H^2_{\mathbb{C}^n}$ denotes the block Toeplitz operator with symbol Φ defined by

$$T_\Phi F := P_n(\Phi F) \quad \text{for } F \in H^2_{\mathbb{C}^n},$$

where P_n is the orthogonal projection of $L^2_{\mathbb{C}^n}$ onto $H^2_{\mathbb{C}^n}$. A block Hankel operator with symbol $\Phi \in L^\infty_{M_n}$ is the operator $H_\Phi : H^2_{\mathbb{C}^n} \to H^2_{\mathbb{C}^n}$ defined by

$$H_\Phi F := J_n P_n^\perp(\Phi F) \quad \text{for } F \in H^2_{\mathbb{C}^n},$$

where $P_n^\perp$ is the orthogonal projection of $L^2_{\mathbb{C}^n}$ onto $(H^2_{\mathbb{C}^n})^\perp$ and J_n denotes the unitary operator on $L^2_{\mathbb{C}^n}$ given by $J_n(F)(z) := \overline{z} I_n F(\overline{z})$ for $F \in L^2_{\mathbb{C}^n}$ (where I_n is the $n \times n$ identity matrix). If we set $H^2_{\mathbb{C}^n} = H^2 \oplus \cdots \oplus H^2$ then we see that

$$T_\Phi = \begin{bmatrix} T_{\varphi_{11}} & \dots & T_{\varphi_{1n}} \\ & \vdots & \\ T_{\varphi_{n1}} & \dots & T_{\varphi_{nn}} \end{bmatrix} \quad \text{and} \quad H_\Phi = \begin{bmatrix} H_{\varphi_{11}} & \dots & H_{\varphi_{1n}} \\ & \vdots & \\ H_{\varphi_{n1}} & \dots & H_{\varphi_{nn}} \end{bmatrix},$$

where

$$\Phi = \begin{bmatrix} \varphi_{11} & \dots & \varphi_{1n} \\ & \vdots & \\ \varphi_{n1} & \dots & \varphi_{nn} \end{bmatrix} \in L^\infty_{M_n}.$$

For $\Phi \in L^\infty_{M_n}$, we write

$$\widetilde{\Phi}(z) := \Phi^*(\overline{z}). \tag{6}$$

For $\Phi \in L^\infty_{M_n}$, we also write

$$\Phi_+ := P_n\Phi \in H^2_{M_n} \quad \text{and} \quad \Phi_- := \left(P_n^\perp \Phi\right)^* \in H^2_{M_n}.$$

Thus we can write $\Phi = \Phi_-^* + \Phi_+$. However, it will often be convenient to allow the constant term in Φ_-. When this is the case, $\Phi_-(0)^*$ will not be zero; however, we will still ensure that $\Phi(0) = \Phi_+(0) + \Phi_-(0)^*$.

A matrix-valued function $\Theta \in H^\infty_{M_{n\times m}}$ $(= H^\infty \otimes M_{n\times m})$ is called *inner* if Θ is isometric a.e. on $\mathbb{T}$. The following basic relations can be easily derived:

$$T^*_\Phi = T_{\Phi^*}, \quad H^*_\Phi = H_{\widetilde{\Phi}} \quad (\Phi \in L^\infty_{M_n});$$

$$T_{\Phi\Psi} - T_\Phi T_\Psi = H^*_{\Phi^*} H_\Psi \quad (\Phi, \Psi \in L^\infty_{M_n}); \tag{7}$$

$$H_\Phi T_\Psi = H_{\Phi\Psi}, \quad H_{\Psi\Phi} = T^*_{\widetilde{\Psi}} H_\Phi \quad (\Phi \in L^\infty_{M_n}, \Psi \in H^\infty_{M_n}). \tag{8}$$

For a matrix-valued function $\Phi = [\phi_{ij}] \in L^\infty_{M_n}$, we say that Φ is of *bounded type* if each entry ϕ_{ij} is of bounded type and that Φ is *rational* if each entry ϕ_{ij} is a rational function.

For a matrix-valued function $\Phi \in H^2_{M_{n\times r}}$, we say that $\Delta \in H^2_{M_{n\times m}}$ is a *left inner divisor* of Φ if Δ is an inner matrix function such that $\Phi = \Delta A$ for some $A \in H^2_{M_{m\times r}}$ $(m \le n)$. We also say that two matrix functions $\Phi \in H^2_{M_{n\times r}}$ and $\Psi \in H^2_{M_{n\times m}}$ are *left coprime* if the only common left inner divisor of both Φ and Ψ is a unitary constant and that $\Phi \in H^2_{M_{n\times r}}$ and $\Psi \in H^2_{M_{m\times r}}$ are *right coprime* if $\widetilde{\Phi}$ and $\widetilde{\Psi}$ are left coprime. Two matrix functions Φ and Ψ in $H^2_{M_n}$ are said to be *coprime* if they are both left and right coprime. We remark that if $\Phi \in H^2_{M_n}$ is such that $\det \Phi$ is not identically zero then any left inner divisor Δ of Φ is square, i.e., $\Delta \in H^2_{M_n}$. If $\Phi \in H^2_{M_n}$ is such that $\det \Phi$ is not identically zero then we say that $\Delta \in H^2_{M_n}$ is a *right inner divisor* of Φ if $\widetilde{\Delta}$ is a left inner divisor of $\widetilde{\Phi}$.

The *shift* operator S on $H^2_{\mathbb{C}^n}$ is defined by

$$S := \sum_{j=1}^{n} \bigoplus T_z.$$

The following fundamental result known as the Beurling–Lax–Halmos Theorem is useful in the sequel.

The Beurling–Lax–Halmos theorem. *A nonzero subspace M of $H^2_{\mathbb{C}^n}$ is invariant for the shift operator S on $H^2_{\mathbb{C}^n}$ if and only if $M = \Theta H^2_{\mathbb{C}^m}$, where Θ is an inner matrix function in $H^\infty_{M_{n\times m}}$ $(m \le n)$. Furthermore, Θ is unique up to a unitary constant right factor; that is, if $M = \Delta H^2_{\mathbb{C}^r}$ (for Δ an inner function in $H^\infty_{M_{n\times r}}$), then $m = r$ and $\Theta = \Delta W$, where W is a unitary matrix mapping $\mathbb{C}^m$ onto $\mathbb{C}^m$.*

As is customarily done, we say that two matrix-valued functions A and B are *equal* if they are equal up to a unitary constant right factor. Observe by (8) that for $\Phi \in L^\infty_{M_n}$, $H_\Phi S = H_\Phi T_{zI_n} = H_{\Phi\cdot zI_n} = H_{zI_n\cdot\Phi} = T^*_{zI_n} H_\Phi$, which implies that the kernel of a block Hankel operator H_Φ is an invariant subspace of the shift operator on $H^2_{\mathbb{C}^n}$. Thus, if $\ker H_\Phi \ne \{0\}$, then by the Beurling–Lax–Halmos Theorem,

$$\ker H_\Phi = \Theta H^2_{\mathbb{C}^m}$$

for some inner matrix function Θ. We note that Θ need not be a square matrix. For example, let θ_i $(i = 0, 1, 2)$ be a scalar inner function such that θ_1 and θ_2 are coprime and let $q \in L^\infty$ be such that $\ker H_q = \{0\}$. Define

$$\Theta := \frac{1}{\sqrt{2}} \begin{bmatrix} \theta_0\theta_1 \\ \theta_0\theta_2 \end{bmatrix} \quad \text{and} \quad \Phi := \begin{bmatrix} \overline{\theta_0\theta_1} & \overline{\theta_0\theta_2} \\ q\theta_2 & -q\theta_1 \end{bmatrix}.$$

Then a straightforward calculation shows that $\ker H_\Phi = \Theta H^2$ (cf. [GHR, Example 2.9]).

The following result was shown in [GHR, Theorem 2.2].

Theorem 1.1 ([GHR]). *For $\Phi \in L^\infty_{M_n}$, the following statements are equivalent:*
(i) *Φ is of bounded type;*
(ii) $\ker H_\Phi = \Theta H^2_{\mathbb{C}^n}$ *for some square inner matrix function Θ;*
(iii) $\Phi = A\Theta^*$, *where $A \in H^\infty_{M_n}$ and A and Θ are right coprime.*

For an inner matrix function $\Theta \in H^2_{M_n}$, we write

$$\mathcal{H}_\Theta := H^2_{\mathbb{C}^n} \ominus \Theta H^2_{\mathbb{C}^n}.$$

In view of Theorem 1.1, if $\Phi \in L^\infty_{M_n}$ is such that Φ and Φ^* are of bounded type then Φ_+ and Φ_- can be written in the form

$$\Phi_+ = \Theta_1 A^* \quad \text{and} \quad \Phi_- = \Theta_2 B^*, \tag{9}$$

where Θ_1 and Θ_2 are inner, $A, B \in H^2_{M_n}$, Θ_1 and A are right coprime, and Θ_2 and B are right coprime. In (9), $\Theta_1 A^*$ and $\Theta_2 B^*$ will be called *right coprime factorizations* of Φ_+ and Φ_-, respectively.

In general, it is not easy to check the condition "B and Θ are right coprime". But if $\Theta \equiv \theta I_n$ for a finite Blaschke product θ, then we have a tractable criterion (cf. [CHL2, Lemma 3.3]):

$$\Theta \text{ and } B \text{ are right coprime} \iff B(\alpha) \text{ is invertible for each zero } \alpha \text{ of } \theta. \tag{10}$$

1.3. Hyponormality of Toeplitz operators

An elegant and useful theorem of C. Cowen [Co4] characterizes the hyponormality of a Toeplitz operator T_φ by properties of the symbol $\varphi \in L^\infty(\mathbb{T})$. This result makes it possible to answer an algebraic question coming from operator theory – namely, is T_φ hyponormal? – by studying the function φ itself. Normal Toeplitz operators were characterized by a property of their symbol in the early 1960's by A. Brown and P.R. Halmos [BH], and so it is somewhat surprising that 25 years passed before the exact nature of the relationship between the symbol $\varphi \in L^\infty$ and the positivity of the self-commutator $[T^*_\varphi, T_\varphi]$ was understood (via Cowen's Theorem). As Cowen notes in his survey paper [Co3], the intensive study of subnormal Toeplitz operators in the 1970s and early 80s is one explanation for the relatively late appearance of the sequel to the Brown-Halmos work. The characterization of hyponormality via Cowen's Theorem requires one to solve a certain functional equation in the unit ball of H^∞.

Cowen's theorem ([Co4], [NT]). *For each $\varphi \in L^\infty$, let*

$$\mathcal{E}(\varphi) \equiv \{k \in H^\infty : \ ||k||_\infty \le 1 \text{ and } \varphi - k\overline{\varphi} \in H^\infty\}.$$

Then T_φ is hyponormal if and only if $\mathcal{E}(\varphi)$ is nonempty.

Cowen's Theorem has been used in [CHL1], [CL1], [CL2], [FL], [Gu1], [Gu2], [GS], [HKL1], [HKL2], [HL1], [HL2], [HL3], [Le], [NT] and [Zhu], which have been devoted to the study of hyponormality for Toeplitz operators on H^2. Particular attention has been paid to Toeplitz operators with polynomial symbols, rational symbols, and bounded type symbols [HL2], [HL3], [CHL1]. However, the case of arbitrary symbol φ, though solved in principle by Cowen's theorem, is in practice very complicated. Indeed, it may not even be possible to find tractable necessary and sufficient condition for the hyponormality of T_φ in terms of the Fourier coefficients of the symbol φ unless certain assumptions are made about φ. To date, tractable criteria for the cases of trigonometric polynomial symbols (resp. rational symbols) were derived from a Carathéodory–Schur interpolation problem ([Zhu]) (resp. a tangential Hermite–Fejér interpolation problem ([Gu1]) or the classical Hermite–Fejér interpolation problem ([HL3])). Very recently, a tractable and explicit criterion on the hyponormality of Toeplitz operators having bounded type symbols was established via the triangularization theorem for compressions of the shift operator ([CHL1]).

When one studies the hyponormality (also, normality and subnormality) of the Toeplitz operator T_φ one may, without loss of generality, assume that $\varphi(0) = 0$; this is because hyponormality is invariant under translation by scalars.

In 2006, Gu, Hendricks and Rutherford [GHR] characterized the hyponormality of block Toeplitz operators in terms of their symbols. Their characterization for hyponormality of block Toeplitz operators T_Φ resembles Cowen's Theorem except for an additional condition which is trivially satisfied in the scalar case – the normality of the symbol, i.e., $\Phi^*\Phi = \Phi\Phi^*$.

Theorem 1.2 (Hyponormality of block Toeplitz operators, Gu–Hendricks–Rutherford [GHR]). *For each $\Phi \in L^\infty_{M_n}$, let*

$$\mathcal{E}(\Phi) := \left\{K \in H^\infty_{M_n} : \ ||K||_\infty \le 1 \ \text{ and } \ \Phi - K\Phi^* \in H^\infty_{M_n}\right\}.$$

Then T_Φ is hyponormal if and only if Φ is normal and $\mathcal{E}(\Phi)$ is nonempty.

In [GHR], the normality of block Toeplitz operator T_Φ was also characterized in terms of the symbol Φ, under a "determinant" assumption on the symbol Φ.

Theorem 1.3 (Normality of block Toeplitz operators, Gu–Hendricks–Rutherford [GHR]). *Let $\Phi \equiv \Phi_+ + \Phi_-^*$ be normal. If $\det\Phi_+$ is not identically zero then*

$$\begin{aligned} T_\Phi \text{ is normal} \Longleftrightarrow \Phi_+ - \Phi_+(0) \\ = \left(\Phi_- - \Phi_-(0)\right) U \ \text{ for some constant unitary matrix } U. \end{aligned} \tag{11}$$

Until now, tractable criteria for the hyponormality of block Toeplitz operators T_Φ with matrix-valued trigonometric polynomials, rational functions or bounded type functions Φ have been established via interpolation problems or the so-called triangularization theorem for compressions of the shift operator ([GHR], [HL4], [HL5], [CHL1]).

1.4. Halmos' Problem 5

In view of the preceding argument, it is natural and significant to elucidate the subnormality of Toeplitz operators. In 1970, P.R. Halmos addressed a problem on subnormality of Toeplitz operators acting on H^2, the so-called Halmos' Problem 5 in his lectures "Ten problems in Hilbert space" [Hal1]:

Halmos' Problem 5. Is every subnormal Toeplitz operator either normal or analytic?

A Toeplitz operator T_φ is called *analytic* if $\varphi \in H^\infty$. Any analytic Toeplitz operator is easily seen to be subnormal: indeed, M_φ is a normal extension of T_φ, where M_φ is the normal operator of multiplication by φ on L^2. Thus the question is natural because the two classes, the normal and analytic Toeplitz operators, are well understood and are subnormal. In the 1970's, interesting partial (affirmative) answers appeared. Thus, when in 1979 Halmos wrote a report on progress on his ten problems (cf. [Hal2]), he stated that "some very good mathematics had gone into that answer" on Problem 5. He then conjectured that the future of Problem 5 was hopeful in the affirmative direction. However, in 1984, Halmos' Problem 5 was answered in the negative by C. Cowen and J. Long [CoL]: they found an analytic function ψ for which $T_{\psi+\alpha\overline{\psi}}$ $(0 < \alpha < 1)$ is subnormal – in fact, this Toeplitz operator is unitarily equivalent to a subnormal weighted shift W_β with weight sequence $\beta \equiv \{\beta_n\}$, where $\beta_n = (1-\alpha^{2n+2})^{\frac{1}{2}}$ for $n = 0, 1, 2, \ldots$. A similar result was independently obtained by S. Sun ([Sun1], [Sun2], [Sun3]). Unfortunately, these constructions do not provide an intrinsic connection between subnormality and the theory of Toeplitz operators.

Until now researchers have been unable to characterize subnormal Toeplitz operators in terms of their symbols. In fact it may not even be possible to find tractable necessary and sufficient condition for the subnormality of T_φ in terms of their symbols unless certain assumptions are made about φ. On the other hand, surprisingly, as C. Cowen notes in [Co2], some analytic Toeplitz operators are unitarily equivalent to non-analytic Toeplitz operators; i.e., the analyticity of Toeplitz operators is not invariant under unitary equivalence. In this sense, we might ask whether Cowen and Long's non-analytic subnormal Toeplitz operator is unitarily equivalent to an analytic Toeplitz operator. It was shown in [CHL2] that Cowen and Long's non-analytic subnormal Toeplitz operator T_φ is not unitarily equivalent to any analytic Toeplitz operator. Consequently, even if we interpret "*is*" in Halmos' Problem 5 as "*is up to unitary equivalence*," the answer to Halmos' Problem 5 is still negative. Thus we would like to reformulate Halmos' Problem 5 as follows:

Halmos' Problem 5 reformulated. Which Toeplitz operators are subnormal?

Directly connected with Halmos' Problem 5 is the following question:

$$\text{Which subnormal Toeplitz operators are normal or analytic?} \tag{12}$$

In 1976, M.B. Abrahamse proved that the answer to Halmos' question is affirmative for Toeplitz operators with bounded type symbols ([Ab]):

Abrahamse's theorem ([Ab, theorem]). *Let $\varphi \in L^\infty$ be such that φ or $\overline{\varphi}$ is of bounded type. If T_φ is hyponormal and* $\ker[T_\varphi^*, T_\varphi]$ *is invariant under T_φ then T_φ is normal or analytic.*

Consequently, if $\varphi \in L^\infty$ is such that φ or $\overline{\varphi}$ is of bounded type, then every subnormal Toeplitz operator must be either normal or analytic. Partial answers to question (12) have been obtained by many authors (cf. [AIW], [Co2], [CoL], [CHL1], [CHL2], [CL1], [CL2], [CL3], [ItW], [NT]). More generally, we are interested in the following question:

$$\text{Which subnormal block Toeplitz operators are normal or analytic?} \tag{13}$$

Question (13) is more difficult to answer, in comparison with the scalar-valued case. Indeed, Abrahamse's Theorem does not hold for block Toeplitz operators (even with matrix-valued trigonometric polynomial symbol): For instance, if

$$\Phi := \begin{bmatrix} z+\overline{z} & 0 \\ 0 & z \end{bmatrix},$$

then

$$T_\Phi = \begin{bmatrix} U_+ + U_+^* & 0 \\ 0 & U_+ \end{bmatrix} \quad (U_+ := \text{the unilateral shift on } H^2)$$

is neither normal nor analytic although T_Φ is evidently subnormal.

Recall that an operator $T \in \mathcal{B}(\mathcal{H})$ is said to be *quasinormal* if T commutes with T^*T and is said to be *pure* if it has no nonzero reducing subspace on which it is normal. It is well known that quasinormal $\Rightarrow$ subnormal. On the other hand, in [ItW], it was shown that every quasinormal Toeplitz operator is either normal or analytic, i.e., the answer to the Halmos' Problem 5 is affirmative for quasinormal Toeplitz operators. However, this is not true for the cases of matrix-valued symbols: indeed, if

$$\Phi \equiv \begin{bmatrix} \overline{z} & \overline{z}+2z \\ \overline{z}+2z & \overline{z} \end{bmatrix}. \tag{14}$$

then T_Φ is quasinormal, but it is neither normal nor analytic. Since

$$T_\Phi = \begin{bmatrix} U_+^* & U_+^*+2U_+ \\ U_+^*+2U_+ & U_+^* \end{bmatrix},$$

it follows that if $W := \frac{1}{\sqrt{2}}\left[\begin{smallmatrix} 1 & -1 \\ 1 & 1 \end{smallmatrix}\right]$, then W is unitary and

$$W^*T_\Phi W = 2\begin{bmatrix} U_+^* + U_+ & 0 \\ 0 & -U_+ \end{bmatrix},$$

which says that T_Φ is unitarily equivalent to a direct sum of the normal operator $2(U_+^* + U_+)$ and the analytic Toeplitz operator $-2U_+$. This phenomenon is not an accident. Indeed, very recently, in [CHKL], it was shown that every pure quasinormal operator with finite rank self-commutator is unitarily equivalent to a Toeplitz operator with a matrix-valued analytic rational symbol and (as a corollary) that every pure quasinormal Toeplitz operator with a matrix-valued rational symbol is unitarily equivalent to an analytic Toeplitz operator.

Also, in [CHKL], the following theorem was obtained:

Theorem 1.4 (Abrahamse's theorem for matrix-valued rational symbols, [CHKL]). *Let $\Phi \equiv \Phi_-^* + \Phi_+ \in L^\infty_{M_n}$ be a matrix-valued rational function. Thus in view of* (9), *we may write*

$$\Phi_- = \Theta B^* \quad (\textit{right coprime factorization}).$$

Assume that Θ has an inner divisor of the form θI_n, where θ is a nonconstant inner function. If

(i) T_Φ *is hyponormal;*
(ii) $\ker[T_\Phi^*, T_\Phi]$ *is invariant for* T_Φ,

then T_Φ is normal. Hence in particular, if T_Φ is subnormal then T_Φ is normal.

Theorem 1.4 may fail if we drop the assumption "Θ has a nonconstant diagonal-constant inner divisor." To see this, consider the matrix-valued function in (14):

$$\Phi \equiv \begin{bmatrix} \overline{z} & \overline{z}+2z \\ \overline{z}+2z & \overline{z} \end{bmatrix}.$$

We thus have

$$\Phi_- = \begin{bmatrix} z & z \\ z & z \end{bmatrix} = \left(\frac{1}{\sqrt{2}}\begin{bmatrix} 1 & z \\ -1 & z \end{bmatrix}\right)\left(\frac{1}{\sqrt{2}}\begin{bmatrix} 0 & 2 \\ 0 & 2 \end{bmatrix}\right)^*,$$

where

$$\Theta \equiv \frac{1}{\sqrt{2}}\begin{bmatrix} 1 & z \\ -1 & z \end{bmatrix} \text{ and } B \equiv \frac{1}{\sqrt{2}}\begin{bmatrix} 0 & 2 \\ 0 & 2 \end{bmatrix} \text{ are right coprime.}$$

As we saw in the preceding, T_Φ is quasinormal, and hence subnormal. But clearly, T_Φ is neither normal nor analytic. Here we note that Θ does not have any nonconstant diagonal inner divisor of the form θI_n with a nonconstant inner function θ.

1.5. A special subnormal Toeplitz completion

Given a partially specified operator matrix with some known entries, the problem of finding suitable operators to complete the given partial operator matrix so that the resulting matrix satisfies certain given properties is called a *completion problem.* Dilation problems are special cases of completion problems: in other words, the dilation of T is a completion of the partial operator matrix $\left[\begin{smallmatrix} T & ? \\ ? & ? \end{smallmatrix}\right]$. A *partial block Toeplitz matrix* is simply an $n \times n$ matrix some of whose entries are specified Toeplitz operators and whose remaining entries are unspecified. A *subnormal completion* of a partial operator matrix is a particular specification of the unspecified entries resulting in a subnormal operator. In particular, to avoid the triviality, we are interested in the cases whose diagonal entries are specified. For example, if ω is a finite Blaschke product, then T_ω is evidently subnormal, so that $\left[\begin{smallmatrix} T_\omega & 0 \\ 0 & T_\omega \end{smallmatrix}\right]$ is itself subnormal. On the other hand,

$$\begin{bmatrix} T_\omega & 1 - T_\omega T_{\overline{\omega}} \\ 0 & T_{\overline{\omega}} \end{bmatrix} \quad (\omega \text{ is a finite Blaschke product}) \tag{15}$$

is a subnormal (even unitary) completion of the 2×2 partial operator matrix

$$\begin{bmatrix} T_\omega & ? \\ ? & T_{\overline{\omega}} \end{bmatrix}.$$

A *subnormal Toeplitz completion* of a partial block Toeplitz matrix is a subnormal completion whose unspecified entries are Toeplitz operators. Then the following question comes up at once: Does there exist a subnormal Toeplitz completion of $\left[\begin{smallmatrix} T_\omega & ? \\ ? & T_{\overline{\omega}} \end{smallmatrix}\right]$? Evidently, (15) is not such a completion. To answer this question, let

$$\Phi \equiv \begin{bmatrix} \omega & \varphi \\ \psi & \overline{\omega} \end{bmatrix} \qquad (\varphi, \psi \in L^\infty).$$

If T_Φ is hyponormal then by Theorem 1.2, Φ should be normal. Thus a straightforward calculation shows that

$$|\varphi| = |\psi| \quad \text{and} \quad \overline{\omega}(\varphi + \overline{\psi}) = \omega(\varphi + \overline{\psi}),$$

which implies that $\varphi = -\overline{\psi}$. Thus a direct calculation shows that

$$[T_\Phi^*, T_\Phi] = \begin{bmatrix} * & * \\ * & T_\omega T_{\overline{\omega}} - 1 \end{bmatrix},$$

which is not positive semi-definite because $T_\omega T_{\overline{\omega}} - 1$ is not. Therefore, there are no hyponormal Toeplitz completions of $\begin{bmatrix} T_\omega & ? \\ ? & T_{\overline{\omega}} \end{bmatrix}$. The following question seems to be more difficult: Does there exist a subnormal Toeplitz completion of

$$\begin{bmatrix} T_{\overline{\omega}} & ? \\ ? & T_{\overline{\omega}} \end{bmatrix} \qquad (\omega \text{ is a finite Blaschke product}) \ ?$$

Special cases of this question were successfully considered in [CHL1] and [CHL3]. In the next section, we consider a subnormal Toeplitz completion problem.

2. Subnormal Toeplitz completions

In this section we consider the following:

Problem A. Complete the unspecified *rational* Toeplitz operators of the partial block Toeplitz matrix

$$G := \begin{bmatrix} T_{\overline{\omega_1}} & ? \\ ? & T_{\overline{\omega_2}} \end{bmatrix} \qquad (\omega_1 \text{ and } \omega_2 \text{ are finite Blaschke products}) \tag{16}$$

to make G subnormal.

To answer Problem A, we need several auxiliary lemmas. We write

$$b_\alpha(z) := \frac{z - \alpha}{1 - \overline{\alpha} z} \quad (\alpha \in \mathbb{D}).$$

We begin with:

Lemma 2.1. *Suppose* $\varphi, \psi \in L^\infty$. *Then*

$$[T^*_{\varphi \circ b_\alpha}, T_{\psi \circ b_\alpha}] \cong [T^*_\varphi, T_\psi] \quad (\cong \textit{ denotes unitary equivalence}).$$

In particular, $T_{\varphi \circ b_\alpha}$ *is hyponormal if and only if* T_φ *is hyponormal.*

Proof. By a well-known fact due to C. Cowen [Co1, Theorem 1], there exists a unitary operator V such that

$$T_{\varphi \circ b_\alpha} = V^* T_\varphi V \quad \text{and} \quad T_{\psi \circ b_\alpha} = V^* T_\psi V.$$

We thus have $[T^*_{\varphi \circ b_\alpha}, T_{\psi \circ b_\alpha}] = V^*[T^*_\varphi, T_\psi]V$, which gives the result. □

Lemma 2.2. *Let* $\varphi, \psi \in L^\infty$ *be rational functions and let* ω_1 *and* ω_2 *be finite Blaschke products. If*

$$\Phi := \begin{bmatrix} \overline{\omega_1} & \varphi \\ \psi & \overline{\omega_2} \end{bmatrix}$$

is such that T_Φ *is hyponormal then* $\omega_1 = \omega_2$.

Proof. We first observe $(b_\alpha \circ b_{-\alpha})(z) = z$. Thus, in view of Lemma 2.1 we may assume that $\omega_1(0) = 0$. Then this lemma follows from a slight variation of the proof of [CHL3, Theorem 4.2], in which $\omega_1 = b_\alpha$ and $\omega_2 = b_\beta$. □

In view of Lemma 2.2, for the problem (16), it suffices to consider the case

$$\Phi := \begin{bmatrix} \overline{\omega} & \varphi \\ \psi & \overline{\omega} \end{bmatrix} \quad (\varphi, \psi \in L^\infty \text{ are rational; } \omega \text{ is a finite Blaschke product}))$$

Lemma 2.3. *Suppose* $\Phi := \Phi_-^* + \Phi_+ \in L^\infty_{M_n}$ *is a matrix-valued rational function. Then we may write* (*cf.* [CHL3, Lemma 3.1])

$$\Phi_+ = A^* \Delta_0 \Delta \quad \textit{and} \quad \Phi_- = B^* \Delta,$$

where $\Delta_0 \Delta \equiv \theta I_n$ *with an inner function* θ, B *and* Δ *are left coprime and* $A, B \in H^2_{M_n}$. *If* $\ker[T^*_\Phi, T_\Phi]$ *is invariant under* T_Φ *and* $K \in \mathcal{E}(\Phi)$, *then*

$$\operatorname{cl}\operatorname{ran} H_{A\Delta^*} \subseteq \ker(I - T_{\widetilde{K}} T^*_{\widetilde{K}}).$$

(*For the definition of* $\widetilde{K}$, *see* (6).)

Proof. This follows from formula (16) in [CHL2], together with a careful analysis that the proof of (16) in [CHL2] does not employ the diagonal-*constant*-ness of Δ. □

Lemma 2.4. *Let* $\Phi \equiv \Phi_-^* + \Phi_+ \in L^\infty_{M_n}$ *be a matrix-valued rational function such that*

$$\Phi_- := \begin{bmatrix} \omega & \psi_- \\ \varphi_- & \omega \end{bmatrix},$$

where ω *is a finite Blaschke product of the form*

$$\omega = \prod_{i=1}^{p} b_i^{q_i} \quad \left(b_i(z) := \frac{z - \alpha_i}{1 - \overline{\alpha_i} z} \ (\alpha_i \neq \alpha_j \textit{ if } i \neq j) \textit{ and } q_i \geq 1 \right).$$

If

$$\Phi_- = \Theta B^* \quad \textit{(right coprime factorization)},$$

then Θ *has an inner divisor of the form* $b_i I_2$ *for some* $i = 1, 2, \ldots, p$, *except in the following two cases:*

(i) $m_i + n_i = 2q_i$ *for all* $i = 1, 2, \ldots, p$;
(ii) $m_{i_0} + n_{i_0} > 2q_{i_0}$ *and* $m_{i_0} n_{i_0} = 0$ *for some* i_0,

in the representation

$$\varphi_- \equiv \theta_0 \overline{a} = \Big(\prod_{i=1}^p b_i^{m_i}\Big)\theta_0' \overline{a} \quad \textit{and} \quad \psi_- \equiv \theta_1 \overline{b} = \Big(\prod_{i=1}^p b_i^{n_i}\Big)\theta_1' \overline{b} \quad \left(\begin{array}{c}\textit{coprime}\\ \textit{factorizations}\end{array}\right)$$

($m_i, n_i = 0, 1, \ldots$ *and* $(\theta_0'\theta_1')(\alpha_i) \neq 0$ *for all* $i = 1, 2, \ldots, p$).

Proof. By Theorem 1.1, $\ker H_{\Phi_-^*} = \Theta H^2_{\mathbb{C}^2}$. We observe that for $f, g \in H^2$,

$$\Phi_-^* \begin{bmatrix} f \\ g \end{bmatrix} \in H^2_{\mathbb{C}^2} \iff \begin{bmatrix} \overline{\omega} & \overline{\theta_0 a} \\ \overline{\theta_1 b} & \overline{\omega} \end{bmatrix} \begin{bmatrix} f \\ g \end{bmatrix} \in H^2_{\mathbb{C}^2},$$

which implies that if $\begin{bmatrix} f \\ g \end{bmatrix} \in \ker H_{\Phi_-^*}$, then

$$\Big(\prod_{i=1}^p \overline{b_i}^{q_i}\Big) f + \Big(\prod_{i=1}^p \overline{b_i}^{m_i}\Big)\overline{\theta_0'} a g \in H^2 \quad \text{and} \quad \Big(\prod_{i=1}^p \overline{b_i}^{n_i}\Big)\overline{\theta_1'} b f + \Big(\prod_{i=1}^p \overline{b_i}^{q_i}\Big) g \in H^2. \tag{17}$$

We split the proof into two cases.

Case 1 $(0 \leq m_{i_0} + n_{i_0} < 2q_{i_0}$ **for some** $i_0 = 1, 2, \ldots, d)$. In this case, $n_{i_0} < q_{i_0}$ or $m_{i_0} < q_{i_0}$. Suppose that $m_{i_0} < q_{i_0}$. Then by the first statement of (17) we have

$$\Big(\prod_{i \neq i_0} \overline{b_i}^{q_i - m_i}\Big)\overline{b_{i_0}}^{q_{i_0} - m_{i_0}} \theta_0' f \in H^2,$$

which implies that $f = b_{i_0}^{q_{i_0} - m_{i_0}} f_1$ for some $f_1 \in H^2$. In turn, by the second statement of (17) we have

$$\Big(\prod_{i \neq i_0} \overline{b_i}^{n_i}\Big)\overline{b_{i_0}}^{m_{i_0} + n_{i_0} - q_{i_0}} \overline{\theta_1'} b f_1 + \Big(\prod_{i=1}^p \overline{b_i}^{q_i}\Big) g \in H^2.$$

Thus if $m_{i_0} + n_{i_0} - q_{i_0} \leq 0$, then $g = b_{i_0}^{q_{i_0}} g_1$ for some $g_1 \in H^2$ and if instead $m_{i_0} + n_{i_0} - q_{i_0} > 0$, then

$$\overline{b_{i_0}}^{2q_{i_0} - m_{i_0} - n_{i_0}} \Big(\prod_{i \neq i_0} \overline{b_i}^{q_i - n_i}\Big)\theta_1' g \in H^2,$$

which implies that $g = b_{i_0}^{2q_{i_0} - m_{i_0} - n_{i_0}} g_2$ for some $g_2 \in H^2$. Therefore $b_{i_0} I_2$ is an inner divisor of Θ.

If instead $n_{i_0} < q_{i_0}$ then the same argument as the above gives that $b_{i_0} I_2$ is an inner divisor of Θ.

Case 2 $(m_{i_0}+n_{i_0}>2q_{i_0}$ **and** $m_{i_0}n_{i_0}\neq 0$ **for some** $i_0)$.

(a) Suppose $m_{i_0}\geq q_{i_0}+1$. If $\begin{bmatrix} f\\ g\end{bmatrix}\in\ker H_{\Phi_-^*}$, then by the first statement of (17) we have

$$\Big(\prod_{i\neq i_0}\overline{b_i}^{m_i-q_i}\Big)\overline{b_{i_0}}^{m_{i_0}-q_{i_0}}\overline{\theta_0'}ag\in H^2,$$

which implies that $g=b_{i_0}^{m_{i_0}-q_{i_0}}g_1$ for some $g_1\in H^2$. In turn, by the second statement of (17) we have

$$\Big(\prod_{i\neq i_0}\overline{b_i}^{n_i}\Big)\overline{b_{i_0}}^{n_{i_0}}\overline{\theta_1'}bf+\Big(\prod_{i\neq i_0}\overline{b_i}^{q_i}\Big)\overline{b_{i_0}}^{2q_{i_0}-m_{i_0}}g_1\in H^2.$$

Thus if $2q_{i_0}\leq m_{i_0}$, then $f=b_{i_0}^{n_{i_0}}f_1$ for some $f_1\in H^2$ and if instead $2q_{i_0}>m_{i_0}$, then

$$\Big(\prod_{i\neq i_0}\overline{b_i}^{n_i-q_i}\Big)\overline{b_{i_0}}^{m_{i_0}+n_{i_0}-2q_{i_0}}\overline{\theta_1'}bf\in H^2,$$

which implies that $f=b_{i_0}^{m_{i_0}+n_{i_0}-2q_{i_0}}f_2$ for some $f_2\in H^2$. Therefore $b_{i_0}I_2$ is an inner divisor of Θ.

(b) Suppose $m_{i_0}<q_{i_0}+1$. Then $n_{i_0}\geq q_{i_0}+1$ and the same argument as the Case 2(a) gives that $b_{i_0}I_2$ is an inner divisor of Θ.

From Case 1 and Case 2, we can conclude that Θ has an inner divisor of the form b_iI_2 for some $i=1,2,\dots,p$ except the cases $m_i+n_i=2q_i$ for all $i=1,2,\dots,p$ and $m_{i_0}+n_{i_0}>2q_{i_0}$ with $m_{i_0}n_{i_0}=0$ for some i_0. This completes the proof. □

Lemma 2.5. *Let $\Phi\equiv\Phi_-^*+\Phi_+\in L^\infty_{M_n}$ be a matrix-valued rational function such that*

$$\Phi_-:=\begin{bmatrix}\omega & \varphi_-\\ \psi_- & \omega\end{bmatrix},$$

where ω is a finite Blaschke product of the form

$$\omega=\prod_{i=1}^p b_i^{q_i}\quad\Big(b_i(z):=\frac{z-\alpha_i}{1-\overline{\alpha_i}z},\ q_i\geq 1\Big),$$

$$\varphi_-\equiv\theta_0\overline{a}=\Big(\prod_{i=1}^p b_i^{m_i}\Big)\theta_0'\overline{a}\quad\textit{and}\quad\psi_-\equiv\theta_1\overline{b}=\Big(\prod_{i=1}^p b_i^{n_i}\Big)\theta_1'\overline{b}\quad\Big(\begin{matrix}\textit{coprime}\\ \textit{factorizations}\end{matrix}\Big)$$

$(m_i,n_i=0,1,\dots$ and $(\theta_0'\theta_1')(\alpha_i)\neq 0$ for all $i=1,2,\dots,p)$. If $\alpha_{i_0}=0$, $m_{i_0}>2q_{i_0}$ and $n_{i_0}=0$ for some i_0, then

$$\ker H_{\Phi_-^*}\subseteq\frac{1}{\sqrt{|\alpha|^2+1}}\begin{bmatrix} z^{m_{i_0}-q_{i_0}}\theta_0' & -\overline{\alpha}z^{m_{i_0}-q_{i_0}+1}\theta_0'\\ \alpha\theta_1' & z\theta_1'\end{bmatrix}H^2_{\mathbb{C}^2}\quad\Big(\alpha:=-\frac{a'(0)}{\theta_1''(0)}\Big),$$

where

$$a':=\Big(\prod_{i\neq i_0}b_i^{M_i-m_i}\Big)a\quad\textit{and}\quad\theta_1'':=\Big(\prod_{i\neq i_0}b_i^{M_i-q_i}\Big)\theta_1'$$

$(M_i:=\max(m_i,q_i)$ for $i\neq i_0)$.

Proof. Observe that for $f, g \in H^2$,

$$\Phi_-^* \begin{bmatrix} f \\ g \end{bmatrix} \in H^2_{\mathbb{C}^2} \iff \begin{bmatrix} \overline{\omega} & \overline{\theta_1 b} \\ \overline{\theta_0 a} & \overline{\omega} \end{bmatrix} \begin{bmatrix} f \\ g \end{bmatrix} \in H^2_{\mathbb{C}^2},$$

which implies that if $\begin{bmatrix} f \\ g \end{bmatrix} \in \ker H_{\Phi_-^*}$, then

$$\Big(\prod_{i=1}^p \overline{b_i}^{q_i}\Big) f + \Big(\prod_{i=1}^p \overline{b_i}^{n_i}\Big) \overline{\theta_1'} b g \in H^2 \quad \text{and} \quad \Big(\prod_{i=1}^p \overline{b_i}^{m_i}\Big) \overline{\theta_0'} a f + \Big(\prod_{i=1}^p \overline{b_i}^{q_i}\Big) g \in H^2. \tag{18}$$

It follows from the first statement of (18) that $g = \theta_1' g_1$ for some $g_1 \in H^2$. In turn

$$\Big(\prod_{i=1}^p \overline{b_i}^{q_i}\Big) f + \Big(\prod_{i=1}^p \overline{b_i}^{n_i}\Big) b g_1 \in H^2.$$

Since $n_{i_0} = 0$, we have $f = z^{q_{i_0}} f_1$ for some $f_1 \in H^2$. Thus, by the second statement of (18) we have

$$\Big(\prod_{i \neq i_0} \overline{b_i}^{m_i}\Big) \overline{z}^{m_{i_0} - q_{i_0}} \overline{\theta_0'} a f_1 + \Big(\prod_{i=1}^p \overline{b_i}^{q_i}\Big) g \in H^2, \tag{19}$$

so that

$$\Big(\prod_{i \neq i_0} \overline{b_i}^{m_i - q_i}\Big) \overline{z}^{m_{i_0} - 2q_{i_0}} \overline{\theta_0'} a f_1 \in H^2.$$

Since $m_{i_0} > 2q_{i_0}$, it follows that $f_1 = \theta_0' z^{m_{i_0} - 2q_{i_0}} f_2$ for some $f_2 \in H^2$. Thus, by (19) we have

$$\Big(\prod_{i \neq i_0} \overline{b_i}^{m_i}\Big) \overline{z}^{q_{i_0}} a f_2 + \Big(\prod_{i=1}^p \overline{b_i}^{q_i}\Big) \theta_1' g_1 \in H^2. \tag{20}$$

Then it follows from (20) that

$$\Big(\prod_{i \neq i_0} b_i^{M_i - m_i}\Big) a f_2 + \Big(\prod_{i \neq i_0} b_i^{M_i - q_i}\Big) \theta_1' g_1 \in z^{q_{i_0}} H^2. \tag{21}$$

Write

$$a' := \Big(\prod_{i \neq i_0} b_i^{M_i - m_i}\Big) a \quad \text{and} \quad \theta_1'' := \Big(\prod_{i \neq i_0} b_i^{M_i - q_i}\Big) \theta_1'.$$

Then we have $a'(0) \neq 0$ and $\theta_1''(0) \neq 0$, and by (21) we have

$$g_1(0) = \alpha f_2(0) \quad \Big(\alpha := -\frac{a'(0)}{\theta_1''(0)}\Big).$$

Therefore, we have

$$\begin{bmatrix} f \\ g \end{bmatrix} \in \ker H_{\Phi_-^*} \Longrightarrow f = z^{m_{i_0} - q_{i_0}} \theta_0' f_2, \; g = \theta_1' g_1, \text{ and } g_1(0) = \alpha f_2(0). \tag{22}$$

Put

$$\Omega := \frac{1}{\sqrt{|\alpha|^2+1}} \begin{bmatrix} z^{m_{i_0}-q_{i_0}}\theta_0' & -\overline{\alpha} z^{m_{i_0}-q_{i_0}+1}\theta_0' \\ \alpha\theta_1' & z\theta_1' \end{bmatrix}.$$

Then Ω is inner, and for $h_1, h_2 \in H^2$,

$$\begin{aligned} \Omega \begin{bmatrix} h_1 \\ h_2 \end{bmatrix} &= \frac{1}{\sqrt{|\alpha|^2+1}} \begin{bmatrix} z^{m_{i_0}-q_{i_0}}\theta_0' h_1 - \overline{\alpha} z^{m_{i_0}-q_{i_0}+1}\theta_0' h_2 \\ \alpha\theta_1' h_1 + z\theta_1' h_2 \end{bmatrix} \\ &= \frac{1}{\sqrt{|\alpha|^2+1}} \begin{bmatrix} z^{m_{i_0}-q_{i_0}}\theta_0'(h_1 - \overline{\alpha} z h_2) \\ \theta_1'(\alpha h_1 + z h_2) \end{bmatrix}. \end{aligned}$$

Since $(\alpha h_1 + z h_2)(0) = \alpha h_1(0) = \alpha(h_1 - \overline{\alpha} z h_2)(0)$, it follows from (22) that

$$\ker H_{\Phi_-^*} \subseteq \Omega H^2_{\mathbb{C}^2},$$

which gives the result. □

To answer Problem A, we recall ([CHL2, Lemma 3.2]) that if $\Phi \equiv \Phi_-^* + \Phi_+ \in L^\infty_{M_n}$ is such that Φ and Φ^* are of bounded type, we may write, as in (9),

$$\Phi_+ = \Theta_1 A^* \quad \text{and} \quad \Phi_- = \Theta B^* \quad \text{(right coprime factorizations).}$$

If T_Φ is hyponormal, then

$$\Theta_1 = \Theta\Theta_0 \quad \text{for some inner matrix function } \Theta_0; \tag{23}$$

in other words, Θ is a left inner divisor of Θ_1.

We are ready for:

Theorem 2.6. *Let $\varphi, \psi \in L^\infty$ be rational functions and consider*

$$G := \begin{bmatrix} T_{\overline{\omega_1}} & T_\varphi \\ T_\psi & T_{\overline{\omega_2}} \end{bmatrix} \quad (\omega_i \text{ is a finite Blaschke product for } i = 1, 2). \tag{24}$$

Then the following statements are equivalent:

1. *G is normal;* 2. *G is subnormal;* 3. *G is 2-hyponormal;*
4. *G is hyponormal and $\ker[G^*, G]$ is invariant for G;*
5. *$\omega_1 = \omega_2 =: \omega$ and the following condition holds:*

$$\varphi = e^{i\delta_1}\omega + \zeta \quad \text{and} \quad \psi = e^{i\delta_2}\varphi \quad (\zeta \in \mathbb{C};\ \delta_1, \delta_2 \in [0, 2\pi)), \tag{25}$$

except in the following case:

$$m_i + n_i = 2q_i \quad \text{for some } i = 1, 2, \ldots, p, \tag{26}$$

in the representation

$$\omega := \prod_{i=1}^p b_i^{q_i} \quad \left(b_i(z) := \frac{z - \alpha_i}{1 - \overline{\alpha_i} z},\ q_i \geq 1 \right),$$

$$\varphi_- \equiv \theta_0 \overline{a} = \left(\prod_{i=1}^p b_i^{m_i} \right) \theta_0' \overline{a} \quad \text{and} \quad \psi_- \equiv \theta_1 \overline{b} = \left(\prod_{i=1}^p b_i^{n_i} \right) \theta_1' \overline{b} \quad \left(\begin{matrix} \text{coprime} \\ \text{factorizations} \end{matrix} \right)$$

($m_i, n_i = 0, 1, \ldots$ and $(\theta_0'\theta_1')(\alpha_i) \neq 0$ for all $i = 1, 2, \ldots, p$).

Proof. Clearly, (1) $\Rightarrow$ (2) and (2) $\Rightarrow$ (3). Also (3) $\Rightarrow$ (4) is evident because $\ker[T^*, T]$ is invariant under T for every 2-hyponormal operator $T \in \mathcal{B}(\mathcal{H})$ (cf. [CL2]). Moreover, (5) $\Rightarrow$ (1) follows from a straightforward calculation.

(4) $\Rightarrow$ (5): By Lemma 2.2, $\omega_1 = \omega_2 =: \omega$. Thus we may write

$$\Phi \equiv \begin{bmatrix} \overline{\omega} & \varphi \\ \psi & \overline{\omega} \end{bmatrix} \equiv \Phi_-^* + \Phi_+ = \begin{bmatrix} \omega & \psi_- \\ \varphi_- & \omega \end{bmatrix}^* + \begin{bmatrix} 0 & \varphi_+ \\ \psi_+ & 0 \end{bmatrix}$$

and assume that T_Φ is hyponormal and $\ker[T_\Phi, T_\Phi^*]$ is invariant for T_Φ. Since, by Theorem 1.2, Φ is normal, we have

$$|\varphi| = |\psi|, \tag{27}$$

and also there exists a function $K \equiv \left[\begin{smallmatrix} k_1 & k_2 \\ k_3 & k_4 \end{smallmatrix}\right] \in H^\infty_{M_2}$ such that $\Phi_-^* - K\Phi_+^* \in H^2_{M_2}$, i.e.,

$$\begin{bmatrix} \overline{\omega} & \overline{\varphi_-} \\ \overline{\psi_-} & \overline{\omega} \end{bmatrix} - \begin{bmatrix} k_1 & k_2 \\ k_3 & k_4 \end{bmatrix} \begin{bmatrix} 0 & \overline{\psi_+} \\ \overline{\varphi_+} & 0 \end{bmatrix} \in H^2_{M_2},$$

which implies that φ_+ and ψ_+ are not identically zero and hence $\det \Phi_+$ is not identically zero.

We now split the proof into three cases.

Case 1 ($\boldsymbol{m_{i_0} = n_{i_0} = 0}$ **for some** $\boldsymbol{i_0}$). In this case, by Lemma 2.4 and Theorem 1.4, we can conclude that T_Φ is normal. Since $\det \Phi_+$ is not identically zero, it follows from Theorem 1.3 that $\Phi_+ - \Phi_- U \in M_n(\mathbb{C})$ for some constant unitary matrix $U \equiv \left[\begin{smallmatrix} c_1 & c_2 \\ c_3 & c_4 \end{smallmatrix}\right]$. We observe

$$\begin{aligned} \Phi_+ - \Phi_- U \in M_n(\mathbb{C}) &\Longleftrightarrow \begin{bmatrix} 0 & \varphi_+ \\ \psi_+ & 0 \end{bmatrix} - \begin{bmatrix} \omega & \theta_1 \overline{b} \\ \theta_0 \overline{a} & \omega \end{bmatrix} \begin{bmatrix} c_1 & c_2 \\ c_3 & c_4 \end{bmatrix} \in M_n(\mathbb{C}) \\ &\Longrightarrow \begin{cases} c_1\omega + c_3\theta_1\overline{b} = \xi_1 \\ c_4\omega + c_2\theta_0\overline{a} = \xi_2 \\ \varphi_+ = c_2\omega + c_4\theta_1\overline{b} + \xi_3 \\ \psi_+ = c_3\omega + c_1\theta_0\overline{a} + \xi_4 \end{cases} \qquad (\xi_i \in \mathbb{C} \text{ for } i = 1, \dots, 4), \end{aligned} \tag{28}$$

which gives

$$\overline{c_1} H_{\overline{\omega}} = -\overline{c_3} H_{\overline{\theta_1 b}} \quad \text{and} \quad \overline{c_4} H_{\overline{\omega}} = -\overline{c_2} H_{\overline{\theta_0 a}}\,. \tag{29}$$

Thus if $c_1 \neq 0$ then $c_3 \neq 0$ and hence $\omega = \theta_1$, which is a contradiction because $\omega(\alpha_{i_0}) = 0$, but $\theta_1(\alpha_{i_0}) \neq 0$. Thus $c_1 = 0$ and similarly, $c_4 = 0$. Since U is unitary, it follows that $|c_2| = |c_3| = 1$, and hence $\theta_1 \overline{b}$ and $\theta_0 \overline{a}$ are constants. Thus, again by (28), we have

$$\varphi = \varphi_+ = e^{i\delta_1}\omega + \beta_1 \quad \text{and} \quad \psi = \psi_+ = e^{i\delta_2}\omega + \beta_2 \quad (\delta_1, \delta_2 \in [0, 2\pi);\ \beta_1, \beta_2 \in \mathbb{C}).$$

Since $|\varphi| = |\psi|$, it follows that

$$\varphi = e^{i\delta_1}\omega + \zeta \quad \text{and} \quad \psi = e^{i\delta_2}\varphi \quad (\delta_1, \delta_2 \in [0, 2\pi);\ \zeta \in \mathbb{C})).$$

Case 2 ((i) $\boldsymbol{0 < m_{i_0} + n_{i_0} < 2q_{i_0}}$; **or (ii)** $\boldsymbol{m_{i_0} + n_{i_0} > 2q_{i_0}}$ ($\boldsymbol{m_{i_0} n_{i_0} \neq 0}$) **for some** $\boldsymbol{i_0}$). In this case, by Lemma 2.4 and Theorem 1.4, we can conclude that

T_Φ is normal. By case assumption, we have $m_{i_0} \neq q_{i_0}$ or $n_{i_0} \neq q_{i_0}$. Suppose that $m_{i_0} \neq q_{i_0}$. Then $\omega \neq \theta_0$, and hence by (29) we have $c_2 = c_4 = 0$. Therefore U is not unitary, a contradiction.

If instead $n_{i_0} \neq q_{i_0}$ then the same argument as above gives that U is not unitary, a contradiction. Thus this case cannot occur.

Case 3 $(\boldsymbol{m_i + n_i > 2q_i}$ $(\boldsymbol{m_i n_i = 0})$ **for all** $\boldsymbol{i = 1, \dots, p})$**.** Fix i_0 $(1 \leq i_o \leq p)$; we may, without loss of generality, assume that $n_{i_0} = 0$ (and hence, $m_{i_0} > 2q_{i_0}$). By Lemma 2.1, we may also assume that $b_{i_0} = z$. It follows from Theorem 1.2 that there exists a matrix function $K \equiv \left[\begin{smallmatrix} k_1 & k_2 \\ k_3 & k_4 \end{smallmatrix}\right] \in \mathcal{E}(\Phi)$, so that

$$\begin{bmatrix} \overline{\omega} & \overline{\varphi_-} \\ \overline{\psi_-} & \overline{\omega} \end{bmatrix} - \begin{bmatrix} k_1 & k_2 \\ k_3 & k_4 \end{bmatrix} \begin{bmatrix} 0 & \overline{\psi_+} \\ \overline{\varphi_+} & 0 \end{bmatrix} \in H^2_{M_2},$$

which implies that

$$\begin{cases} \overline{\omega} - k_2\overline{\varphi_+} \in H^2, & \overline{\theta}_1 b - k_4\overline{\varphi_+} \in H^2 \\ \overline{\omega} - k_3\overline{\psi_+} \in H^2, & \overline{\theta}_0 a - k_1\overline{\psi_+} \in H^2. \end{cases} \tag{30}$$

Since $||K||_\infty \leq 1$ and hence $||k_i||_\infty \leq 1$ for each $i = 1, \dots, 4$, the following Toeplitz operators are all hyponormal (by Cowen's Theorem):

$$T_{\overline{\omega}+\varphi_+}, \ T_{\overline{\theta}_1 b+\varphi_+}, \ T_{\overline{\omega}+\psi_+}, \ T_{\overline{\theta}_0 a+\psi_+}. \tag{31}$$

Put $M_i := \max(m_i, q_i)$ and $N_i := \max(n_i, q_i)$. Then by (31) and a scalar-valued version of (23), we can see that

$$\varphi_+ = z^{q_{i_0}} \prod_{i \neq i_0} b_i^{N_i} \theta_1' \theta_3 \overline{d} \ \text{ and } \ \psi_+ = z^{m_{i_0}} \prod_{i \neq i_0} b_i^{M_i} \theta_0' \theta_2 \overline{c} \qquad \text{(coprime factorizations)},$$

where θ_2 and θ_3 are finite Blaschke products. Thus, in particular, $c(0) \neq 0$ and $d(0) \neq 0$. Thus, by (30), we can see that

$$k_3(0) = 0 \quad \text{and} \quad k_4(0) = 0 : \tag{32}$$

indeed, in (30),

$$\begin{aligned} \overline{\omega} - k_3\overline{\psi_+} \in H^2 &\Longrightarrow \overline{z}^{q_{i_0}} \prod_{i \neq i_0} \overline{b_i}^{q_i} - k_3 \overline{z}^{m_{i_0}} \prod_{i \neq i_0} \overline{b_i}^{M_i} \overline{\theta_0'} \overline{\theta_2} c \in H^2 \\ &\Longrightarrow z^{m_{i_0} - q_{i_0}} \prod_{i \neq i_0} b_i^{M_i - q_i} \theta_0' \theta_2 - k_3 c \in z^{m_{i_0}} H^2 \\ &\Longrightarrow k_3(0) = 0 \quad (\text{since } m_{i_0} > 2q_{i_0}) \end{aligned}$$

and

$$\begin{aligned} \overline{\theta}_1 b - k_4\overline{\varphi_+} \in H^2 &\Longrightarrow \Big(\prod_{i=1}^{d} \overline{b_i}^{n_i}\Big) \overline{\theta_1'} b - k_4 \overline{z}^{q_{i_0}} \prod_{i \neq i_0} \overline{b_i}^{N_i} \overline{\theta_1'} \overline{\theta_3} d \in H^2 \\ &\Longrightarrow z^{q_{i_0}} \Big(\prod_{i \neq i_0} b_i^{N_i - n_i}\Big) \theta_3 b - k_4 d \in z^{q_{i_0}} H^2 \\ &\Longrightarrow k_4(0) = 0\,, \end{aligned}$$

which proves (32). Write

$$\theta_2 = z^{l_2}\theta_2' \quad \text{and} \quad \theta_3 = z^{l_3}\theta_3' \quad (\theta_2'(0) \neq 0,\ \theta_3'(0) \neq 0).$$

Then we can write

$$\Phi_+ = \begin{bmatrix} 0 & z^{q_{i_0}+l_3}\prod_{i\neq i_0} b_i^{N_i}\theta_1'\theta_3'\overline{d} \\ z^{m_{i_0}+l_2}\prod_{i\neq i_0} b_i^{M_i}\theta_0'\theta_2'\overline{c} & 0 \end{bmatrix}.$$

On the other hand, write

$$a' := \Big(\prod_{i\neq i_0} b_i^{M_i-m_i}\Big)a, \quad \theta_1'' := \Big(\prod_{i\neq i_0} b_i^{M_i-q_i}\Big)\theta_1'$$

and

$$\alpha := -\frac{a'(0)}{\theta_1''(0)} \quad \text{and} \quad \nu := \frac{1}{\sqrt{|\alpha|^2+1}}.$$

Note that

$$\widetilde{\Phi_-} = \begin{bmatrix} \widetilde{\omega} & \widetilde{\theta}_0\overline{\widetilde{a}} \\ \widetilde{\theta}_1\overline{\widetilde{b}} & \widetilde{\omega} \end{bmatrix}.$$

Since $\widetilde{\Phi_-^*}$ is of bounded type, it follows from Theorem 1.1 that there exists a square inner matrix function Δ such that $\ker H_{\widetilde{\Phi_-^*}} = \widetilde{\Delta}H^2_{\mathbb{C}^2}$ and

$$\widetilde{\Phi_-^*} = \widetilde{B}\widetilde{\Delta}^* \quad \text{(right coprime factorization).}$$

Thus, by Lemma 2.5 we have

$$\ker H_{\widetilde{\Phi_-^*}} = \widetilde{\Delta}H^2_{\mathbb{C}_2} \subseteq \widetilde{\Omega}H^2_{\mathbb{C}^2} \quad \text{and} \quad \Phi_- = B^*\Delta \quad \text{(left coprime factorization)}, \tag{33}$$

where

$$\Omega = \nu\begin{bmatrix} z^{m_{i_0}-q_{i_0}}\theta_0' & \alpha\theta_1' \\ -\overline{\alpha}z^{m_{i_0}-q_{i_0}+1}\theta_0' & z\theta_1' \end{bmatrix}.$$

Since $\widetilde{\Delta}H^2_{\mathbb{C}_2} \subseteq \widetilde{\Omega}H^2_{\mathbb{C}^2}$, it follows that $\widetilde{\Omega}$ is a left inner divisor of $\widetilde{\Delta}$. Thus, we can write

$$\widetilde{\Delta} = \widetilde{\Omega}\widetilde{\Omega}_1 \ \text{ for some } \Omega_1, \ \text{ so that } \ \Delta = \Omega_1\Omega.$$

We suppose that $q_{i_0}+l_3 \leq m_{i_0}+l_2$ and write $r := (m_{i_0}+l_2)-(q_{i_0}+l_3) \geq 0$. Then there exist finite Blaschke products θ_4 and θ_5 with $\theta_i(0) \neq 0$ $(i = 4, 5)$ such that

$$\Phi_+ = \prod_{i\neq i_0} b_i^{\max(M_i,N_i)}(z^{m_{i_0}+l_2}\theta_1'\theta_3'\theta_0'\theta_2')I_2\begin{bmatrix} 0 & \theta_5\theta_1'\theta_3'c \\ z^r\theta_4\theta_0'\theta_2'd & 0 \end{bmatrix}^* \equiv (\theta I_2)A^*,$$

where $\theta := \prod_{i\neq i_0} b_i^{\max(M_i,N_i)}(z^{m_{i_0}+l_2}\theta_1'\theta_3'\theta_0'\theta_2')$. Since $H_{A\Delta^*} = H_{A\Omega^*\Omega_1^*}$, it follows that

$$\text{ran}H_{A\Delta^*} \supseteq \text{ran}H_{A\Omega^*}. \tag{34}$$

Observe that

$$A\Omega^* = \nu \begin{bmatrix} 0 & \theta_5\theta_1'\theta_3'c \\ z^r\theta_4\theta_0'\theta_2'd & 0 \end{bmatrix} \begin{bmatrix} z^{m_{i_0}-q_{i_0}}\theta_0' & \alpha\theta_1' \\ -\overline{\alpha}z^{m_{i_0}-q_{i_0}+1}\theta_0' & z\theta_1' \end{bmatrix}^* = \nu \begin{bmatrix} \overline{\alpha}\theta_5\theta_3'c & \overline{z}\theta_5\theta_3'c \\ z^{r-m_{i_0}+q_{i_0}}\theta_4\theta_2'd & -\alpha z^{r-m_{i_0}+q_{i_0}-1}\theta_4\theta_2'd \end{bmatrix}.$$

If $r \le m_{i_0} - q_{i_0}$, then we have

$$H_{A\Omega^*}\begin{bmatrix} 0 \\ z^{m_{i_0}-q_{i_0}-r} \end{bmatrix} = \nu \begin{bmatrix} H_{\overline{z}}(z^{m_{i_0}-q_{i_0}-r}\theta_5\theta_3'c) \\ -\alpha H_{\overline{z}}(\theta_4\theta_2'd) \end{bmatrix}.$$

Since $(\theta_4\theta_2'd)(0) \neq 0$, it follows from Lemma 2.3, (33) and (34) that

$$\begin{bmatrix} \beta \\ 1 \end{bmatrix} \in \text{cl ran}\, H_{A\Omega^*} \subseteq \text{cl ran}\, H_{A\Delta^*} \subseteq \ker(I - T_{\widetilde{K}}T_{\widetilde{K}}^*) \quad \text{for some } \beta \in \mathbb{C}. \tag{35}$$

It thus follows from (32) and (35) that

$$\begin{bmatrix} \beta \\ 1 \end{bmatrix} = T_{\widetilde{K}}T_{\widetilde{K}}^*\begin{bmatrix} \beta \\ 1 \end{bmatrix} = \begin{bmatrix} T_{\widetilde{k}_1} & T_{\widetilde{k}_3} \\ T_{\widetilde{k}_2} & T_{\widetilde{k}_4} \end{bmatrix}\begin{bmatrix} T_{\overline{\widetilde{k}_1}} & T_{\overline{\widetilde{k}_2}} \\ T_{\overline{\widetilde{k}_3}} & T_{\overline{\widetilde{k}_4}} \end{bmatrix}\begin{bmatrix} \beta \\ 1 \end{bmatrix} = \begin{bmatrix} T_{\widetilde{k_1}} & T_{\widetilde{k_3}} \\ T_{\widetilde{k_2}} & T_{\widetilde{k_4}} \end{bmatrix}\begin{bmatrix} (\beta k_1(0) + k_2(0)) \\ 0 \end{bmatrix} = \begin{bmatrix} \widetilde{k_1}(\beta k_1(0) + k_2(0)) \\ \widetilde{k_2}(\beta k_1(0) + k_2(0)) \end{bmatrix},$$

which implies that k_1 is a constant and k_2 is a nonzero constant. Again by (30),

$$\begin{aligned} \overline{\omega} - k_2\overline{\varphi_+} \in H^2 &\Longrightarrow \omega\overline{z}^{q_{i_0}}\prod_{i\neq i_0}\overline{b_i}^{N_i}\overline{\theta_1'\theta_3}d \in H^2 \\ &\Longrightarrow q_i \ge n_i \ (i \neq i_0) \text{ and } \overline{\theta_1'\theta_3}d \in H^2 \\ &\Longrightarrow n_i = 0 \ (i \neq i_0) \text{ and } \theta_1'\theta_3 = 1, \end{aligned} \tag{36}$$

where the last implication follows from the observation that if $n_i \neq 0$ then by the case assumption, $m_i = 0$ and hence, $2q_i < n_i \le q_i$, a contradiction. We thus have $n_i = 0$ for all $i = 1, \dots, p$. Since $\theta_1' = 1$, it follows that

$$\theta_1 = 1 \text{ and hence, } \psi_- = 0.$$

In turn, $m_i > 2q_i$ for all $i = 1, \dots, p$, so that θ_0 is nonconstant, and hence $\varphi_- = \theta_0\overline{a} \neq 0$. Since by (30), $\overline{\theta_0}a - k_1\overline{\psi_+} \in H^2$, it follows that $k_1 \neq 0$. We thus have

$$\begin{aligned} \overline{\theta_0}a - k_1\overline{\psi_+} \in H^2 &\Longrightarrow \overline{\theta_0}a - k_1\overline{z}^{m_{i_0}}\prod_{i\neq i_0}\overline{b_i}^{M_i}\overline{\theta_0'\theta_2}c \in H^2 \\ &\Longrightarrow \prod_{i=1}^{p}\overline{b_i}^{m_i}a - k_1\overline{z}^{m_{i_0}}\prod_{i\neq i_0}\overline{b_i}^{M_i}\overline{\theta_2}c \in H^2 \\ &\Longrightarrow \overline{\theta_2}c \in H^2 \quad (\text{because } m_i > 2q_i \text{ and hence, } M_i = m_i) \\ &\Longrightarrow \theta_2 = 1. \end{aligned} \tag{37}$$

Therefore, we have

$$\varphi_+ = z^q \prod_{i\neq i_0} b_i^{N_i}\theta_1'\theta_3\overline{d} = \omega\overline{d} \quad (q := q_{i_0}) \quad \text{and} \quad \psi_+ = z^{m_{i_0}} \prod_{i\neq i_0} b_i^{M_i}\theta_0'\theta_2\overline{c} = \theta_0\overline{c}.$$

Since by (27), $|\varphi| = |\psi|$, we have

$$|\omega\overline{d} + \overline{\theta_0}a| = |\varphi_+ + \overline{\varphi_-}| = |\psi_+| = |\theta_0\overline{c}| \quad (\text{where } a \in \mathcal{H}_{\theta_0},\ d \in \mathcal{H}_{z\omega},\ c \in \mathcal{H}_{z\theta_0}),$$

which implies

$$\omega\theta_0(\omega\overline{d} + \overline{\theta_0}a)(\overline{\omega}d + \theta_0\overline{a}) = \omega\theta_0 c\overline{c},$$

so that

$$ad = z\Big((\theta_0\overline{c})(\overline{z}\omega)c - (\theta_0\overline{d})(\overline{z}\omega)d - (\theta_0\overline{a})(\theta_0\overline{d})(\overline{z}\omega^2) - (\theta_0\overline{a})(\overline{z}\omega)a\Big). \tag{38}$$

Since $a \in \mathcal{H}_{\theta_0}$, $c \in \mathcal{H}_{z\theta_0}$, $d \in \mathcal{H}_{z\omega}$ and $m_i \geq 2q_i$ for all $i = 1, \dots, p$, it follows that $\theta_0\overline{a} \in H^2$, $\theta_0\overline{c} \in H^2$ and $\theta_0\overline{d} = \left(\prod_{i=1}^p b_i^{m_i}\right)\theta_0'\overline{d} = \left(\prod_{i=1}^p b_i^{m_i - q_i}\theta_0'\right)(\omega\overline{d}) \in H^2$. Thus, (38) implies that $ad = zh$ for some $h \in H^2$, and hence $(ad)(0) = 0$, a contradiction. Therefore this case cannot occur.

If instead $r > m_{i_0} - q_{i_0}$, then the same argument as before leads to a contradiction. Moreover, by the same argument as in the case $q_{i_0} + l_3 \leq m_{i_0} + l_2$, the case $q_{i_0} + l_3 > m_{i_0} + l_2$ cannot occur either.

Therefore, Case 3 cannot occur. This proves the implication (4) ⇒ (5).

This completes the proof. □

Remark 2.7. From the proof of Theorem 2.6 we can see that if G is given by (24) then G is subnormal if and only if G is normal, except in the case (26). However we need not expect that the exceptional case (26) implies normality of G. For example, if

$$\Phi := \begin{bmatrix} \overline{\omega} & \overline{\omega} + 2\omega \\ \overline{\omega} + 2\omega & \overline{\omega} \end{bmatrix} \quad (\omega \text{ is a finite Blaschke product})$$

then T_Φ satisfies the case (26) (where $m_i = n_i = q_i$ and $a = b = \theta_0' = \theta_1' = 1$). A straightforward calculation shows that T_Φ is not normal. Since

$$T_\Phi = \begin{bmatrix} T_{\overline{\omega}} & T_{\overline{\omega}} + 2T_\omega \\ T_{\overline{\omega}} + 2T_\omega & T_{\overline{\omega}} \end{bmatrix},$$

it follows that if $W := \frac{1}{\sqrt{2}}\left[\begin{smallmatrix} 1 & -1 \\ 1 & 1 \end{smallmatrix}\right]$, then W is unitary and

$$W^* T_\Phi W = 2\begin{bmatrix} T_{\overline{\omega}} + T_\omega & 0 \\ 0 & -T_\omega \end{bmatrix},$$

which says that T_Φ is unitarily equivalent to a direct sum of the normal operator $2(T_{\overline{\omega}} + T_\omega)$ and the analytic Toeplitz operator $-2T_\omega$. From this viewpoint, we might conjecture that every subnormal rational Toeplitz operator is unitarily equivalent to a direct sum of a normal operator and an analytic Toeplitz operator. However we have been unable to settle this conjecture.

Acknowledgment

The authors are deeply indebted to the referee for many helpful comments that helped improved the presentation and mathematical content of the paper.

References

[Ab] M.B. Abrahamse, *Subnormal Toeplitz operators and functions of bounded type*, Duke Math. J. **43** (1976), 597–604.

[AIW] I. Amemiya, T. Ito, and T.K. Wong, *On quasinormal Toeplitz operators*, Proc. Amer. Math. Soc. **50**(1975), 254–258.

[At] A. Athavale *On joint hyponormality of operators*, Proc. Amer. Math. Soc. **103**(1988), 417–423.

[Br] J. Bram, *Subnormal operators*, Duke Math. J. **22**(1955), 75–94.

[BH] A. Brown and P.R. Halmos, *Algebraic properties of Toeplitz operators*, J. Reine Angew. Math. **213**(1963/1964), 89–102.

[Con] J.B. Conway, The Theory of Subnormal Operators, *Math. Surveys and Monographs*, **36**(1991), Amer. Math. Soc. Providence, Rhode Island.

[CoS] J.B. Conway and W. Szymanski, *Linear combination of hyponormal operators*, Rocky Mountain J. Math. **18**(1988), 695–705.

[Co1] C.C. Cowen, *On equivalence of Toeplitz operators*, J. Operator Theory **7**(1982), 167–172.

[Co2] C.C. Cowen, *More subnormal Toeplitz operators*, J. Reine Angew. Math. **367**(1986), 215–219.

[Co3] C.C. Cowen, *Hyponormal and subnormal Toeplitz operators*, Surveys of Some Recent Results in Operator Theory, I (J.B. Conway and B.B. Morrel, eds.), Pitman Research Notes in Mathematics, Volume 171, Longman, 1988, pp. (155–167).

[Co4] C.C. Cowen, *Hyponormality of Toeplitz operators*, Proc. Amer. Math. Soc. **103**(1988), 809–812.

[CoL] C.C. Cowen and J. Long, *Some subnormal Toeplitz operators*, J. Reine Angew. Math. **351**(1984), 216–220.

[Cu1] R.E. Curto, *Fredholm and invertible n-tuples of operators. The deformation problem*, Trans. Amer. Math. Soc. **266**(1981), 129–159

[Cu2] R.E. Curto, *Quadratically hyponormal weighted shifts*, Integral Equations Operator Theory, **13**(1990), 49–66.

[CuF1] R.E. Curto and L.A. Fialkow, *Recursiveness, positivity, and truncated moment problems*, Houston J. Math. **17**(1991), 603–635.

[CuF2] R.E. Curto and L.A. Fialkow, *Recursively generated weighted shifts and the subnormal completion problem*, Integral Equations Operator Theory, **17**(1993), 202–246.

[CuF3] R.E. Curto and L.A. Fialkow, *Recursively generated weighted shifts and the subnormal completion problem II*, Integral Equations Operator Theory, **18**(1994), 369–426.

[CHKL] R.E. Curto, I.S. Hwang, D. Kang and W.Y. Lee, *Subnormal and quasinormal Toeplitz operator with matrix-valued rational symbols*, Adv. Math. **255**(2014), 562–585.

[CHL1] R.E. Curto, I.S. Hwang and W.Y. Lee, *Hyponormality and subnormality of block Toeplitz operators*, Adv. Math. **230**(2012), 2094–2151.

[CHL2] R.E. Curto, I.S. Hwang and W.Y. Lee, *Which subnormal Toeplitz operators are either normal or analytic ?*, J. Funct. Anal. **263(8)**(2012), 2333–2354.

[CHL3] R.E. Curto, I.S. Hwang and W.Y. Lee, *Abrahamse's Theorem for matrix-valued symbols and subnormal Toeplitz completions*, (preprint 2012) (arXiv: 1301.6901,2013).

[CLL] R.E. Curto, S.H. Lee and W.Y. Lee, *Subnormality and 2-hyponormality for Toeplitz operators*, Integral Equations Operator Theory, **44**(2002), 138–148.

[CL1] R.E. Curto and W.Y. Lee, *Joint hyponormality of Toeplitz pairs*, Memoirs Amer. Math. Soc. **712**, Amer. Math. Soc., Providence, 2001.

[CL2] R.E. Curto and W.Y. Lee, *Towards a model theory for 2–hyponormal operators*, Integral Equations Operator Theory **44**(2002), 290–315.

[CL3] R.E. Curto and W.Y. Lee, *Subnormality and k-hyponormality of Toeplitz operators: A brief survey and open questions*, Operator Theory and Banach Algebras (Rabat, 1999), 73–81, Theta, Bucharest, 2003.

[CMX] R.E. Curto, P.S. Muhly and J. Xia, *Hyponormal pairs of commuting operators*, Contributions to Operator Theory and Its Applications (Mesa, AZ, 1987) (I. Gohberg, J.W. Helton and L. Rodman, eds.), Operator Theory: Advances and Applications, vol. 35, Birkhäuser, Basel–Boston, 1988, 1–22.

[CP1] R.E. Curto and M. Putinar, *Existence of non-subnormal polynomially hyponormal operators*, Bull. Amer. Math. Soc. (N.S.), **25**(1991), 373–378.

[CP2] R.E. Curto and M. Putinar, *Nearly subnormal operators and moment problems*, J. Funct. Anal. **115**(1993), 480–497.

[DPY] R.G. Douglas, V.I. Paulsen, and K. Yan, *Operator theory and algebraic geometry* Bull. Amer. Math. Soc. (N.S.) **20**(1989), 67–71.

[FL] D.R. Farenick and W.Y. Lee, *Hyponormality and spectra of Toeplitz operators*, Trans. Amer. Math. Soc. **348**(1996), 4153–4174.

[Gu1] C. Gu, *A generalization of Cowen's characterization of hyponormal Toeplitz operators*, J. Funct. Anal. **124**(1994), 135–148.

[Gu2] C. Gu, *On a class of jointly hyponormal Toeplitz operators*, Trans. Amer. Math. Soc. **354**(2002), 3275–3298.

[GHR] C. Gu, J. Hendricks and D. Rutherford, *Hyponormality of block Toeplitz operators*, Pacific J. Math. **223**(2006), 95–111.

[GS] C. Gu and J.E. Shapiro, *Kernels of Hankel operators and hyponormality of Toeplitz operators*, Math. Ann. **319**(2001), 553–572.

[Hal1] P.R. Halmos, *Ten problems in Hilbert space*, Bull. Amer. Math. Soc. **76**(1970), 887–933.

[Hal2] P.R. Halmos, *Ten years in Hilbert space*, Integral Equations Operator Theory **2**(1979), 529–564.

[HKL1] I.S. Hwang, I.H. Kim and W.Y. Lee, *Hyponormality of Toeplitz operators with polynomial symbols*, Math. Ann. **313(2)**(1999), 247–261.

[HKL2] I.S. Hwang, I.H. Kim and W.Y. Lee, *Hyponormality of Toeplitz operators with polynomial symbols: An extremal case*, Math. Nach. **231**(2001), 25–38.

[HL1] I.S. Hwang and W.Y. Lee, *Hyponormality of trigonometric Toeplitz operators*, Trans. Amer. Math. Soc. **354**(2002), 2461–2474.

[HL2] I.S. Hwang and W.Y. Lee, *Hyponormality of Toeplitz operators with rational symbols*, Math. Ann. **335** (2006), 405–414.

[HL3] I.S. Hwang and W.Y. Lee, *Hyponormal Toeplitz operators with rational symbols*, J. Operator Theory **56**(2006), 47–58.

[HL4] I.S. Hwang and W.Y. Lee, *Block Toeplitz Operators with rational symbols*, J. Phys. A: Math. Theor. **41(18)**(2008), 185207.

[HL5] I.S. Hwang and W.Y. Lee, *Block Toeplitz Operators with rational symbols (II)*, J. Phys. A: Math. Theor. **41(38)**(2008), 385206.

[Le] W.Y. Lee, *Cowen sets for Toeplitz operators with finite rank selfcommutators*, J. Operator Theory **54(2)**(2005), 301–307.

[ItW] T. Ito and T.K. Wong, *Subnormality and quasinormality of Toeplitz operators*, Proc. Amer. Math. Soc. **34**(1972), 157–164.

[McCP] S. McCullough and V. Paulsen, *A note on joint hyponormality*, Proc. Amer. Math. Soc. **107**(1989), 187–195.

[NT] T. Nakazi and K. Takahashi, *Hyponormal Toeplitz operators and extremal problems of Hardy spaces*, Trans. Amer. Math. Soc. **338**(1993), 753–769.

[Ni] N.K. Nikolskii, Treatise on the shift operator, *Springer*, New York, 1986.

[Sun1] S. Sun, *On hyponormal weighted shift*, Chinese Ann. Math. Ser. B **5** (1984), no. 1, 101–108. (A Chinese summary appears in Chinese Ann. Math. Ser. A **5** (1984), no. 1, 124.)

[Sun2] S. Sun, *On Toeplitz operators in the θ-class*, Sci. Sinica Ser. A **28** (1985), no. 3, 235–241.

[Sun3] S. Sun, *On hyponormal weighted shift, II*, Chinese Ann. Math. Ser. B **6** (1985), no. 3, 359–361. (A Chinese summary appears in Chinese Ann. Math. Ser. A **6** (1985), no. 4, 516.)

[Zhu] K. Zhu, *Hyponormal Toeplitz operators with polynomial symbols*, Integral Equations Operator Theory **21**(1996), 376–381.

Raúl E. Curto
Department of Mathematics, University of Iowa
Iowa City, IA 52242, USA
e-mail: `raul-curto@uiowa.edu`

In Sung Hwang
Department of Mathematics, Sungkyunkwan University
Suwon 440-746, Korea
e-mail: `ihwang@skku.edu`

Woo Young Lee
Department of Mathematics, Seoul National University
Seoul 151-742, Korea
e-mail: `wylee@snu.ac.kr`

Operator Theory:
Advances and Applications, Vol. 240, 111–135

Generalized Repeated Interaction Model and Transfer Functions

Santanu Dey and Kalpesh J. Haria

Abstract. Using a scheme involving a lifting of a row contraction we introduce a toy model of repeated interactions between quantum systems. In this model there is an outgoing Cuntz scattering system involving two wandering subspaces. We associate to this model an input/output linear system which leads to a transfer function. This transfer function is a multi-analytic operator, and we show that it is inner if we assume that the system is observable. Finally it is established that transfer functions coincide with characteristic functions of associated liftings.

Mathematics Subject Classification (2010). Primary 47A13; Secondary 47A20, 46L53, 47A48, 47A40, 81R15.

Keywords. Repeated interaction, quantum system, multivariate operator theory, row contraction, contractive lifting, outgoing Cuntz scattering system, transfer function, multi-analytic operator, input-output formalism, linear system, observability, scattering theory, characteristic function.

1. Introduction

In page 287 of the article [9] the author has commented the following while comparing [9] with [4, 5]: In [4] a row contraction $\underline{A}$ on a Hilbert space $\mathcal{H}$ with a one-dimensional eigenspace is considered and the theory of minimal isometric dilations is used. The characteristic function introduced in [5] is a multi-analytic operator associated to a lifting and the ergodic case is studied in detail in [4]. In [9] minimality is not considered but one starts with an interaction U (which is a unitary operator) in a scheme similar to [4] and obtains a multi-analytic operator which represents the transfer function of an input-output system associated with the interaction. It is expected that the scheme developed [9] is more directly applicable to physical models. In the setting of [5] the assumption of a one-dimensional eigenspace is dropped and the theory is much more general in another direction. A further integration of these schemes in the future may help to remove unnecessarily restrictive assumptions of the toy model considered in [9] and lead to the study of other and of more realistic models.

This paper achieves some of these objectives. In the model of repeated interactions between quantum systems, also called a noncommutative Markov chain, studied in [9] (cf. [8]) for given three Hilbert spaces $\mathcal{H}, \mathcal{K}$ and $\mathcal{P}$ with unit vectors $\Omega^{\mathcal{H}}, \Omega^{\mathcal{K}}$ and $\Omega^{\mathcal{P}}$ an *interaction* is defined to be a unitary operator $U : \mathcal{H} \otimes \mathcal{K} \to \mathcal{H} \otimes \mathcal{P}$ such that

$$U(\Omega^{\mathcal{H}} \otimes \Omega^{\mathcal{K}}) = \Omega^{\mathcal{H}} \otimes \Omega^{\mathcal{P}}. \tag{1.1}$$

Define $\mathcal{K}_\infty := \bigotimes_{i=1}^{\infty} \mathcal{K}$ and $\mathcal{P}_\infty := \bigotimes_{i=1}^{\infty} \mathcal{P}$ as infinite tensor products of Hilbert spaces with distinguished unit vectors. We denote mth copy of $\mathcal{K}$ in $\mathcal{K}_\infty$ by $\mathcal{K}_m$ and set $\mathcal{K}_{[m,n]} := \mathcal{K}_m \otimes \cdots \otimes \mathcal{K}_n$ for $m \leq n$. Similar notations are also used with respect to $\mathcal{P}$. The repeated interaction is defined as

$$U(n) := U_n \dots U_1 : \mathcal{H} \otimes \mathcal{K}_\infty \to \mathcal{H} \otimes \mathcal{P}_{[1,n]} \otimes \mathcal{K}_{[n+1,\infty)}$$

where U_i's are copies of U on the factors $\mathcal{H} \otimes \mathcal{K}_i$ of the infinite tensor products and U_i's leaves other factors fixed. Equation (1.1) tells us that the tensor product of the vacuum vectors $\Omega^{\mathcal{H}}, \Omega^{\mathcal{K}}$ (along with $\Omega^{\mathcal{P}}$) represents a state of the coupled system which is not affected by the interaction U. This entire setting represents interactions of an atom with light beams or fields. In particular $\Omega^{\mathcal{H}}$ in [9] is thought of as the vacuum state of an atom, and $\Omega^{\mathcal{K}}$ and $\Omega^{\mathcal{P}}$ as a state indicating the absence of photons.

In the generalized repeated interaction model that we introduce in this article we use a pair of unitaries to encode the interactions instead of one unitary as follows:
Let $\tilde{\mathcal{H}}$ be a (closed) subspace of $\mathcal{H}$, and $U : \mathcal{H} \otimes \mathcal{K} \to \mathcal{H} \otimes \mathcal{P}$ and $\tilde{U} : \tilde{\mathcal{H}} \otimes \mathcal{K} \to \tilde{\mathcal{H}} \otimes \mathcal{P}$ be two unitaries such that

$$U(\tilde{h} \otimes \Omega^{\mathcal{K}}) = \tilde{U}(\tilde{h} \otimes \Omega^{\mathcal{K}}) \text{ for all } \tilde{h} \in \tilde{\mathcal{H}}. \tag{1.2}$$

We fix $\{\epsilon_1, \dots, \epsilon_d\}$ to be an orthonormal basis of $\mathcal{P}$. The equation (1.2) is the analog of the equation (1.1) for our model and thus our model can be used for the setting where a quantum system interacts with a stream of copies of another quantum system in such a way that there is no backaction (so we get a Markovian type of dynamics) and such that there is a certain kind of subprocess. In the model of [9] the vacuum state $\Omega^{\mathcal{H}}$ of an atom plays an important role. For a model describing interaction of a quantum system with a stream of copies of another quantum sytem we need that the computations do not involve any fixed unit vector $\Omega^{\mathcal{H}}$ and we are able to achieve this in our model by using a pair of unitaries. Instead of $\Omega^{\mathcal{H}}$ we now have a kind of subprocess, described by $\tilde{U}$, which can be treated on the same level as the full process, described by U.

The main condition imposed on the unitary $U : \mathcal{H} \otimes \mathcal{K} \to \mathcal{H} \otimes \mathcal{P}$ in order to get a generalized interaction model is that $U(\tilde{\mathcal{H}} \otimes \Omega^{\mathcal{K}}) \subset \tilde{\mathcal{H}} \otimes \mathcal{P}$ (cf. Proposition 3.1 of [10] for an interesting consequence of this assumption). We can then define $\tilde{U}$ restricted to $\tilde{\mathcal{H}} \otimes \Omega^{\mathcal{K}}$ as U restricted to $\tilde{\mathcal{H}} \otimes \Omega^{\mathcal{K}}$, and assume that $\mathcal{H} \otimes \mathcal{P}$ is big enough to allow a unitary extension $\tilde{U} : \tilde{\mathcal{H}} \otimes \mathcal{K} \to \tilde{\mathcal{H}} \otimes \mathcal{P}$. The focus of the study done here, as also in [9], is to bring out that certain multi-analytic operators of the multivariate operator theory are associated to noncommutative Markov chains and related models, and that these operators can be exploited as powerful tools. These operators occur as central objects in various context such as in the systems theory related works (cf. [3]) and noncommutative multivariable operator theory related works (cf. [14], [15]).

A tuple $\underline{T} = (T_1, \ldots, T_d)$ of operators T_i's on a common Hilbert space $\mathcal{L}$ is called a *row contraction* if $\sum_{i=1}^d T_i T_i^* \leq I$. In particular if $\sum_{i=1}^d T_i T_i^* = I$, then the tuple $\underline{T} = (T_1, \ldots, T_d)$ is called *coisometric.* We introduce the notation $\tilde{\Lambda}$ for the free semigroup with generators $1, \ldots, d$. Suppose $T_1, \ldots, T_d \in \mathcal{B}(\mathcal{L})$ for a Hilbert space $\mathcal{L}$. If $\alpha \in \tilde{\Lambda}$ is the word $\alpha_1 \ldots \alpha_n$ with length $|\alpha| = n$, where each $\alpha_j \in \{1, \ldots, d\}$, then T_α denote $T_{\alpha_1} \ldots T_{\alpha_n}$. For the empty word $\emptyset$ we define $|\emptyset| = 0$ and $T_\emptyset = I$.

The unitary $U : \mathcal{H} \otimes \mathcal{K} \to \mathcal{H} \otimes \mathcal{P}$ from our model can be decomposed as

$$U(h \otimes \Omega^{\mathcal{K}}) = \sum_{j=1}^{d} E_j^* h \otimes \epsilon_j \text{ for } h \in \mathcal{H}, \tag{1.3}$$

where E_j's are some operators in $\mathcal{B}(\mathcal{H})$, for $j = 1, \ldots, d$. Likewise there exist some operators C_j's in $\mathcal{B}(\tilde{\mathcal{H}})$ such that

$$\tilde{U}(\tilde{h} \otimes \Omega^{\mathcal{K}}) = \sum_{j=1}^{d} C_j^* \tilde{h} \otimes \epsilon_j \text{ for } \tilde{h} \in \tilde{\mathcal{H}}. \tag{1.4}$$

Observe that $\sum_{j=1}^d E_j E_j^* = I$ and $\sum_{j=1}^d C_j C_j^* = I$, i.e., $\underline{E}$ and $\underline{C}$ are coisometric tuples. By equation (1.2)

$$E_j^* \tilde{h} = C_j^* \tilde{h} \text{ for all } \tilde{h} \in \tilde{\mathcal{H}}, j = 1, \ldots, d.$$

We recall from [5] that such tuple $\underline{E} = (E_1, \ldots, E_d)$ is called a *lifting* of $\underline{C} = (C_1, \ldots, C_d)$.

From a physicist perspective our model is a Markovian approximation of the repeated interaction between a quantum system and a stream of copies of another quantum system in such a way that there is no backaction. The change of an observable $X \in \mathcal{B}(\mathcal{H})$ until time n, compressed to $\mathcal{H}$, is written as

$$Z_n(X) := P_{\mathcal{H}} U(n)^* (X \otimes I) U(n)|_{\mathcal{H}}.$$

From equation (1.3) it follows that $Z_n(X) = Z^n(X)$ where $Z(X) = \sum_{i=1}^d E_i X E_i^* : B(\mathcal{H}) \to B(\mathcal{H})$ and Z is called the *transition operator* of the noncommutative Markov chain.

In Section 2 we develop our generalized repeated interaction model and obtain a coisometric operator which intertwines between the minimal isometric dilations of $\underline{E}$ and $\underline{C}$, and which will be crucial for the further investigation in this article. Using this an outgoing Cuntz scattering system in the sense of [3] is constructed for our model in Section 3. Popescu introduced the minimal isometric dilation in

[13] and the characteristic function in [14] of a row contraction, and systematically developed an extensive theory of row contractions (cf. [16], [17]). We use some of the concepts from Popescu's theory in this work.

For the outgoing Cuntz scattering system in Section 4 we give a $\tilde{\Lambda}$-linear system with an input-output formalism. A multi-analytic operator appears here as the transfer function and in the next section we show that this transfer function can be derived from the intertwining coisometry of Section 2. In the scattering interpretation of the transfer function this now mediates between two processes. This together with a nice product formula obtained in Proposition 2.1 tells us that this identification of transfer function is a reminiscent of the scattering operator construction using wave operators in Lax–Phillips scattering theory [12], equation (1.5) (cf. [18]), with one of the processes moving forward combined with the other moving backward. In [20] and [7] there are other approaches to transfer functions. Several works on transfer functions and on quantum systems using linear system theory can be found in recent theoretical physics and control theory surveys.

In Section 5 we investigate in regard to our model what the notion of observability implies for the scattering theory and the theory of liftings. Some techniques used here are similar to those of scattering theory of noncommutative Markov chains introduced in [11]. Characteristic functions for liftings, introduced in [5], are multi-analytic operators which classify certain class of liftings. Our model generalizes the setting of [9], and a comparison is done in Section 6 between the transfer function of our model and the characteristic function for the associated lifting using the series expansion of the transfer function obtained in Section 4. As a consequence mathematically generalized interaction models get firmly linked into the theory of functional models.

2. A generalised repeated interaction model

We begin with three Hilbert spaces $\mathcal{H}, \mathcal{K}$ and $\mathcal{P}$ with unit vectors $\Omega^{\mathcal{K}} \in \mathcal{K}$ and $\Omega^{\mathcal{P}} \in \mathcal{P}$, and unitaries U and $\tilde{U}$ as in equation (1.2). In $\mathcal{K}_\infty = \bigotimes_{i=1}^{\infty} \mathcal{K}$ and $\mathcal{P}_\infty = \bigotimes_{i=1}^{\infty} \mathcal{P}$ define $\Omega_\infty^{\mathcal{K}} := \bigotimes_{i=1}^{\infty} \Omega^{\mathcal{K}}$ and $\Omega_\infty^{\mathcal{P}} := \bigotimes_{i=1}^{\infty} \Omega^{\mathcal{P}}$ respectively. We denote mth copy of $\Omega^{\mathcal{K}}$ in $\Omega_\infty^{\mathcal{K}}$ by $\Omega_m^{\mathcal{K}}$ and in terms of this we introduce the notation $\Omega_{[m,n]}^{\mathcal{K}} := \Omega_m^{\mathcal{K}} \otimes \cdots \otimes \Omega_n^{\mathcal{K}}$. Identify $\mathcal{K}_{[m,n]}$ with $\Omega_{[1,m-1]}^{\mathcal{K}} \otimes \mathcal{K}_{[m,n]} \otimes \Omega_{[n+1,\infty)}^{\mathcal{K}}$, $\mathcal{H}$ with $\mathcal{H} \otimes \Omega_\infty^{\mathcal{K}}$ as a subspace of $\mathcal{H} \otimes \mathcal{K}_\infty$ and $\tilde{\mathcal{H}}$ with $\tilde{\mathcal{H}} \otimes \Omega_\infty^{\mathcal{K}}$ as a subspace of $\tilde{\mathcal{H}} \otimes \mathcal{K}_\infty$. Similar notations with respect to $\mathcal{P}$ are also used. For simplicity we assume that d is finite but all the results here can be derived also for $d = \infty$.

Associate a row contraction $\underline{E}$ to the unitary U as in equation (1.3) and define isometries

$$\hat{V}_j^E(h \otimes \eta) := U^*(h \otimes \epsilon_j) \otimes \eta \text{ for } j = 1, \ldots, d,$$

on the elementary tensors $h \otimes \eta \in \mathcal{H} \otimes \mathcal{K}_\infty$ and extend it linearly to obtain $\hat{V}_j^E \in \mathcal{B}(\mathcal{H} \otimes \mathcal{K}_\infty)$ for $j = 1, \ldots, d$. We recall that a lifting $\underline{T} = (T_1, \ldots, T_d)$ of any row contraction $\underline{S} = (S_1, \ldots, S_d)$ is called its isometric dilation if T_i's are isometries with orthogonal ranges. It can be easily verified that $\widehat{\underline{V}}^E = (\widehat{V}_1^E, \dots, \widehat{V}_d^E)$ on the space $\mathcal{H} \otimes \mathcal{K}_\infty$ is an isometric dilation of $\underline{E} = (E_1, \dots, E_d)$. If $h \in \mathcal{H}$ and $k_1 \in \mathcal{K}$, then there exist $h_i \in \mathcal{H}$ for $i = 1, \dots, d$ such that $U^*(\sum_{i=1}^d h_i \otimes \epsilon_i) = h \otimes k_1$ because U is a unitary. This implies

$$\sum_{i=1}^{d} \widehat{V}_i^E (h_i \otimes \Omega_\infty^{\mathcal{K}}) = h \otimes k_1 \otimes \Omega_{[2,\infty)}^{\mathcal{K}}.$$

In addition if $k_2 \in \mathcal{K}$, then

$$\sum_{i=1}^{d} \widehat{V}_i^E (h_i \otimes k_2 \otimes \Omega_{[2,\infty)}^{\mathcal{K}}) = U^*(\sum_{i=1}^{d} h_i \otimes \epsilon_i) \otimes k_2 \otimes \Omega_{[3,\infty)}^{\mathcal{K}} = h \otimes k_1 \otimes k_2 \otimes \Omega_{[3,\infty)}^{\mathcal{K}}.$$

By induction we conclude that

$$\mathcal{H} \otimes \mathcal{K}_\infty = \overline{\operatorname{span}}\{\widehat{V}_\alpha^E (h \otimes \Omega_\infty^{\mathcal{K}}) : h \in \mathcal{H}, \alpha \in \tilde{\Lambda}\},$$

i.e., $\widehat{\underline{V}}^E$ is the minimal isometric dilation of $\underline{E}$. Note that the minimal isometric dilation is unique up to unitary equivalence (cf. [13]).

Similarly, associate a row contraction $\underline{C}$ to the unitary $\tilde{U}$ as in equation (1.4) and define isometries

$$\widehat{V}_j^C (\tilde{h} \otimes \eta) := \tilde{U}^*(\tilde{h} \otimes \epsilon_j) \otimes \eta \text{ for } j = 1, \dots, d \tag{2.1}$$

on the elementary tensors $\tilde{h} \otimes \eta \in \tilde{\mathcal{H}} \otimes \mathcal{K}_\infty$ and extend it linearly to obtain $\widehat{V}_j^C \in \mathcal{B}(\tilde{\mathcal{H}} \otimes \mathcal{K}_\infty)$ for $j = 1, \dots, d$. The tuple $\widehat{\underline{V}}^C = (\widehat{V}_1^C, \dots, \widehat{V}_d^C)$ on the space $\tilde{\mathcal{H}} \otimes \mathcal{K}_\infty$ is the minimal isometric dilation of $\underline{C} = (C_1, \dots, C_d)$. Recall that

$$U_m : \mathcal{H} \otimes \mathcal{K}_\infty \to \mathcal{H} \otimes \mathcal{K}_{[1,m-1]} \otimes \mathcal{P}_m \otimes \mathcal{K}_{[m+1,\infty)}$$

is nothing but the operator which acts as U on $\mathcal{H} \otimes \mathcal{K}_m$ and fixes other factors of the infinite tensor products. Similarly, we define $\tilde{U}_m$ using $\tilde{U}$.

Proposition 2.1. *Let $P_n := P_{\tilde{\mathcal{H}}} \otimes I_{\mathcal{P}_{[1,n]}} \otimes I_{\mathcal{K}_{[n+1,\infty)}} \in \mathcal{B}(\mathcal{H} \otimes \mathcal{P}_{[1,n]} \otimes \mathcal{K}_{[n+1,\infty)})$ for $n \in \mathbb{N}$. Then*

$$sot - \lim_{n\to\infty} \tilde{U}_1^* \dots \tilde{U}_n^* P_n U_n \dots U_1$$

exists and this limit defines a coisometry $\widehat{W} : \mathcal{H} \otimes \mathcal{K}_\infty \to \tilde{\mathcal{H}} \otimes \mathcal{K}_\infty$. Its adjoint $\widehat{W}^* : \tilde{\mathcal{H}} \otimes \mathcal{K}_\infty \to \mathcal{H} \otimes \mathcal{K}_\infty$ is given by

$$\widehat{W}^* = sot - \lim_{n\to\infty} U_1^* \dots U_n^* \tilde{U}_n \dots \tilde{U}_1.$$

Here sot stands for the strong operator topology.

Proof. At first we construct the adjoint $\widehat{W}^*$. For that consider the dense subset $\bigcup_{m\geq 1} \tilde{\mathcal{H}} \otimes \mathcal{K}_{[1,m]}$ of $\tilde{\mathcal{H}} \otimes \mathcal{K}_\infty$ and let an arbitrary simple tensor element of this dense subset be $\tilde{h} \otimes k_1 \otimes \dots \otimes k_\ell \otimes \Omega_{[\ell+1,\infty)}^{\mathcal{K}}$ for some $\ell \in \mathbb{N}, \tilde{h} \in \tilde{\mathcal{H}}$ and $k_i \in \mathcal{K}_i$. Set $a_p = U_1^* \dots U_p^* \tilde{U}_p \dots \tilde{U}_1 (\tilde{h} \otimes k_1 \otimes \dots \otimes k_\ell \otimes \Omega_{[\ell+1,\infty)}^{\mathcal{K}})$ for $p \in \mathbb{N}$. Since $U(\tilde{h} \otimes \Omega^{\mathcal{K}}) =$

$\tilde{U}(\tilde{h}\otimes\Omega^{\mathcal{K}})$ for all $\tilde{h}\in\tilde{\mathcal{H}}$, we have $a_\ell = a_{\ell+n}$ for all $n\in\mathbb{N}$. Therefore we deduce that

$$\lim_{n\to\infty} U_1^*\dots U_n^*\tilde{U}_n\dots\tilde{U}_1(\tilde{h}\otimes k_1\otimes\cdots\otimes k_\ell\otimes\Omega^{\mathcal{K}}_{[\ell+1,\infty)})$$

exists. Because U and $\tilde{U}$ are unitaries, we obtain an isometric extension $\widehat{W}^*$ to the whole of $\tilde{\mathcal{H}}\otimes\mathcal{K}_\infty$. Thus its adjoint is a coisometry $\widehat{W}:\mathcal{H}\otimes\mathcal{K}_\infty\to\tilde{\mathcal{H}}\otimes\mathcal{K}_\infty$.

Now we will derive the limit form for $\widehat{W}$ as claimed in the statement of the proposition. If $h\otimes\eta\in\mathcal{H}\otimes\mathcal{K}_{[1,k]}$, $\tilde{h}\otimes\tilde{\eta}\in\tilde{\mathcal{H}}\otimes\mathcal{K}_{[1,n]}$ and $k\le n$, then

$$\begin{aligned}\langle\widehat{W}(h\otimes\eta),\tilde{h}\otimes\tilde{\eta}\rangle &= \langle h\otimes\tilde{\eta},\widehat{W}^*(\tilde{h}\otimes\tilde{\eta})\rangle\\ &= \langle h\otimes\eta, U_1^*\dots U_n^*\tilde{U}_n\dots\tilde{U}_1(\tilde{h}\otimes\tilde{\eta})\rangle\\ &= \langle\tilde{U}_1^*\dots\tilde{U}_n^*P_nU_n\dots U_1(h\otimes\eta),\tilde{h}\otimes\tilde{\eta}\rangle.\end{aligned}$$

Consequently $\widehat{W} = sot-\lim_{n\to\infty}\tilde{U}_1^*\dots\tilde{U}_n^*P_nU_n\dots U_1$ on a dense subset and therefore it can be extended to the whole of $\mathcal{H}\otimes\mathcal{K}_\infty$. □

Observe that

$$\widehat{W}^*(\tilde{h}\otimes\Omega^{\mathcal{K}}_\infty) = \tilde{h}\otimes\Omega^{\mathcal{K}}_\infty \text{ for all } \tilde{h}\in\tilde{\mathcal{H}}. \tag{2.2}$$

Next we show that this coisometry $\widehat{W}$ intertwines between $\widehat{V}_j^E$ and $\widehat{V}_j^C$ for all $j=1,\dots,d$. For $j=1,\dots,d$, define

$$\begin{aligned}S_j:\mathcal{H}\otimes\mathcal{K}_\infty &\to \mathcal{H}\otimes\mathcal{P}_1\otimes\mathcal{K}_{[2,\infty)},\\ h\otimes\eta &\mapsto h\otimes\epsilon_j\otimes\eta.\end{aligned}$$

The following are immediate:

(1) $S_j^*(h\otimes p_1\otimes\eta) = \langle\epsilon_j,p_1\rangle(h\otimes\eta)$ for $(h\otimes p_1\otimes\eta)\in\mathcal{H}\otimes\mathcal{P}_1\otimes\mathcal{K}_{[2,\infty)}$.
(2) $\widehat{V}_j^E(h\otimes\eta) = U_1^*S_j(h\otimes\eta)$ for $h\otimes\eta\in\mathcal{H}\otimes\mathcal{K}_\infty$.
(3) $\widehat{V}_j^C(\tilde{h}\otimes\eta) = \tilde{U}_1^*S_j(\tilde{h}\otimes\eta)$ for $\tilde{h}\otimes\eta\in\tilde{\mathcal{H}}\otimes\mathcal{K}_\infty$.

Proposition 2.2. *If $\widehat{W}$ is as in Proposition 2.1, then*

$$\widehat{W}\widehat{V}_j^E = \widehat{V}_j^C\widehat{W},\qquad \widehat{V}_j^E\widehat{W}^* = \widehat{W}^*\widehat{V}_j^C \text{ for all } j=1,\dots,d.$$

Proof. If $h\in\mathcal{H},\eta\in\mathcal{K}_\infty,\tilde{h}\in\tilde{\mathcal{H}}$ and $k_i\in\mathcal{K}_i$, then by the three observations that were noted preceding this proposition we obtain for $j=1,\dots,d$

$$\begin{aligned}&\langle\,\widehat{W}\widehat{V}_j^E(h\otimes\eta),\ \tilde{h}\otimes k_1\otimes\cdots\otimes k_\ell\otimes\Omega^{\mathcal{K}}_{[\ell+1,\infty)}\,\rangle\\ &\quad= \langle\, U^*(h\otimes\epsilon_j)\otimes\eta,\ U_1^*\dots U_\ell^*\tilde{U}_\ell\dots\tilde{U}_1(\tilde{h}\otimes k_1\otimes\cdots\otimes k_\ell\otimes\Omega^{\mathcal{K}}_{[\ell+1,\infty)})\,\rangle.\end{aligned}$$

Substituting $\tilde{U}(\tilde{h}\otimes k_1) = \sum_i\tilde{h}^{(i)}\otimes k_1^{(i)}$ where $\tilde{h}^{(i)}\in\tilde{\mathcal{H}}$ and $k_1^{(i)}\in\mathcal{K}$ we obtain

$$\begin{aligned}&\langle\,\widehat{W}\widehat{V}_j^E(h\otimes\eta),\ \tilde{h}\otimes k_1\otimes\cdots\otimes k_\ell\otimes\Omega^{\mathcal{K}}_{[\ell+1,\infty)}\,\rangle\\ &\quad= \Big\langle\, h\otimes\epsilon_j\otimes\eta,\ U_2^*\dots U_\ell^*\tilde{U}_\ell\dots\tilde{U}_2\Big(\sum\nolimits_i(\tilde{h}^{(i)}\otimes k_1^{(i)})\\ &\qquad\otimes k_2\otimes\cdots\otimes k_\ell\otimes\Omega^{\mathcal{K}}_{[\ell+1,\infty)}\Big)\,\Big\rangle\end{aligned}$$

$$
\begin{aligned}
&= \sum_i \langle \epsilon_j,\ k_1^{(i)} \rangle \ \langle\ h \otimes \eta,\ \widehat{W}^*(\tilde{h}^{(i)} \otimes k_2 \otimes \cdots \otimes k_\ell \otimes \Omega^{\mathcal{K}}_{[\ell+1,\infty)})\ \rangle \\
&= \langle\ \widehat{W}(h \otimes \eta),\ S_j^* \tilde{U}_1(\tilde{h} \otimes k_1 \otimes \cdots \otimes k_\ell \otimes \Omega^{\mathcal{K}}_{[\ell+1,\infty)})\ \rangle \\
&= \langle\ \tilde{U}_1^* S_j \widehat{W}((h \otimes \eta),\ \tilde{h} \otimes k_1 \otimes \cdots \otimes k_\ell \otimes \Omega^{\mathcal{K}}_{[\ell+1,\infty)}\ \rangle \\
&= \langle\ \widehat{V}_j^C \widehat{W}(h \otimes \eta),\ \tilde{h} \otimes k_1 \otimes \cdots \otimes k_\ell \otimes \Omega^{\mathcal{K}}_{[\ell+1,\infty)}\ \rangle.
\end{aligned}
$$

Hence $\widehat{W}\widehat{V}_j^E = \widehat{V}_j^C \widehat{W}$ for all $j = 1, \ldots, d$. To obtain the other equation of the proposition we again use the last two of the three observations as follows: For $j = 1, \ldots, d$

$$
\begin{aligned}
&\widehat{W}^* \widehat{V}_j^C(\tilde{h} \otimes k_1 \otimes \cdots \otimes k_\ell \otimes \Omega^{\mathcal{K}}_{[\ell+1,\infty)}) \\
&\quad = \widehat{W}^* \tilde{U}_1^*(\tilde{h} \otimes \epsilon_j \otimes k_1 \otimes \cdots \otimes k_\ell \otimes \Omega^{\mathcal{K}}_{[\ell+2,\infty)}) \\
&\quad = U_1^* U_2^* \ldots U_{\ell+1}^* \tilde{U}_{\ell+1} \ldots \tilde{U}_2 \tilde{U}_1 \tilde{U}_1^*(\tilde{h} \otimes \epsilon_j \otimes k_1 \otimes \cdots \otimes k_\ell \otimes \Omega^{\mathcal{K}}_{[\ell+2,\infty)}) \\
&\quad = U_1^* U_2^* \ldots U_{\ell+1}^* \tilde{U}_{\ell+1} \ldots \tilde{U}_2 S_j(\tilde{h} \otimes k_1 \otimes \cdots \otimes k_\ell \otimes \Omega^{\mathcal{K}}_{[\ell+1,\infty)}) \\
&\quad = U_1^* S_j U_1^* \ldots U_\ell^* \tilde{U}_\ell \ldots \tilde{U}_1(\tilde{h} \otimes k_1 \otimes \cdots \otimes k_\ell \otimes \Omega^{\mathcal{K}}_{[\ell+1,\infty)}) \\
&\quad = \widehat{V}_j^E \widehat{W}^*(\tilde{h} \otimes k_1 \otimes \cdots \otimes k_\ell \otimes \Omega^{\mathcal{K}}_{[\ell+1,\infty)}).
\end{aligned}
$$
□

Further define

$$
\begin{aligned}
(\mathcal{H} \otimes \mathcal{K}_\infty)^\circ &:= (\mathcal{H} \otimes \mathcal{K}_\infty) \ominus (\tilde{\mathcal{H}} \otimes \Omega^{\mathcal{K}}_\infty), \\
(\tilde{\mathcal{H}} \otimes \mathcal{K}_\infty)^\circ &:= (\tilde{\mathcal{H}} \otimes \mathcal{K}_\infty) \ominus (\tilde{\mathcal{H}} \otimes \Omega^{\mathcal{K}}_\infty) \quad \text{and} \quad \mathcal{H}^\circ := \mathcal{H} \ominus \tilde{\mathcal{H}}.
\end{aligned} \tag{2.3}
$$

Let $\sum_{i=1}^k \xi_i \otimes \eta_i \in (\mathcal{H} \otimes \mathcal{K}_\infty)^\circ$ and $\tilde{h} \in \tilde{\mathcal{H}}$. Then for $j = 1, \ldots, d$

$$
\begin{aligned}
\left\langle \widehat{V}_j^E \left(\sum_i \xi_i \otimes \eta_i \right), \tilde{h} \otimes \Omega^{\mathcal{K}}_\infty \right\rangle &= \left\langle \sum_i U^*(\xi_i \otimes \epsilon_j) \otimes \eta_i, \tilde{h} \otimes \Omega^{\mathcal{K}}_\infty \right\rangle \\
&= \left\langle \sum_i \xi_i \otimes \epsilon_j \otimes \eta_i, \tilde{U}(\tilde{h} \otimes \Omega^{\mathcal{K}}_1) \otimes \Omega^{\mathcal{K}}_{[2,\infty)} \right\rangle = 0
\end{aligned}
$$

because $\tilde{U}$ maps into $\tilde{\mathcal{H}} \otimes \mathcal{P}$ and $\sum_{i=1}^k \xi_i \otimes \eta_i \perp \tilde{\mathcal{H}} \otimes \Omega^{\mathcal{K}}$. Therefore $\widehat{V}_j^E(\mathcal{H} \otimes \mathcal{K}_\infty)^\circ \subset (\mathcal{H} \otimes \mathcal{K}_\infty)^\circ$ for $j = 1, \ldots, d$. Similarly $\widehat{V}_j^C(\tilde{\mathcal{H}} \otimes \mathcal{K}_\infty)^\circ \subset (\tilde{\mathcal{H}} \otimes \mathcal{K}_\infty)^\circ$ for $j = 1, \ldots, d$. Set $V_j^E := \widehat{V}_j^E|_{(\mathcal{H} \otimes \mathcal{K}_\infty)^\circ}$ and $V_j^C := \widehat{V}_j^C|_{(\tilde{\mathcal{H}} \otimes \mathcal{K}_\infty)^\circ}$ for $j = 1, \ldots, d$. If we define

$$
W^* := \widehat{W}^*|_{(\tilde{\mathcal{H}} \otimes \mathcal{K}_\infty)^\circ},
$$

then by equation (2.2) it follows that $W^* \in \mathcal{B}((\tilde{\mathcal{H}} \otimes \mathcal{K}_\infty)^\circ, (\mathcal{H} \otimes \mathcal{K}_\infty)^\circ)$. The operator W^* is an isometry because it is a restriction of an isometry and W, the adjoint of W^*, is the restriction of $\widehat{W}$ to $(\mathcal{H} \otimes \mathcal{K}_\infty)^\circ$, i.e., $W = \widehat{W}|_{(\mathcal{H} \otimes \mathcal{K}_\infty)^\circ}$.

Remark 2.3. It follows that

$$
W V_j^E = V_j^C W
$$

for $j = 1, \ldots, d$.

3. Outgoing Cuntz scattering systems

In this section we aim to construct an outgoing Cuntz scattering system (cf. [3]) for our model. This will assist us in the next section to work with an input-output formalism and to associate a transfer function to the model.

Following are some notions from the multivariable operator theory:

Definition 3.1. Suppose $\underline{T} = (T_1, \dots, T_d)$ is a row contraction where $T_i \in \mathcal{B}(\mathcal{L})$.

(1) If T_i's are isometries with orthogonal ranges, then the tuple $\underline{T} = (T_1, \dots, T_d)$ is called a *row isometry*.
(2) If $\overline{\text{span}}_{j=1,\dots,d} T_j\mathcal{L} = \mathcal{L}$ and $\underline{T} = (T_1, \dots, T_d)$ is a row isometry, then $\underline{T}$ is called a *row unitary*.
(3) If there exist a subspace $\mathcal{E}$ of $\mathcal{L}$ such that $\mathcal{L} = \bigoplus_{\alpha\in\tilde{\Lambda}} T_\alpha\mathcal{E}$ and $\underline{T} = (T_1, \dots, T_d)$ is a row isometry, then $\underline{T}$ is called a *row shift* and $\mathcal{E}$ is called a *wandering subspace* of $\mathcal{L}$ w.r.t. $\underline{T}$.

Definition 3.2. A collection $(\mathcal{L}, \underline{V} = (V_1, \dots, V_d), \mathcal{G}_*^+, \mathcal{G})$ is called an outgoing Cuntz scattering system (cf. [3]), if $\underline{V}$ is a row isometry on the Hilbert space $\mathcal{L}$, and $\mathcal{G}_*^+$ and $\mathcal{G}$ are subspaces of $\mathcal{L}$ such that

(1) for $\mathcal{E}_* := \mathcal{L} \ominus \overline{\text{span}}_{j=1,\dots,d} V_j\mathcal{L}$, the tuple $\underline{V}|\mathcal{G}_*^+$ is a row shift where $\mathcal{G}_*^+ = \bigoplus_{\alpha\in\tilde{\Lambda}} V_\alpha\mathcal{E}_*$.
(2) there exist $\mathcal{E} := \mathcal{G} \ominus \overline{\text{span}}_{j=1,\dots,d} V_j\mathcal{G}$ with $\mathcal{G} = \bigoplus_{\alpha\in\tilde{\Lambda}} V_\alpha\mathcal{E}$, i.e., $\underline{V}|_{\mathcal{G}}$ is a row shift.

In the above definition the part (1) is the Wold decomposition (cf. [13]) of the row isometry $\underline{V}$ and therefore $\mathcal{G}_*^+$ can be derived from $\underline{V}$. But $\mathcal{G}_*^+$ is included in the data because it helps in describing the scattering phenomenon. We continue using the notations from the previous section. $\widehat{V}_j^E$'s are isometries with orthogonal ranges and because $(\epsilon_j)_{j=1}^d$ is an orthonormal basis of $\mathcal{P}$, we have

$$\overline{\text{span}}_{j=1,\dots,d} \widehat{V}_j^E(\mathcal{H} \otimes \mathcal{K}_\infty) = \mathcal{H} \otimes \mathcal{K}_\infty.$$

Thus $\widehat{\underline{V}}^E$ is a row unitary on $\mathcal{H}\otimes\mathcal{K}_\infty$. Now using the fact that $V_j^E = \widehat{V}_j^E|_{(\mathcal{H}\otimes\mathcal{K}_\infty)^\circ}$ we infer that V_j^E's are isometries with orthogonal ranges. Therefore $\underline{V}^E$ is a row isometry on $(\mathcal{H} \otimes \mathcal{K}_\infty)^\circ$.

Proposition 3.3. *If* $\mathcal{Y} := \tilde{\mathcal{H}} \otimes (\Omega_1^{\mathcal{K}})^\perp \otimes \Omega_{[2,\infty)}^{\mathcal{K}} \subset \tilde{\mathcal{H}} \otimes \mathcal{K}_\infty$, *then*

$$W^*\mathcal{Y} \perp \overline{\text{span}}_{j=1,\dots,d}\ V_j^E(\mathcal{H} \otimes \mathcal{K}_\infty)^\circ.$$

Proof. By Proposition 2.1 it is easy to see that

$$W^*\mathcal{Y} = U_1^*\tilde{U}_1\mathcal{Y} \subset \mathcal{H} \otimes \mathcal{K}_1 \otimes \Omega_{[2,\infty)}^{\mathcal{K}}. \tag{3.1}$$

Let $\tilde{h}_i \in \tilde{\mathcal{H}}$ and $k_i \perp \Omega_1^{\mathcal{K}}$ for $i = 1, \ldots, n$, i.e., $\sum_i \tilde{h}_i \otimes k_i \otimes \Omega^{\mathcal{K}}_{[2,\infty)} \in \mathcal{Y}$. For $\sum_k h_k \otimes \eta_k \in (\mathcal{H} \otimes \mathcal{K}_\infty)^\circ$ with $h_k \in \mathcal{H}$ and $\eta_k \in \mathcal{K}_\infty$

$$\begin{aligned}
&\langle\, W^*(\textstyle\sum_i \tilde{h}_i \otimes k_i \otimes \Omega^{\mathcal{K}}_{[2,\infty)}), V_j^E(\sum_k h_k \otimes \eta_k)\,\rangle \\
&= \langle\, U^*\tilde{U}(\textstyle\sum_i \tilde{h}_i \otimes k_i) \otimes \Omega^{\mathcal{K}}_{[2,\infty)}, \sum_k U^*(h_k \otimes \epsilon_j) \otimes \eta_k \,\rangle \\
&= \langle\, \tilde{U}(\textstyle\sum_i \tilde{h}_i \otimes k_i) \otimes \Omega^{\mathcal{K}}_{[2,\infty)}, \sum_k h_k \otimes \epsilon_j \otimes \eta_k \,\rangle = 0.
\end{aligned}$$

The last equality holds because $\sum_k h_k \otimes \eta_k \perp \tilde{\mathcal{H}} \otimes \Omega^{\mathcal{K}}_\infty$.

Thus $W^*\mathcal{Y} \perp \overline{\operatorname{span}}_{j=1,\ldots,d} V_j^E(\mathcal{H} \otimes \mathcal{K}_\infty)^\circ$. □

The following Proposition gives an explicit description of the Wold decomposition of $\underline{V}^E$:

Proposition 3.4. *If $\mathcal{Y}$ is defined as in the previous proposition, then $W^*\mathcal{Y}$ is a wandering subspace of $\underline{V}^E$, i.e., $V_\alpha^E(W^*\mathcal{Y}) \perp V_\beta^E(W^*\mathcal{Y})$ whenever $\alpha, \beta \in \tilde{\Lambda}$, $\alpha \neq \beta$, and*

$$W^*\mathcal{Y} = (\mathcal{H} \otimes \mathcal{K}_\infty)^\circ \ominus \overline{\operatorname{span}}_{j=1,\ldots,d} V_j^E(\mathcal{H} \otimes \mathcal{K}_\infty)^\circ.$$

Proof. By Proposition 3.3 it is immediate that $V_\alpha^E(W^*\mathcal{Y}) \perp V_\beta^E(W^*\mathcal{Y})$ whenever $\alpha, \beta \in \tilde{\Lambda}$, $\alpha \neq \beta$ and $W^*\mathcal{Y} \subset (\mathcal{H} \otimes \mathcal{K}_\infty)^\circ \ominus \overline{\operatorname{span}}_{j=1,\ldots,d} V_j^E(\mathcal{H} \otimes \mathcal{K}_\infty)^\circ$. The only thing that remains to be shown is that

$$(\mathcal{H} \otimes \mathcal{K}_\infty)^\circ \ominus \overline{\operatorname{span}}_{j=1,\ldots,d} V_j^E(\mathcal{H} \otimes \mathcal{K}_\infty)^\circ \subset W^*\mathcal{Y}.$$

Let $x \in (\mathcal{H} \otimes \mathcal{K}_\infty)^\circ \ominus \overline{\operatorname{span}}_{j=1,\ldots,d} V_j^E(\mathcal{H} \otimes \mathcal{K}_\infty)^\circ$. Write down the decomposition of x as $x_1 \oplus x_2$ w.r.t. $W^*\mathcal{Y} \oplus (W^*\mathcal{Y})^\perp$. So $x - x_1 = x_2$ is orthogonal to both $\overline{\operatorname{span}}_{j=1,\ldots,d} V_j^E(\mathcal{H} \otimes \mathcal{K}_\infty)^\circ$ and $W^*\mathcal{Y}$. Now we show that if any element in $(\mathcal{H} \otimes \mathcal{K}_\infty)^\circ$ is orthogonal to $\overline{\operatorname{span}}_{j=1,\ldots,d} V_j^E(\mathcal{H} \otimes \mathcal{K}_\infty)^\circ$ and $W^*\mathcal{Y}$, then it is the zero vector. Let x_0 be such an element. Because $x_0 \in (\mathcal{H} \otimes \mathcal{K}_\infty)^\circ$ and $x_0 \perp W^*\mathcal{Y}$,

$$x_0 \perp U^*(\tilde{\mathcal{H}} \otimes \epsilon_j) \otimes \Omega^{\mathcal{K}}_{[2,\infty)}$$

for $j = 1, \ldots, d$. This implies $x_0 \perp \overline{\operatorname{span}}_{j=1,\ldots,d} \widehat{V}_j^E(\tilde{\mathcal{H}} \otimes \Omega^{\mathcal{K}}_\infty)$. We also know that

$$x_0 \perp \overline{\operatorname{span}}_{j=1,\ldots,d} V_j^E(\mathcal{H} \otimes \mathcal{K}_\infty)^\circ (= \overline{\operatorname{span}}_{j=1,\ldots,d} \widehat{V}_j^E(\mathcal{H} \otimes \mathcal{K}_\infty)^\circ).$$

Therefore

$$x_0 \perp \overline{\operatorname{span}}_{j=1,\ldots,d} \widehat{V}_j^E(\mathcal{H} \otimes \mathcal{K}_\infty).$$

Since $\underline{\widehat{V}}^E$ is a row unitrary, $x_0 \perp \mathcal{H} \otimes \mathcal{K}_\infty$. So $x_0 = 0$ and hence $x = x_1 \in W^*\mathcal{Y}$. We conclude that $(\mathcal{H} \otimes \mathcal{K}_\infty)^\circ \ominus \overline{\operatorname{span}}_{j=1,\ldots,d} V_j^E(\mathcal{H} \otimes \mathcal{K}_\infty)^\circ \subset W^*\mathcal{Y}$. □

Proposition 3.5. *If $\mathcal{E} := \mathcal{H} \otimes (\Omega_1^{\mathcal{K}})^\perp \otimes \Omega^{\mathcal{K}}_{[2,\infty)} \subset (\mathcal{H} \otimes \mathcal{K}_\infty)^\circ$, then $V_\alpha^E \mathcal{E} \perp V_\beta^E \mathcal{E}$ whenever $\alpha, \beta \in \tilde{\Lambda}, \alpha \neq \beta$ and $(\mathcal{H} \otimes \mathcal{K}_\infty)^\circ = \mathcal{H}^\circ \oplus \bigoplus_{\alpha \in \tilde{\Lambda}} V_\alpha^E \mathcal{E}$.*

Proof. If $|\alpha| = |\beta|$ and $\alpha \neq \beta$, then it is easy to see that $V_\alpha^E \mathcal{E} \perp V_\beta^E \mathcal{E}$ because ranges of V_j^E's are mutually orthogonal. If $|\alpha| \neq |\beta|$ (without loss of generality we can assume that $|\alpha| > |\beta|$), then by taking the inner product at the tensor factor $\mathcal{K}_{|\alpha|+1}$ we obtain $V_\alpha^E \mathcal{E} \perp V_\beta^E \mathcal{E}$.

To prove the second part of the proposition, observe that for $n \in \mathbb{N}$,

$$\begin{aligned}
&\mathcal{H} \otimes \mathcal{K}_{[1,n]} \otimes \Omega^{\mathcal{K}}_{[n+1,\infty]} \\
&\quad = (\mathcal{H} \otimes \Omega^{\mathcal{K}}_\infty) \oplus (\mathcal{H} \otimes (\Omega^{\mathcal{K}}_1)^\perp \otimes \Omega^{\mathcal{K}}_{[2,\infty)}) \oplus (\mathcal{H} \otimes \mathcal{K}_1 \otimes \\
&\qquad (\Omega^{\mathcal{K}}_2)^\perp \otimes \Omega^{\mathcal{K}}_{[3,\infty)}) \oplus \cdots \oplus \ (\mathcal{H} \otimes \mathcal{K}_{[1,n-1]} \otimes (\Omega^{\mathcal{K}}_n)^\perp \otimes \Omega^{\mathcal{K}}_{[n+1,\infty)}) \\
&\quad = (\tilde{\mathcal{H}} \otimes \Omega^{\mathcal{K}}_\infty) \oplus (\mathcal{H}^\circ \otimes \Omega^{\mathcal{K}}_\infty) \oplus \mathcal{E} \oplus \bigoplus_{j=1}^{d} V_j^E \mathcal{E} \oplus \cdots \oplus \bigoplus_{|\alpha|=n-1}^{d} V_\alpha^E \mathcal{E}.
\end{aligned}$$

Taking $n \to \infty$ we have the following:

$$\mathcal{H} \otimes \mathcal{K}_\infty = (\tilde{\mathcal{H}} \otimes \Omega^{\mathcal{K}}_\infty) \oplus (\mathcal{H}^\circ \otimes \Omega^{\mathcal{K}}_\infty) \oplus \bigoplus_{\alpha \in \tilde{\Lambda}} V_\alpha^E \mathcal{E}.$$

Since $(\mathcal{H} \otimes \mathcal{K}_\infty)^\circ = (\mathcal{H} \otimes \mathcal{K}_\infty) \ominus (\tilde{\mathcal{H}} \otimes \Omega^{\mathcal{K}}_\infty)$, it follows that

$$(\mathcal{H} \otimes \mathcal{K}_\infty)^\circ = \mathcal{H}^\circ \oplus \bigoplus_{\alpha \in \tilde{\Lambda}} V_\alpha^E \mathcal{E}. \qquad \square$$

We sum up Propositions 3.3, 3.4 and 3.5 in the following theorem:

Theorem 3.6. *For a generalized repeated interaction model involving unitaries U and $\tilde{U}$ as before set $\mathcal{Y} := \tilde{\mathcal{H}} \otimes (\Omega^{\mathcal{K}}_1)^\perp \otimes \Omega^{\mathcal{K}}_{[2,\infty)}$ and $\mathcal{E} := \mathcal{H} \otimes (\Omega^{\mathcal{K}}_1)^\perp \otimes \Omega^{\mathcal{K}}_{[2,\infty)}$. If $\mathcal{E}_* := W^* \mathcal{Y}$, $\mathcal{G}_*^+ := \bigoplus_{\alpha \in \tilde{\Lambda}} V_\alpha^E \mathcal{E}_*$ and $\mathcal{G} := \bigoplus_{\alpha \in \tilde{\Lambda}} V_\alpha^E \mathcal{E}$, then the collection*

$$((\mathcal{H} \otimes \mathcal{K}_\infty)^\circ, \underline{V}^E = (V_1^E, \ldots, V_d^E), \mathcal{G}_*^+, \mathcal{G})$$

is an outgoing Cuntz scattering system such that $(\mathcal{H} \otimes \mathcal{K}_\infty)^\circ = \mathcal{H}^\circ \oplus \mathcal{G}$.

Remark 3.7. Applying arguments similar to those used for proving the second part of the Proposition 3.5 one can prove the following:

$$(\tilde{\mathcal{H}} \otimes \mathcal{K}_\infty)^\circ = \bigoplus_{\alpha \in \tilde{\Lambda}} V_\alpha^C \mathcal{Y}.$$

We refer the reader to Proposition 3.1 of [10] for a result in a similar direction.

4. $\tilde{\Lambda}$-linear systems and transfer functions

We would demonstrate that the outgoing Cuntz scattering system $((\mathcal{H} \otimes \mathcal{K}_\infty)^\circ$, $\underline{V}^E = (V_1^E, \ldots, V_d^E)$, $\mathcal{G}_*^+, \mathcal{G})$ from Theorem 3.6 has interesting relations with a generalization of the linear systems theory that is associated to our interaction model. For a given model involving unitaries U and $\tilde{U}$ as before, let us define the input space as

$$\mathcal{U} := \mathcal{E} = \mathcal{H} \otimes (\Omega_1^{\mathcal{K}})^{\perp} \otimes \Omega_{[2,\infty)}^{\mathcal{K}} \subset (\mathcal{H} \otimes \mathcal{K}_{\infty})^{\circ}$$

and the output space as

$$\mathcal{Y} = \tilde{\mathcal{H}} \otimes (\Omega_1^{\mathcal{K}})^{\perp} \otimes \Omega_{[2,\infty)}^{\mathcal{K}} \subset (\tilde{\mathcal{H}} \otimes \mathcal{K}_{\infty})^{\circ}.$$

Here we assume that a quantum system $\mathcal{A}$ interacts with a stream of copies of another quantum system $\mathcal{B}$ and we assume $\mathcal{H}$ is the (quantum mechanical) Hilbert space of $\mathcal{A}$. Let $\mathcal{K}_i$ be the Hilbert space of a part of a stream of copies of $\mathcal{B}$ at time i immediately before the interaction with $\mathcal{A}$. Let the Hilbert space $\mathcal{P}_i$ be that the part of a stream of copies of $\mathcal{B}$ at time i immediately after the interaction with $\mathcal{A}$. $\Omega^{\mathcal{K}}$ and $\Omega^{\mathcal{P}}$ denote states indicating that no copy of quantum system $\mathcal{B}$ is present and so no interaction is taking place at time i. Then $\eta \in \mathcal{U} = \mathcal{H} \otimes (\Omega_1^{\mathcal{K}})^{\perp} \otimes \Omega_{[2,\infty)}^{\mathcal{K}} \subset \mathcal{H} \otimes \mathcal{K}_{\infty}$ represents a vector state with copies of quantum system $\mathcal{B}$ arriving at time 1 and stimulating an interaction between the stream of copies of $\mathcal{A}$ and $\mathcal{B}$, but no further copy of $\mathcal{B}$ arriving at later times. But some activity is induced which goes on for a longer period.

Note that $\mathcal{H} \otimes \mathcal{K} = \mathcal{H} \oplus \mathcal{U}$ and $\tilde{\mathcal{H}} \otimes \mathcal{K} = \tilde{\mathcal{H}} \oplus \mathcal{Y}$. So U maps $\mathcal{H} \oplus \mathcal{U}$ onto $\mathcal{H} \otimes \mathcal{P}$ and $\tilde{U}$ maps $\tilde{\mathcal{H}} \oplus \mathcal{Y}$ onto $\tilde{\mathcal{H}} \otimes \mathcal{P}$. Using unitaries U and $\tilde{U}$ we define $F_j : \mathcal{H} \to \mathcal{U}$ and $D_j : \tilde{\mathcal{H}} \to \mathcal{Y}$ for $j = 1, \ldots, d$ by

$$\sum_{j=1}^{d} F_j^* \eta \otimes \epsilon_j := U(0 \oplus \eta), \; \sum_{j=1}^{d} D_j^* y \otimes \epsilon_j := \tilde{U}(0 \oplus y) \text{ for } \eta \in \mathcal{U} \text{ and } y \in \mathcal{Y}. \quad (4.1)$$

Combining equation (4.1) with equations (1.3) and (1.4) we have for $h \in \mathcal{H}$, $\eta \in \mathcal{U}$, $\tilde{h} \in \tilde{\mathcal{H}}$ and $y \in \mathcal{Y}$

$$U(h \oplus \eta) = \sum_{j=1}^{d} (E_j^* h + F_j^* \eta) \otimes \epsilon_j, \quad (4.2)$$

$$\tilde{U}(\tilde{h} \oplus y) = \sum_{j=1}^{d} (C_j^* \tilde{h} + D_j^* y) \otimes \epsilon_j \quad (4.3)$$

respectively. Using equation (4.3) it can be checked that

$$\tilde{U}^*(\tilde{h} \otimes \epsilon_j) = C_j \tilde{h} \oplus D_j \tilde{h} \text{ for } \tilde{h} \in \tilde{\mathcal{H}}; j = 1, \ldots, d. \quad (4.4)$$

Let us define

$$\tilde{C} := \sum_{j=1}^{d} D_j P_{\tilde{\mathcal{H}}} E_j^* : \mathcal{H} \to \mathcal{Y}, \; \tilde{D} := \sum_{j=1}^{d} D_j P_{\tilde{\mathcal{H}}} F_j^* : \mathcal{U} \to \mathcal{Y}$$

where $P_{\tilde{\mathcal{H}}}$ is the orthogonal projection onto $\tilde{\mathcal{H}}$. It follows that

$$P_{\mathcal{Y}} \tilde{U}^* P_1 U(h \oplus \eta) = \tilde{C} h + \tilde{D} \eta \quad (4.5)$$

where $h \in \mathcal{H}, \eta \in \mathcal{U}, P_1$ is as in Proposition 2.1 and $P_{\mathcal{Y}}$ is the orthogonal projection onto $\mathcal{Y}$.

Define a *colligation* of operators (cf. [3]) using the operators E_j^*'s, F_j^*'s, $\tilde{C}$ and $\tilde{D}$ by

$$\mathcal{C}_{U,\tilde{U}} := \begin{pmatrix} E_1^* & F_1^* \\ \vdots & \vdots \\ E_d^* & F_d^* \\ \tilde{C} & \tilde{D} \end{pmatrix} : \mathcal{H} \oplus \mathcal{U} \to \bigoplus_{j=1}^{d} \mathcal{H} \oplus \mathcal{Y}.$$

From the colligation $\mathcal{C}_{U,\tilde{U}}$ we get the following $\tilde{\Lambda}$-*linear system* $\sum_{U,\tilde{U}}$:

$$x(j\alpha) = E_j^* x(\alpha) + F_j^* u(\alpha), \tag{4.6}$$

$$y(\alpha) = \tilde{C} x(\alpha) + \tilde{D} u(\alpha) \tag{4.7}$$

where $j = 1, \ldots, d$ and $\alpha, j\alpha$ are words in $\tilde{\Lambda}$, and

$$x : \tilde{\Lambda} \to \mathcal{H}, \; u : \tilde{\Lambda} \to \mathcal{U}, \; y : \tilde{\Lambda} \to \mathcal{Y}.$$

If $x(\emptyset)$ and u are known, then using $\sum_{U,\tilde{U}}$ we can compute x and y recursively. Such a $\tilde{\Lambda}$-linear system is also called a noncommutative Fornasini–Marchesini system in [1] in reference to [6].

Let $z = (z_1, \ldots, z_d)$ be a d-tuple of formal noncommuting indeterminates. Define the Fourier transforms of x, u and y as

$$\hat{x}(z) = \sum_{\alpha \in \tilde{\Lambda}} x(\alpha) z^\alpha, \; \hat{u}(z) = \sum_{\alpha \in \tilde{\Lambda}} u(\alpha) z^\alpha, \; \hat{y}(z) = \sum_{\alpha \in \tilde{\Lambda}} y(\alpha) z^\alpha$$

respectively where $z^\alpha = z_{\alpha_n} \ldots z_{\alpha_1}$ for $\alpha = \alpha_n \ldots \alpha_1 \in \tilde{\Lambda}$. Assuming that z-variables commute with the coefficients the input-output relation

$$\hat{y}(z) = \Theta_{U,\tilde{U}}(z) \hat{u}(z)$$

can be obtained on setting $x(\emptyset) := 0$ where

$$\Theta_{U,\tilde{U}}(z) := \sum_{\alpha \in \tilde{\Lambda}} \Theta_{U,\tilde{U}}^{(\alpha)} z^\alpha := \tilde{D} + \tilde{C} \sum_{\beta \in \tilde{\Lambda}, j=1,\ldots,d} (E_{\bar{\beta}})^* F_j^* z^{\beta j}. \tag{4.8}$$

Here $\bar{\beta} = \beta_1 \ldots \beta_n$ is the reverse of $\beta = \beta_n \ldots \beta_1 \in \tilde{\Lambda}$ and $\Theta_{U,\tilde{U}}^{(\alpha)}$ maps $\mathcal{U}$ to $\mathcal{Y}$. The formal noncommutative power series $\Theta_{U,\tilde{U}}$ is called the *transfer function* associated to the unitaries U and $\tilde{U}$. The transfer function is a mathematical tool for encoding the evolution of a $\tilde{\Lambda}$-linear system. For $y(\alpha) \in \mathcal{Y}$ with $\sum_{\alpha \in \tilde{\Lambda}} \|y(\alpha)\|^2 < \infty$, any series $\sum_{\alpha \in \tilde{\Lambda}} y(\alpha) z^\alpha$ stands for a series converging to an element of $\ell^2(\tilde{\Lambda}, \mathcal{Y})$.

Theorem 4.1. *The map* $M_{\Theta_{U,\tilde{U}}} : \ell^2(\tilde{\Lambda}, \mathcal{U}) \to \ell^2(\tilde{\Lambda}, \mathcal{Y})$ *defined by*

$$M_{\Theta_{U,\tilde{U}}} \hat{u}(z) := \Theta_{U,\tilde{U}}(z) \hat{u}(z)$$

is a contraction.

Proof. Observe that $P_{\mathcal{Y}}\tilde{U}^*P_1U(\tilde{h}\otimes\Omega_\infty^{\mathcal{K}})=0$ for all $\tilde{h}\in\tilde{\mathcal{H}}$. Consider another colligation which is defined as follows:

$$\mathcal{C}^\circ_{U,\tilde{U}}:=\begin{pmatrix} E_1^{*\circ} & F_1^{*\circ}\\ \vdots & \vdots\\ E_d^{*\circ} & F_d^{*\circ}\\ \tilde{C}^\circ & \tilde{D}\end{pmatrix}:\mathcal{H}^\circ\oplus\mathcal{U}\to\bigoplus_{j=1}^{d}\mathcal{H}^\circ\oplus\mathcal{Y}$$

where $E_j^{*\circ}:=P_{\mathcal{H}^\circ}E_j^*|_{\mathcal{H}^\circ}:\mathcal{H}^\circ\to\mathcal{H}^\circ$, $F_j^{*\circ}:=P_{\mathcal{H}^\circ}F_j^*:\mathcal{U}\to\mathcal{H}^\circ$ and $\tilde{C}^\circ:=\tilde{C}|_{\mathcal{H}^\circ}:\mathcal{H}^\circ\to\mathcal{Y}$ for $j=1,\dots,d$. Recall that $\mathcal{H}^\circ$ and $(\mathcal{H}\otimes\mathcal{K}_\infty)^\circ$ were defined in equation array (2.3). Consider the outgoing Cuntz scattering system $((\mathcal{H}\otimes\mathcal{K}_\infty)^\circ,\underline{V}^E=(V_1^E,\dots,V_d^E),\mathcal{G}_*^+,\mathcal{G})$, with $(\mathcal{H}\otimes\mathcal{K}_\infty)^\circ=\mathcal{H}^\circ\oplus\mathcal{G}$, constructed by us in Theorem 3.6. In Chapter 5.2 of [3] it is shown that there is an associated unitary colligation

$$\begin{pmatrix} \hat{E}_1 & \hat{F}_1\\ \vdots & \vdots\\ \hat{E}_d & \hat{F}_d\\ \hat{M} & \hat{N}\end{pmatrix}:\mathcal{H}^\circ\oplus\mathcal{E}\to\bigoplus_{j=1}^{d}\mathcal{H}^\circ\oplus\mathcal{E}_* \tag{4.9}$$

such that $(\hat{E}_j,\hat{F}_j)=P_{\mathcal{H}^\circ}(V_j^E)^*|_{\mathcal{H}^\circ\oplus\mathcal{E}}$ and $(\hat{M},\hat{N})=P_{\mathcal{E}_*}|_{\mathcal{H}^\circ\oplus\mathcal{E}}$. Recall that $\mathcal{E}$ and $\mathcal{E}_*$ were introduced in Proposition 3.5 and Theorem 3.6 respectively. From equations (4.2) and (4.5) we observe that $(E_j^{*\circ},F_j^{*\circ})=P_{\mathcal{H}^\circ\otimes\epsilon_j}U|_{\mathcal{H}^\circ\oplus\mathcal{E}}$ (identifying $\mathcal{H}^\circ$ with $\mathcal{H}^\circ\otimes\epsilon_j$) and $(\tilde{C}^\circ,\tilde{D})=P_{\mathcal{Y}}\tilde{U}^*P_1U|_{\mathcal{H}^\circ\oplus\mathcal{E}}$. Using these observations we obtain the following relations:

$$\begin{aligned} U^*(E_j^{*\circ},F_j^{*\circ}) &= U^*P_{\mathcal{H}^\circ\otimes\epsilon_j}U|_{\mathcal{H}^\circ\oplus\mathcal{E}}=P_{U^*(\mathcal{H}^\circ\otimes\epsilon_j)}|_{\mathcal{H}^\circ\oplus\mathcal{E}}=P_{V_j^E\mathcal{H}^\circ}|_{\mathcal{H}^\circ\oplus\mathcal{E}}\\ &= V_j^EP_{\mathcal{H}^\circ}(V_j^E)^*|_{\mathcal{H}^\circ\oplus\mathcal{E}}=V_j^E(\hat{E}_j,\hat{F}_j) \end{aligned}\tag{4.10}$$

for $j=1,\dots,d$ and

$$\begin{aligned} U^*\tilde{U}(\tilde{C}^\circ,\tilde{D}) &= U^*\tilde{U}P_{\mathcal{Y}}\tilde{U}^*P_1U|_{\mathcal{H}^\circ\oplus\mathcal{E}}=U^*P_{\tilde{U}\mathcal{Y}}P_1U|_{\mathcal{H}^\circ\oplus\mathcal{E}}=U^*P_{\tilde{U}\mathcal{Y}}U|_{\mathcal{H}^\circ\oplus\mathcal{E}}\\ &= P_{U^*\tilde{U}\mathcal{Y}}|_{\mathcal{H}^\circ\oplus\mathcal{E}}=P_{W^*\mathcal{Y}}|_{\mathcal{H}^\circ\oplus\mathcal{E}}\quad\text{(by equation (3.1))}\\ &= P_{\mathcal{E}_*}|_{\mathcal{H}^\circ\oplus\mathcal{E}}=(\hat{M},\hat{N}). \end{aligned}\tag{4.11}$$

Let $\hat{u}(z)=\sum_{\alpha\in\tilde{\Lambda}}u(\alpha)z^\alpha\in\ell^2(\tilde{\Lambda},\mathcal{U})$ with $u(\alpha)\in\mathcal{U}$ such that $\sum_{\alpha\in\tilde{\Lambda}}\|u(\alpha)\|^2<\infty$. We would prove that

$$\|M_{\Theta_{U,\tilde{U}}}\hat{u}(z)\|^2\le\|\hat{u}(z)\|^2.$$

Define $x:\tilde{\Lambda}\to\mathcal{H}$ by equation (4.6) such that $x(\emptyset)=0$. Further, define $x^\circ(\alpha):=P_{\mathcal{H}^\circ}x(\alpha)$ for all $\alpha\in\tilde{\Lambda}$. Now applying the projection $P_{\mathcal{H}^\circ}$ to relation (4.6) on both sides and using the fact $\tilde{\mathcal{H}}$ is invariant under E_j^* for $j=1,\dots,d$ we obtain the following relation:

$$x^\circ(j\alpha)=E_j^{*\circ}x^\circ(\alpha)+F_j^{*\circ}u(\alpha)\text{ for all }\alpha\in\tilde{\Lambda},\ j=1,\dots,d. \tag{4.12}$$

Because $P_{\mathcal{Y}}\tilde{U}^* P_1 U_1(\tilde{h} \otimes \Omega_\infty^{\mathcal{K}}) = 0$ for all $\tilde{h} \in \tilde{\mathcal{H}}$ we conclude by equation (4.5) that

$$\tilde{C}\tilde{h} = 0 \text{ for } \tilde{h} \in \tilde{\mathcal{H}}. \tag{4.13}$$

This implies

$$\tilde{C}x(\alpha) = \tilde{C}^\circ x^\circ(\alpha) \text{ for all } \alpha \in \tilde{\Lambda}. \tag{4.14}$$

Define $y : \tilde{\Lambda} \to \mathcal{Y}$ by

$$y(\alpha) := \tilde{C}x(\alpha) + \tilde{D}u(\alpha) \tag{4.15}$$

for all $\alpha \in \tilde{\Lambda}$. Recall that the input-output relation stated just before the theorem is

$$\hat{y}(z) = \sum_{\alpha \in \tilde{\Lambda}} y(\alpha) z^\alpha = \Theta_{U,\tilde{U}}(z)\hat{u}(z)(= M_{\Theta_{U,\tilde{U}}}\hat{u}(z)).$$

Using the unitary colligation given in equation (4.9) we have

$$\begin{aligned}
\|x^\circ(\alpha)\|^2 + \|u(\alpha)\|^2 &= \sum_{j=1}^d \|\hat{E}_j x^\circ(\alpha) + \hat{F}_j u(\alpha)\|^2 + \|\hat{M}x^\circ(\alpha) + \hat{N}u(\alpha)\|^2 \\
&= \sum_{j=1}^d \|E_j^{*\circ} x^\circ(\alpha) + F_j^{*\circ} u(\alpha)\|^2 + \|\tilde{C}^\circ x^\circ(\alpha) + \tilde{D}u(\alpha)\|^2 \\
&= \sum_{j=1}^d \|x^\circ(j\alpha)\|^2 + \|\tilde{C}x(\alpha) + \tilde{D}u(\alpha)\|^2 \\
&= \sum_{j=1}^d \|x^\circ(j\alpha)\|^2 + \|y(\alpha)\|^2
\end{aligned}$$

for all $\alpha \in \tilde{\Lambda}$. In the above calculation equations (4.10), (4.11), (4.12), (4.14) and (4.15) respectively have been used. This gives us

$$\|u(\alpha)\|^2 - \|y(\alpha)\|^2 = \sum_{j=1}^d \|x^\circ(j\alpha)\|^2 - \|x^\circ(\alpha)\|^2$$

for all $\alpha \in \tilde{\Lambda}$. Summing over all $\alpha \in \tilde{\Lambda}$ with $|\alpha| \leq n$ and using the fact that $x^\circ(\emptyset) = 0$ we obtain

$$\sum_{|\alpha| \leq n} \|u(\alpha)\|^2 - \sum_{|\alpha| \leq n} \|y(\alpha)\|^2 = \sum_{|\alpha| = n+1} \|x^\circ(\alpha)\|^2 \geq 0 \text{ for all } n \in \mathbb{N}.$$

Therefore

$$\sum_{|\alpha| \leq n} \|y(\alpha)\|^2 \leq \sum_{|\alpha| \leq n} \|u(\alpha)\|^2 \text{ for all } n \in \mathbb{N}.$$

Finally taking limit $n \to \infty$ both the sides we get that $M_{\Theta_{U,\tilde{U}}}$ is a contraction. □

$M_{\Theta_{U,\tilde{U}}}$ is a *multi-analytic operator* ([15]) (also called *analytic intertwining operator* in [3]) because

$$M_{\Theta_{U,\tilde{U}}}\left(\sum_{\alpha\in\tilde{\Lambda}} u(\alpha)z^\alpha z^j\right) = M_{\Theta_{U,\tilde{U}}}\left(\sum_{\alpha\in\tilde{\Lambda}} u(\alpha)z^\alpha\right) z^j \quad \text{for } j=1,\dots,d,$$

i.e., $M_{\Theta_{U,\tilde{U}}}$ intertwines with right translation. The noncommutative power series $\Theta_{U,\tilde{U}}$ is called the *symbol* of $M_{\Theta_{U,\tilde{U}}}$.

5. Transfer functions, observability and scattering

We would now establish that the transfer function can be derived from the coisometry W of Section 2. In the last section d-tuple $z=(z_1,\dots,z_d)$ of formal noncommuting indeterminates were employed. Treat $(z^\alpha)_{\alpha\in\tilde{\Lambda}}$ as an orthonormal basis of $\ell^2(\tilde{\Lambda},\mathbb{C})$. Assume $\mathcal{Y}$ and $\mathcal{U}$ to be the spaces associated with our model with unitaries U and $\tilde{U}$ as in the last section. It follows from Remark 3.7 that there exist a unitary operator $\tilde{\Gamma}:(\tilde{\mathcal{H}}\otimes\mathcal{K}_\infty)^\circ \to \ell^2(\tilde{\Lambda},\mathcal{Y})$ defined by

$$\tilde{\Gamma}(V^C_\alpha y) := yz^{\bar{\alpha}} \quad \text{for all} \quad \alpha\in\tilde{\Lambda}, y\in\mathcal{Y}.$$

We observe the following intertwining relation:

$$\tilde{\Gamma}(V^C_\alpha y) = (\tilde{\Gamma}y)z^{\bar{\alpha}}. \tag{5.1}$$

Similarly, using Theorem 3.6, we can define a unitary operator $\Gamma:(\mathcal{H}\otimes\mathcal{K}_\infty)^\circ(=(\mathcal{H}^\circ\oplus\mathcal{G}))\to\mathcal{H}^\circ\oplus\ell^2(\tilde{\Lambda},\mathcal{U})$ by

$$\Gamma(\mathring{h}\oplus V^E_\alpha\eta) := \mathring{h}\oplus\eta z^{\bar{\alpha}} \text{ for all } \alpha\in\tilde{\Lambda}$$

where $\mathring{h}\in\mathcal{H}^\circ, \eta\in\mathcal{U}$. In this case the intertwining relation is

$$\Gamma(V^E_\alpha\eta) = (\Gamma\eta)z^{\bar{\alpha}}. \tag{5.2}$$

Using the coisometric operator W, which appears in Remark 2.3, we define Γ_W by the following commutative diagram:

$$\begin{array}{ccc} (\mathcal{H}\otimes\mathcal{K}_\infty)^\circ & \xrightarrow{W} & (\tilde{\mathcal{H}}\otimes\mathcal{K}_\infty)^\circ \\ \Gamma\big\downarrow & & \big\downarrow\tilde{\Gamma} \\ \mathcal{H}^\circ\oplus\ell^2(\tilde{\Lambda},\mathcal{U}) & \xrightarrow{\Gamma_W} & \ell^2(\tilde{\Lambda},\mathcal{Y}), \end{array} \tag{5.3}$$

i.e., $\Gamma_W = \tilde{\Gamma}W\Gamma^{-1}$.

Theorem 5.1. *Γ_W defined by the above commutative diagram satisfies*

$$\Gamma_W|_{\ell^2(\tilde{\Lambda},\,\mathcal{U})} = M_{\Theta_{U,\tilde{U}}}.$$

Proof. Using the intertwining relation $V_j^C W = W V_j^E$ from Remark 2.3, and equations (5.1) and (5.2) we obtain

$$\begin{aligned}\Gamma_W(\eta z^\beta z^j) &= \tilde{\Gamma} W \Gamma^{-1}(\eta z^\beta z^j) = \tilde{\Gamma} W V_j^E V_{\bar{\beta}}^E \eta \\ &= \tilde{\Gamma} V_j^C V_{\bar{\beta}}^C W\eta = (\tilde{\Gamma} W\eta) z^\beta z^j = \Gamma_W(\eta z^\beta) z^j\end{aligned}$$

for $\eta \in \mathcal{U}, \beta \in \tilde{\Lambda}, j = 1, \ldots, d$. Hence, $\Gamma_W|_{\ell^2(\tilde{\Lambda},\, \mathcal{U})}$ is a multi-analytic operator. For computing its symbol we determine $\Gamma_W \eta$ for $\eta \in \mathcal{U}$, where η is identified with $\eta z^\phi \in \ell^2(\tilde{\Lambda}, \mathcal{U})$. For $\alpha = \alpha_{n-1} \ldots \alpha_1 \in \tilde{\Lambda}$ let P_α be the orthogonal projection onto

$$\begin{aligned}&\tilde{\Gamma}^{-1}\{f \in \ell^2(\tilde{\Lambda}, \mathcal{Y}) : f = y z^\alpha \text{ for some } y \in \mathcal{Y}\} \\ &= V_{\bar{\alpha}}^C \mathcal{Y} = \tilde{U}_1^* \ldots \tilde{U}_{n-1}^*(\tilde{\mathcal{H}} \otimes \epsilon_{\alpha_1} \otimes \cdots \otimes \epsilon_{\alpha_{n-1}} \otimes (\Omega_n^{\mathcal{K}})^\perp \otimes \Omega_{[n+1,\infty)}^{\mathcal{K}})\end{aligned}$$

with $\tilde{U}_i$'s as in Proposition 2.1.

Recall that the tuple $\underline{E}$ associated with the unitary U is a lifting of the tuple $\underline{C}$ (associated with the unitary $\tilde{U}$) and so $\underline{E}$ can be written as a block matrix in terms of $\underline{C}$ as follows: $E_j = \begin{pmatrix} C_j & 0 \\ B_j & A_j \end{pmatrix}$ for $j = 1, \ldots, d$ w.r.t. to the decomposition $\mathcal{H} = \tilde{\mathcal{H}} \oplus \mathcal{H}^\circ$ where $\underline{B}$ and $\underline{A}$ are some row contractions. Because $\underline{E}$ is a coisometric lifting of $\underline{C}$ we have

$$\sum_{j=1}^d C_j C_j^* = I \quad \text{and} \quad \sum_{j=1}^d C_j B_j^* = 0$$

(cf. [5]). Now using these relations and equations (4.2), (4.3) and (4.4) it can be easily verified that

$$P_\alpha \tilde{U}_1^* \ldots \tilde{U}_n^* P_n U_n \ldots U_1 \eta = P_\alpha \tilde{U}_1^* \ldots \tilde{U}_m^* P_m U_m \ldots U_1 \eta \text{ for all } m \geq n, \eta \in \mathcal{U}.$$

Using the formula of W from Proposition 2.1 we obtain

$$P_\alpha W\eta = P_\alpha \tilde{U}_1^* \ldots \tilde{U}_n^* P_n U_n \ldots U_1 \eta \text{ for } \eta \in \mathcal{U}.$$

Finally for $\eta \in \mathcal{U}$

$$P_\alpha \tilde{U}_1^* \ldots \tilde{U}_n^* P_n U_n \ldots U_1 \eta = \begin{cases} \tilde{D}\eta & \text{if } n = 1, \alpha = \emptyset, \\ V_{\bar{\alpha}}^C(\tilde{C} E_{\alpha_{n-1}}^* \ldots E_{\alpha_2}^* F_{\alpha_1}^* \eta) & \text{if } n = |\alpha| + 1 \geq 2. \end{cases}$$

This implies for $\eta \in \mathcal{U}$

$$\tilde{\Gamma} W \Gamma^{-1} \eta = \tilde{\Gamma} W \eta = \tilde{D}\eta \oplus \sum_{|\alpha| \geq 1} (\tilde{C} E_{\alpha_{n-1}}^* \ldots E_{\alpha_2}^* F_{\alpha_1}^* \eta) z^\alpha.$$

Comparing this with equation (4.8) we conclude that $\Gamma_W|_{\ell^2(\tilde{\Lambda},\, \mathcal{U})} = M_{\Theta_{U,\tilde{U}}}$. □

Note that the Theorem 4.1 and its proof concern the transfer function of the $\tilde{\Lambda}$-linear system and has nothing to do with the scattering theory. Theorem 5.1, on the other hand, is the scattering theory part in the sense of Lax–Phillips [12]. The same function $M_{\Theta_{U,\tilde{U}}}$ relates the outgoing Fourier representation for a vector in the ambient scattering Hilbert space to the incoming Fourier representation for the same vector. This makes $M_{\Theta_{U,\tilde{U}}}$ the scattering function for the outgoing Cuntz scattering system. We introduce a notion from the linear systems theory for our model:

Definition 5.2. The observability operator $W_0 : \mathcal{H}^\circ \to \ell^2(\tilde{\Lambda}, \mathcal{Y})$ is defined as the restriction of the operator Γ_W to $\mathcal{H}^\circ$, i.e., $W_0 = \Gamma_W|_{\mathcal{H}^\circ}$.

It follows that $W_0\mathring{h} = (\tilde{C}(E_{\tilde{\alpha}})^*\mathring{h})_{\alpha\in\tilde{\Lambda}}$. Popescu has studied the similar types of operators called Poisson kernels in [16].

Definition 5.3. If there exist $k, K > 0$ such that for all $\mathring{h} \in \mathcal{H}^\circ$

$$k\|\mathring{h}\|^2 \leq \sum_{\alpha\in\tilde{\Lambda}} \|\tilde{C}(E_{\tilde{\alpha}})^*\mathring{h}\|^2 = \|W_0\mathring{h}\|^2 \leq K\|\mathring{h}\|^2,$$

then the $\tilde{\Lambda}$-linear system is called (uniformly) observable.

We illustrate below that the notion of observability is closely related to the scattering theory notions of noncommutative Markov chains. Observability of a system for $\dim \mathcal{H} < \infty$ is interpreted as the property of the system that in the absence of $\mathcal{U}$-inputs we can determine the original state $h \in \mathcal{H}^\circ$ of the system from all $\mathcal{Y}$-outputs at all times. Uniform observability is an analog of this for $\dim \mathcal{H} = \infty$.

We extend W_0 to

$$\widehat{W_0} : (\tilde{\mathcal{H}} \oplus \mathcal{H}^\circ)(= \mathcal{H}) \longrightarrow \tilde{\mathcal{H}} \oplus \ell^2(\tilde{\Lambda}, \mathcal{Y})$$

by defining $\widehat{W_0}\tilde{h} := \tilde{h}$ for all $\tilde{h} \in \tilde{\mathcal{H}}$. If W_0 is uniformly observable, then using $\hat{k} = k$ and $\hat{K} = \max\{1, K\}$ the above inequalities can be extended to $\widehat{W_0}$ on $\mathcal{H}$ as

$$\hat{k}\|h\|^2 \leq \|\widehat{W_0}h\|^2 \leq \hat{K}\|h\|^2$$

for all $h \in \mathcal{H}$.

Before stating the main theorem of this section regarding observability we recall from [5] the following: Let $\underline{C}$ be a row contraction on a Hilbert space $\mathcal{H}_C$. The lifting $\underline{E}$ of $\underline{C}$ is called *subisometric* [5] if the minimal isometric dilations $\widehat{\underline{V}}^E$ and $\widehat{\underline{V}}^C$ of $\underline{E}$ and $\underline{C}$ respectively are unitarily equivalent and the corresponding unitary, which intertwines between $\widehat{V}_i^E$ and $\widehat{V}_i^C$ for all $i = 1, 2, \ldots, d$, acts as identity on $\mathcal{H}_C$. Some of the techniques used here are from the scattering theory of noncommutative Markov chains (cf. [11], [8]).

Theorem 5.4. *For any $\tilde{\Lambda}$-linear system associated to a generalized repeated interaction model with unitaries $U, \tilde{U}$ the following statements are equivalent:*

(a) *The system is (uniformly) observable.*
(b) *The observability operator W_0 is isometric.*
(c) *The tuple $\underline{E}$ associated with the unitary U is a subisometric lifting of the tuple $\underline{C}$ (associated with the unitary $\tilde{U}$).*
(d) *$W : (\mathcal{H} \otimes \mathcal{K}_\infty)^\circ \to (\tilde{\mathcal{H}} \otimes \mathcal{K}_\infty)^\circ$ is unitary.*

If one of the above holds, then

(e) *The transfer function* $\Theta_{U,\tilde{U}}$ *is inner, i.e.,* $M_{\Theta_{U,\tilde{U}}} : \ell^2(\tilde{\Lambda},\mathcal{U}) \to \ell^2(\tilde{\Lambda},\mathcal{Y})$ *is isometric.*

If we have additional assumptions, viz. $\dim \mathcal{H} < \infty$ *and* $\dim \mathcal{P} \geq 2$, *then the converse holds, i.e.,* (e) *implies all of* (a), (b), (c) *and* (d).

Proof. Clearly $(d) \Rightarrow (b) \Rightarrow (a)$. We now prove $(a) \Rightarrow (d)$. Because the system is (uniformly) observable there exist $k > 0$ such that for all $\mathring{h} \in \mathcal{H}^\circ$

$$k\|\mathring{h}\|^2 \leq \|W_0\mathring{h}\|^2.$$

Since $\bigcup_{m\geq 1} \mathcal{H} \otimes \mathcal{K}_{[1,m]}$ is a dense subspace of $\mathcal{H} \otimes \mathcal{K}_\infty$, for any $0 \neq \eta \in \mathcal{H} \otimes \mathcal{K}_\infty$ there exist $n \in \mathbb{N}$ and $\eta' \in \mathcal{H} \otimes \mathcal{K}_{[1,n]}$ such that

$$\|\eta - \eta'\| < \frac{\sqrt{k}}{\sqrt{k}+1}\|\eta\|.$$

Let $\eta_0 \in \mathcal{H} \otimes \mathcal{K}_{[1,n]}$. Suppose $U_n \ldots U_1\eta_0 = h_0 \otimes p_0 \otimes \Omega^{\mathcal{K}}_{[n+1,\infty)}$, where $h_0 \in \mathcal{H}$, $p_0 \in \mathcal{P}_{[1,n]}$. Then clearly

$$\lim_{N\to\infty} \|\tilde{U}_1^* \ldots \tilde{U}_n^*\tilde{U}_{n+1}^* \ldots \tilde{U}_N^* P_N U_N \ldots U_{n+1}U_n \ldots U_1\eta_0\| = \|\widehat{W}_0 h_0\|\|p_0\|$$

and thus by Proposition 2.1 it is equal to $\|\widehat{W}\eta_0\|$. Because the system is (uniformly) observable,

$$\|\widehat{W}_0 h_0\|\|p_0\| \geq \sqrt{k}\|h_0\|\|p_0\|.$$

Therefore $\|\widehat{W}\eta_0\|^2 \geq k\|\eta_0\|^2$. However, in general $U_n \ldots U_1\eta_0 = \sum_j h_0^{(j)} \otimes p_0^{(j)} \otimes \Omega^{\mathcal{K}}_{[n+1,\infty)}$ with $h_0^{(j)} \in \mathcal{H}$ and some mutually orthogonal vectors $p_0^{(j)} \in \mathcal{P}_{[1,n]}$. By using the above inequality for each term of the summation and then adding them we find that in general for all $\eta_0 \in \mathcal{H} \otimes \mathcal{K}_{[1,n]}$

$$\|\widehat{W}\eta_0\|^2 \geq k\|\eta_0\|^2.$$

In particular, for $\eta' \in \mathcal{H} \otimes \mathcal{K}_{[1,n]}$ we have the above inequality. Therefore

$$\begin{aligned}\|\widehat{W}\eta\| &\geq \|\widehat{W}\eta'\| - \|\widehat{W}(\eta' - \eta)\| \\ &\geq \sqrt{k}\|\eta'\| - \|\eta - \eta'\| \\ &\geq \sqrt{k}\|\eta\| - (\sqrt{k}+1)\|\eta - \eta'\| > 0.\end{aligned}$$

This implies $\widehat{W}\eta \neq 0$ for all $0 \neq \eta \in \mathcal{H} \otimes \mathcal{K}_\infty$ and hence $\widehat{W}$ is injective. Recall that $\widehat{W}$ is a coisometry and an injective coisometry is unitary. Further, because $\widehat{W}(\tilde{h} \otimes \Omega^{\mathcal{K}}_\infty) = \tilde{h} \otimes \Omega^{\mathcal{K}}_\infty$ for all $\tilde{h} \in \tilde{\mathcal{H}}$ it follows that W is unitary. This establishes $(a) \Rightarrow (d)$ and we have proved $(a) \Leftrightarrow (b) \Leftrightarrow (d)$.

Next we prove $(d) \Leftrightarrow (c)$. Assume that (d) holds. Since W is unitary, clearly $\widehat{W}$ is unitary. We know that $\widehat{W}$ intertwines between the minimal isometric dilations $\underline{\widehat{V}}^E$ and $\underline{\widehat{V}}^C$ of $\underline{E}$ and $\underline{C}$ respectively. Hence $\underline{E}$ is a subisometric lifting of $\underline{C}$.

Conversely, if we assume (c), then by the definition of subisometric lifting there exist a unitary operator

$$\widehat{W}_1 : \mathcal{H} \otimes \mathcal{K}_\infty \longrightarrow \tilde{\mathcal{H}} \otimes \mathcal{K}_\infty$$

which intertwines between $\underline{\widehat{V}}^E$ and $\underline{\widehat{V}}^C$, and $\widehat{W}_1$ acts as an identity on $\tilde{\mathcal{H}} \otimes \Omega^{\mathcal{K}}_\infty$. To prove W is unitary it is enough to prove $\widehat{W}$ is unitary. We show that $\widehat{W} = \widehat{W}_1$. By the definition of the minimal isometric dilation we know that $\tilde{\mathcal{H}} \otimes \mathcal{K}_\infty = \overline{\operatorname{span}}\{\widehat{V}^C_\alpha(\tilde{h} \otimes \Omega^{\mathcal{K}}_\infty) : \tilde{h} \in \tilde{\mathcal{H}}, \alpha \in \tilde{\Lambda}\}$. For $j = 1, \dots, d$ and $\tilde{h} \in \tilde{\mathcal{H}}$, by equation (2.2) and Proposition 2.2,

$$\begin{aligned}\widehat{W}^* \widehat{V}^C_j(\tilde{h} \otimes \Omega^{\mathcal{K}}_\infty) &= \widehat{V}^E_j \widehat{W}^*(\tilde{h} \otimes \Omega^{\mathcal{K}}_\infty) = \widehat{V}^E_j(\tilde{h} \otimes \Omega^{\mathcal{K}}_\infty) \\ &= \widehat{W}_1^* \widehat{V}^C_j \widehat{W}_1(\tilde{h} \otimes \Omega^{\mathcal{K}}_\infty) = \widehat{W}_1^* \widehat{V}^C_j(\tilde{h} \otimes \Omega^{\mathcal{K}}_\infty).\end{aligned}$$

Thus $\widehat{W}^* = \widehat{W}_1^*$ and hence $\widehat{W} = \widehat{W}_1$.

To prove $(d) \Rightarrow (e)$ we at first note that since W is unitary, Γ_W is also unitary. By Theorem 4.2, we have $M_{\Theta_{U,\tilde{U}}} = \Gamma_W|_{\ell^2(\tilde{\Lambda},\mathcal{U})}$. Since a restriction of a unitary operator is an isometry, $M_{\Theta_{U,\tilde{U}}}$ is isometric.

Finally with the additional assumptions $\dim \mathcal{H} < \infty$ and $\dim \mathcal{P} \geq 2$, we show $(e) \Rightarrow (b)$. Define

$$\mathcal{H}_{\text{scat}} := \mathcal{H} \cap \widehat{W}^*(\tilde{\mathcal{H}} \otimes \mathcal{K}_\infty) = \tilde{\mathcal{H}} \oplus \{\mathring{h} \in \mathcal{H}^\circ : \|W_0 \mathring{h}\| = \|\mathring{h}\|\}.$$

Since $\|\widehat{W}_0 h\| = \lim_{n\to\infty} \|\tilde{U}_1 \dots \tilde{U}_n \tilde{P}_n U_n \dots U_1 h\|$ by Proposition 2.1, the following can be easily verified:

$$U(\mathcal{H}_{\text{scat}} \otimes \Omega^{\mathcal{K}}) \subset \mathcal{H}_{\text{scat}} \otimes \mathcal{P}. \tag{5.4}$$

Because $M_{\Theta_{U,\tilde{U}}} = \Gamma_W|_{\ell^2(\tilde{\Lambda},\mathcal{U})}$ is isometric by (e), it can be checked that

$$U(\mathcal{H} \otimes (\Omega^{\mathcal{K}})^\perp) \subset \mathcal{H}_{\text{scat}} \otimes \mathcal{P}. \tag{5.5}$$

Combining equations (5.4) and (5.5) we have

$$U^*((\mathcal{H} \ominus \mathcal{H}_{\text{scat}}) \otimes \mathcal{P}) \subset (\mathcal{H} \ominus \mathcal{H}_{\text{scat}}) \otimes \Omega^{\mathcal{K}}.$$

Since $\dim \mathcal{H} < \infty$ and $\dim \mathcal{P} \geq 2$, we obtain $\mathcal{H} \ominus \mathcal{H}_{\text{scat}} = \{0\}$, i.e., $\mathcal{H} = \mathcal{H}_{\text{scat}}$. This implies W_0 is isometric and hence $(e) \Rightarrow (b)$. □

6. Transfer functions and characteristic functions of liftings

Continuing with the study of our generalized repeated interaction model, from equations (2.1) and (4.4) we obtain

$$\widehat{V}^C_j(\tilde{h} \otimes \Omega^{\mathcal{K}}_\infty) = (C_j \tilde{h} \oplus D_j \tilde{h}) \otimes \Omega^{\mathcal{K}}_{[2,\infty)} \text{ for } \tilde{h} \in \tilde{\mathcal{H}} \text{ and } j = 1, \dots, d. \tag{6.1}$$

Let $D_C := (I - \underline{C}^* \underline{C})^{\frac{1}{2}} : \bigoplus_{i=1}^d \tilde{\mathcal{H}} \to \bigoplus_{i=1}^d \tilde{\mathcal{H}}$ denote the defect operator and $\mathcal{D}_C := \overline{\text{Range } D_C}$. The full Fock space over $\mathbb{C}^d$ $(d \geq 2)$ denoted by $\mathcal{F}$ is

$$\mathcal{F} = \mathbb{C} \oplus \mathbb{C}^d \oplus (\mathbb{C}^d)^{\otimes^2} \oplus \cdots \oplus (\mathbb{C}^d)^{\otimes^m} \oplus \cdots.$$

The vector $e_\emptyset := 1 \oplus 0 \oplus \cdots$ is called the vacuum vector. Let $\{e_1, \dots, e_d\}$ be the standard orthonormal basis of $\mathbb{C}^d$. For $\alpha \in \tilde{\Lambda}$ and $|\alpha| = n$, e_α denote the vector $e_{\alpha_1} \otimes e_{\alpha_2} \otimes \cdots \otimes e_{\alpha_n}$ in the full Fock space $\mathcal{F}$. We recall that Popescu's construction [13] of the minimal isometric dilation $\underline{\tilde{V}}^C = (\tilde{V}_1^C, \dots, \tilde{V}_d^C)$ on $\tilde{\mathcal{H}} \oplus (\mathcal{F} \otimes \mathcal{D}_C)$ of the tuple $\underline{C}$ is

$$\tilde{V}_j^C\Big(\tilde{h} \oplus \sum_{\alpha \in \tilde{\Lambda}} e_\alpha \otimes d_\alpha\Big) = C_j\tilde{h} \oplus \Big[e_\emptyset \otimes (D_C)_j\tilde{h} + e_j \otimes \sum_{\alpha \in \tilde{\Lambda}} e_\alpha \otimes d_\alpha\Big]$$

for $\tilde{h} \in \tilde{\mathcal{H}}$ and $d_\alpha \in \mathcal{D}_C$ where $(D_C)_j\tilde{h} = D_C(0, \dots, \tilde{h}, \dots, 0)$ ($\tilde{h}$ is embedded at the jth component). So

$$\tilde{V}_j^C\tilde{h} = C_j\tilde{h} \oplus (e_\emptyset \otimes (D_C)_j\tilde{h}) \text{ for } \tilde{h} \in \tilde{\mathcal{H}} \text{ and } j = 1, \dots, d. \tag{6.2}$$

From equations (6.1) and (6.2) it follows that

$$\Big\|\sum_{j=1}^d D_j\tilde{h}_j\Big\|^2 = \Big\|\sum_{j=1}^d (D_C)_j\tilde{h}_j\Big\|^2 \tag{6.3}$$

where $\tilde{h}_j \in \tilde{\mathcal{H}}$ for $j = 1, \dots, d$. Let $\Phi_C : \overline{\text{span}}\{D_j\tilde{h} : \tilde{h} \in \tilde{\mathcal{H}}, j = 1, \dots, d\} \to \mathcal{D}_C$ be the unitary given by

$$\Phi_C\Big(\sum_{j=1}^d D_j\tilde{h}_j\Big) = \sum_{j=1}^d (D_C)_j\tilde{h}_j \text{ for } \tilde{h}_j \in \tilde{\mathcal{H}} \text{ and } j = 1, \dots, d.$$

Similarly for E_i's and F_i's obtained from interaction U in equation (4.2) we set $D_E := (I - \underline{E}^*\underline{E})^{\frac{1}{2}} : \bigoplus_{i=1}^d \mathcal{H} \to \bigoplus_{i=1}^d \mathcal{H}$ and $\mathcal{D}_E := \overline{\text{Range } D_E}$, and define another unitary operator $\Phi_E : \overline{\text{span}}\{F_jh : h \in \mathcal{H}, j = 1, \dots, d\} \to \mathcal{D}_E$ by

$$\Phi_E\Big(\sum_{j=1}^d F_jh_j\Big) = \sum_{j=1}^d (D_E)_jh_j \text{ for } h_j \in \mathcal{H} \text{ and } j = 1, \dots, d.$$

The second equation of (4.1) yields

$$\sum_{j=1}^d D_jD_j^*y = y \text{ for } y \in \mathcal{Y}.$$

This implies

$$\overline{\text{span}}\{D_j\tilde{h} : \tilde{h} \in \tilde{\mathcal{H}}, j = 1, \dots, d\} = \mathcal{Y}.$$

Similarly, we can show that $\overline{\text{span}}\{F_jh : h \in \mathcal{H}, j = 1, \dots, d\} = \mathcal{U}$. Thus Φ_C is a unitary from $\mathcal{Y}$ onto $\mathcal{D}_C$ and Φ_E is a unitary from $\mathcal{U}$ onto $\mathcal{D}_E$. As a consequence we have for $i, j = 1, \dots, d$

$$D_j^*D_i = (D_C)_j^*(D_C)_i = \delta_{ij}I - C_j^*C_i, \tag{6.4}$$
$$F_j^*F_i = (D_E)_j^*(D_E)_i = \delta_{ij}I - E_j^*E_i. \tag{6.5}$$

Define unitaries $\tilde{M}_{\Phi_C} : \ell^2(\tilde{\Lambda}, \mathcal{Y}) \to \mathcal{F} \otimes \mathcal{D}_C$ and $\tilde{\Phi}_E : \mathcal{U}z^{\emptyset} \to e_{\emptyset} \otimes \mathcal{D}_E$ by

$$\tilde{M}_{\Phi_C}\left(\sum_{\alpha \in \tilde{\Lambda}} y_\alpha z^\alpha\right) := \sum_{\alpha \in \tilde{\Lambda}} e_{\tilde{\alpha}} \otimes \Phi_C(y_\alpha),$$
$$\tilde{\Phi}_E(uz^{\emptyset}) := e_{\emptyset} \otimes \Phi_E u$$

which would be useful in comparing transfer functions with characteristic functions.

Define $D_{*,A} := (I - \underline{A}\underline{A}^*)^{\frac{1}{2}} : \mathcal{H}^\circ \to \mathcal{H}^\circ$ and $\mathcal{D}_{*,A} := \overline{\text{Range } D_{*,A}}$. Because $\underline{E}$ is a coisometric lifting of $\underline{C}$, using Theorem 2.1 of [5] we conclude that there exist an isometry $\gamma : \mathcal{D}_{*,A} \to \mathcal{D}_C$ with $\gamma D_{*,A} h = \underline{B}^* h$ for all $h \in \mathcal{H}^\circ$. Further, for $h \in \mathcal{H}^\circ$

$$\begin{aligned}
\Phi_C \tilde{C} h &= \Phi_C \sum_{j=1}^{d} D_j P_{\tilde{\mathcal{H}}} E_j^* h = \Phi_C \sum_{j=1}^{d} D_j P_{\tilde{\mathcal{H}}}(B_j^* h \oplus A_j^* h) \\
&= \Phi_C \sum_{j=1}^{d} D_j B_j^* h = \sum_{j=1}^{d} (D_C)_j B_j^* h \\
&= D_C \underline{B}^* h = \underline{B}^* h.
\end{aligned}$$

The last equality holds because for the coisometric tuple $\underline{C}$ the operator D_C is the projection onto $\mathcal{D}_C$ and Range $\underline{B}^* \subset \mathcal{D}_C$. This implies

$$\Phi_C \tilde{C} h = \gamma D_{*,A} h. \tag{6.6}$$

The characteristic function $M_{C,E} : \mathcal{F} \otimes \mathcal{D}_E \to \mathcal{F} \otimes \mathcal{D}_C$ of lifting $\underline{E}$ of $\underline{C}$, which was introduced in [5], and its symbol $\Theta_{C,E}$ has the following expansion: For $i = 1, \dots, d$ and $h \in \tilde{\mathcal{H}}$

$$\Theta_{C,E}(D_E)_i h = e_{\emptyset} \otimes [(D_C)_i h - \gamma D_{*,A} B_i h] - \sum_{|\alpha| \geq 1} e_\alpha \otimes \gamma D_{*,A} (A_\alpha)^* B_i h, \tag{6.7}$$

and for $h \in \mathcal{H}^\circ$

$$\begin{aligned}
\Theta_{C,E}\,(D_E)_i h = &\ - e_{\emptyset} \otimes \gamma D_{*,A} A_i h \\
&+ \sum_{j=1}^{d} e_j \otimes \sum_{\alpha} e_\alpha \otimes \gamma D_{*,A} (A_\alpha)^* (\delta_{ji} I - A_j^* A_i) h.
\end{aligned} \tag{6.8}$$

Theorem 6.1. *Let U and $\tilde{U}$ be unitaries associated with a generalized repeated interaction model, and the lifting $\underline{E}$ of $\underline{C}$ be the corresponding lifting. Then the characteristic function $M_{C,E}$ coincides with the transfer function $\Theta_{U,\tilde{U}}$, i.e.,*

$$\tilde{M}_{\Phi_C} \Theta_{U,\tilde{U}}(z) = \Theta_{C,E} \tilde{\Phi}_E.$$

Proof. If $h \in \mathcal{H}$ and $i = 1, \ldots, d$, then by equation (4.8)

$$\begin{aligned} \tilde{M}_{\Phi_C}\Theta_{U,\tilde{U}}(z)(F_i h z^{\emptyset}) &= \tilde{M}_{\Phi_C}\Big[\tilde{D}\, z^{\emptyset} + \sum_{\beta\in\tilde{\Lambda}, j=1,\ldots,d} \tilde{C}(E_{\bar{\beta}})^* F_j^* z^{\beta j}\Big](F_i h z^{\emptyset}) \\ &= \tilde{M}_{\Phi_C}\Big[\tilde{D}F_i h\, z^{\emptyset} + \sum_{\beta\in\tilde{\Lambda}, j=1,\ldots,d} \tilde{C}(E_{\bar{\beta}})^* F_j^* F_i h\, z^{\beta j}\Big]. \end{aligned} \tag{6.9}$$

Case 1. $h \in \tilde{\mathcal{H}}$:

$$\begin{aligned} \tilde{D}F_i h &= \sum_{j=1}^{d} D_j P_{\tilde{\mathcal{H}}} F_j^* F_i h = \sum_{j=1}^{d} D_j P_{\tilde{\mathcal{H}}}(\delta_{ij} I - E_j^* E_i)h \\ &= D_i h - \Big(\sum_{j=1}^{d} D_j P_{\tilde{\mathcal{H}}} E_j^*\Big) E_i h = D_i h - \tilde{C}E_i h \\ &= D_i h - \tilde{C}(C_i h \oplus B_i h) = D_i h - \tilde{C}B_i h. \end{aligned}$$

Second and last equalities follow from equations (6.5) and (4.13) respectively. By equation (6.5) again we obtain

$$\begin{aligned} &\sum_{\beta\in\tilde{\Lambda}, j=1,\ldots,d} \tilde{C}(E_{\bar{\beta}})^* F_j^* F_i h\, z^{\beta j} \\ &= \sum_{\beta\in\tilde{\Lambda}, j=1,\ldots,d} \tilde{C}(E_{\bar{\beta}})^* (\delta_{ij} I - E_j^* E_i) h\, z^{\beta j} \\ &= \sum_{\beta\in\tilde{\Lambda}} \tilde{C}(E_{\bar{\beta}})^* h\, z^{\beta i} - \sum_{\beta\in\tilde{\Lambda}, j=1,\ldots,d} \tilde{C}(E_{\bar{\beta}})^* E_j^* E_i h\, z^{\beta j} \\ &= - \sum_{\beta\in\tilde{\Lambda}, j=1,\ldots,d} \tilde{C}(E_{\bar{\beta}})^* E_j^* E_i h\, z^{\beta j} \\ &\qquad \text{(because } \tilde{C}(E_{\bar{\beta}})^* h = \tilde{C}(C_{\bar{\beta}})^* h = 0 \text{ by equation (4.13))} \\ &= - \sum_{\beta\in\tilde{\Lambda}, j=1,\ldots,d} \tilde{C}(E_{\bar{\beta}})^* \big((C_j^* C_i + B_j^* B_i) h \oplus A_j^* B_i h\big)\, z^{\beta j} \\ &= - \sum_{\beta\in\tilde{\Lambda}, j=1,\ldots,d} \tilde{C}(A_{\bar{\beta}})^* A_j^* B_i h\, z^{\beta j} \qquad \text{(by equation (4.13))} \\ &= - \sum_{|\alpha|\geq 1} \tilde{C}(A_{\bar{\alpha}})^* B_i h\, z^{\alpha}. \end{aligned}$$

So by equation (6.9) we have for all $i = 1, \ldots, d$ and $h \in \tilde{\mathcal{H}}$

$$\begin{aligned} \tilde{M}_{\Phi_C}\Theta_{U,\tilde{U}}(z)(F_i h z^{\emptyset}) &= \tilde{M}_{\Phi_C}\Big[(D_i h - \tilde{C}B_i h)\, z^{\emptyset} - \sum_{|\alpha|\geq 1} \tilde{C}(A_{\bar{\alpha}})^* B_i h\, z^{\alpha}\Big] \\ &= e_{\emptyset} \otimes \Phi_C(D_i h - \tilde{C}B_i h) - \sum_{|\alpha|\geq 1} e_{\bar{\alpha}} \otimes \Phi_C(\tilde{C}(A_{\bar{\alpha}})^* B_i h) \end{aligned}$$

$$= e_\emptyset \otimes [(D_C)_i h - \gamma D_{*,A} B_i h] - \sum_{|\alpha|\geq 1} e_{\bar{\alpha}} \otimes \gamma D_{*,A}(A_{\bar{\alpha}})^* B_i h.$$

By equation (6.7) it follows that

$$\begin{aligned} \tilde{M}_{\Phi_C} \Theta_{U,\tilde{U}}(z)(F_i h z^\emptyset) &= \Theta_{C,E}(e_\emptyset \otimes (D_E)_i h) \\ &= \Theta_{C,E} \tilde{\Phi}_E (F_i h z^\emptyset). \end{aligned}$$

Case 2. $h \in \mathcal{H}^\circ$:

$$\begin{aligned} \tilde{D} F_i h &= \sum_{j=1}^d D_j P_{\tilde{\mathcal{H}}} F_j^* F_i h = \sum_{j=1}^d D_j P_{\tilde{\mathcal{H}}} (\delta_{ij} I - E_j^* E_i) h \\ &= D_i P_{\tilde{\mathcal{H}}} h - \left(\sum_{j=1}^d D_j P_{\tilde{\mathcal{H}}} E_j^* \right) E_i h = -\tilde{C} A_i h \end{aligned}$$

The second equality follows from equation (6.5). By equations (6.5) and (4.13) again we obtain

$$\begin{aligned} \sum_{\beta \in \tilde{\Lambda}, j=1,\ldots,d} \tilde{C}(E_{\bar{\beta}})^* F_j^* F_i h \; z^{\beta j} &= \sum_{\beta \in \tilde{\Lambda}, j=1,\ldots,d} \tilde{C}(E_{\bar{\beta}})^* (\delta_{ij} I - E_j^* E_i) h \; z^{\beta j} \\ &= \sum_{\beta \in \tilde{\Lambda}, j=1,\ldots,d} \tilde{C}(A_{\bar{\beta}})^* (\delta_{ij} I - A_j^* A_i) h \; z^{\beta j}. \end{aligned}$$

So by equation (6.9) we have for all $i = 1, \ldots, d$ and $h \in \mathcal{H}^\circ$

$$\begin{aligned} &\tilde{M}_{\Phi_C} \Theta_{U,\tilde{U}}(z)(F_i h z^\emptyset) \\ &= \tilde{M}_{\Phi_C} \left[-\tilde{C} A_i h \; z^\emptyset + \sum_{\beta \in \tilde{\Lambda}, j=1,\ldots,d} \tilde{C}(A_{\bar{\beta}})^* (\delta_{ij} I - A_j^* A_i) h \; z^{\beta j} \right] \\ &= -e_\emptyset \otimes \Phi_C(\tilde{C} A_i h) + \sum_{\beta \in \tilde{\Lambda}, j=1,\ldots,d} e_j \otimes e_{\bar{\beta}} \otimes \Phi_C(\tilde{C}(A_{\bar{\beta}})^* (\delta_{ij} I - A_j^* A_i) h) \\ &= -e_\emptyset \otimes \gamma D_{*,A} A_i h + \sum_{\beta \in \tilde{\Lambda}, j=1,\ldots,d} e_j \otimes e_{\bar{\beta}} \otimes \gamma D_{*,A}(A_{\bar{\beta}})^* (\delta_{ij} I - A_j^* A_i) h. \end{aligned}$$

By equation (6.8) it follows that

$$\begin{aligned} \tilde{M}_{\Phi_C} \Theta_{U,\tilde{U}}(z)(F_i h z^\emptyset) &= \Theta_{C,E}(e_\emptyset \otimes (D_E)_i h) \\ &= \Theta_{C,E} \tilde{\Phi}_E (F_i h z^\emptyset). \end{aligned}$$

Hence we conclude that

$$\tilde{M}_{\Phi_C} \Theta_{U,\tilde{U}}(z) = \Theta_{C,E} \tilde{\Phi}_E. \qquad \square$$

The transfer function is a notion affiliated with the input/state/output linear system, while the scattering function is a notion affiliated with the scattering theory in the sense of Lax–Phillips. For our repeated interaction model Theorem 6.1 elucidates that the transfer function is identifiable with the characteristic function of the associated lifting. This establishes a strong connection between a model for quantum systems and the multivariate operator theory. Connections between them were also endorsed in other works like [2], [8], [4] and [10], and this indicates that such approaches to quantum systems using multi-analytic operators are promising.

Acknowledgment

The first author received a support from UKIERI to visit Aberystwyth University, UK in July 2011 which was helpful for this project.

References

[1] J.A. Ball, G. Groenewald, T. Malakorn, *Conservative structured noncommutative multidimensional linear systems. The state space method generalizations and applications*, 179–223, Oper. Theory Adv. Appl., 161, Birkhäuser, Basel (2006).

[2] B.V.R. Bhat, *An index theory for quantum dynamical semigroups*, Trans. Amer. Math. Soc., **348** (1996) 561–583.

[3] J.A. Ball, V. Vinnikov, *Lax–Phillips scattering and conservative linear systems: a Cuntz-algebra multidimensional setting*, Mem. Amer. Math. Soc., **178** (2005).

[4] S. Dey, R. Gohm, *Characteristic functions for ergodic tuples*, Integral Equations and Operator Theory, **58** (2007), 43–63.

[5] S. Dey,; R. Gohm, *Characteristic functions of liftings*, J. Operator Theory, **65** (2011), 17–45.

[6] E. Fornasini,; G. Marchesini, *Doubly-indexed Dynamical Systems: State Space Models and Structural Properties*, Math. Systems Theory, **12** (1978), 59–72.

[7] J. Gough, R. Gohm, Yanagisawa: *Linear Quantum feedback Networks*, Phys. Rev. A, **78** (2008).

[8] R. Gohm, *Noncommutative stationary processes,* Lecture Notes in Mathematics, 1839, Springer-Verlag, Berlin (2004).

[9] R. Gohm, *Non-commutative Markov chains and multi-analytic operators*, J. Math. Anal. Appl., **364** (2010), 275–288.

[10] R. Gohm, *Transfer function for pairs of wandering subspaces*, Spectral theory, mathematical system theory, evolution equations, differential and difference equations, 385–398, Oper. Theory Adv. Appl., 221, Birkhäuser/Springer Basel AG, Basel, (2012).

[11] B. Kümmerer, H. Maassen, *A scattering theory for Markov chains*, Infin. Dimens. Anal. Quantum Probab. Relat. Top. **3** (2000), 161–176.

[12] P.D. Lax, R.S. Phillips, *Scattering theory,* Pure and Applied Mathematics **26** Academic press, New York-London, (1967).

[13] G. Popescu, *Isometric dilations for infinite sequences of noncommuting operators,* Trans. Amer. Math. Soc., **316** (1989), 523–536.

[14] G. Popescu, *Characteristic functions for infinite sequences of noncommuting operators,* J. Operator Theory, **22** (1989), 51–71.

[15] G. Popescu, *Multi-analytic operators on Fock spaces*, Math. Ann., **303** (1995), 31–46.

[16] G. Popescu, *Poisson transforms on some C^*-algebras generated by isometries*, J. Funct. Anal., **161** (1999), 27–61.

[17] G. Popescu, *Free holomorphic functions on the unit ball of* $\mathcal{B}(\mathcal{H})^n$, J. Funct. Anal., **241** (2006), 268–333.

[18] M. Reed, B. Simon, Methods of modern mathematical physics. III. Scattering theory. Academic Press [Harcourt Brace Jovanovich, Publishers], New York-London, (1979).

[19] B. Sz.-Nagy, C. Foias, *Harmonic analysis of operators on Hilbert space,* North–Holland Publ., Amsterdam-Budapest (1970).

[20] M. Yanagisawa, H. Kimura, *Transfer function approach to quantum control, part I: Dynamics of Quantum feedback systems,* IEEE Transactions on Automatic control, **48** (2003), no. 12, 2107–2120.

Santanu Dey and Kalpesh J. Haria
Department of Mathematics
Indian Institute of Technology Bombay
Powai, Mumbai-400076, India
e-mail: `dey@math.iitb.ac.in`
`kalpesh@math.iitb.ac.in`

Operator Theory:
Advances and Applications, Vol. 240, 137–188

Some Remarks on the Spectral Problem Underlying the Camassa–Holm Hierarchy

Fritz Gesztesy and Rudi Weikard

Dedicated with great pleasure to Ludwig Streit on the occasion of his 75th birthday.

Abstract. We study particular cases of left-definite eigenvalue problems

$$A\psi = \lambda B\psi, \quad \text{with } A \geq \varepsilon I \text{ for some } \varepsilon > 0 \text{ and } B \text{ self-adjoint,}$$

but B not necessarily positive or negative definite, applicable, in particular, to the eigenvalue problem underlying the Camassa–Holm hierarchy. In fact, we will treat a more general version where A represents a positive definite Schrödinger or Sturm–Liouville operator T in $L^2(\mathbb{R}; dx)$ associated with a differential expression of the form $\tau = -(d/dx)p(x)(d/dx) + q(x)$, $x \in \mathbb{R}$, and B represents an operator of multiplication by $r(x)$ in $L^2(\mathbb{R}; dx)$, which, in general, is not a weight, that is, it is not nonnegative (or nonpositive) a.e. on $\mathbb{R}$. In fact, our methods naturally permit us to treat certain classes of distributions (resp., measures) for the coefficients q and r and hence considerably extend the scope of this (generalized) eigenvalue problem, without having to change the underlying Hilbert space $L^2(\mathbb{R}; dx)$. Our approach relies on rewriting the eigenvalue problem $A\psi = \lambda B\psi$ in the form $A^{-1/2}BA^{-1/2}\chi = \lambda^{-1}\chi$, $\chi = A^{1/2}\psi$, and a careful study of (appropriate realizations of) the operator $A^{-1/2}BA^{-1/2}$ in $L^2(\mathbb{R}; dx)$.

In the course of our treatment, we review and employ various necessary and sufficient conditions for q to be relatively bounded (resp., compact) and relatively form bounded (resp., form compact) with respect to $T_0 = -d^2/dx^2$ defined on $H^2(\mathbb{R})$. In addition, we employ a supersymmetric formalism which permits us to factor the second-order operator T into a product of two first-order operators familiar from (and inspired by) Miura's transformation linking the KdV and mKdV hierarchy of nonlinear evolution equations. We also treat the case of periodic coefficients q and r, where q may be a distribution and r generates a measure and hence no smoothness is assumed for q and r.

Mathematics Subject Classification (2010). Primary 34B24, 34C25, 34K13, 34L05, 34L40, 35Q58, 47A10, 47A75; Secondary 34B20, 34C10, 34L25, 37K10, 47A63, 47E05.

Keywords. Camassa–Holm hierarchy, left-definite spectral problems, distributional coefficients, Floquet theory, supersymmetric formalism.

1. Introduction

In this paper we are interested in a particular realization of a generalized left-definite spectral problem originally derived from the Camassa–Holm hierarchy of integrable nonlinear evolution equations.

Before specializing to the one-dimensional context at hand, we briefly address the notion of generalized spectral problems associated with operator pencils of the type $A - zB$, $z \in \mathbb{C}$, for appropriate densely defined and closed linear operators A and B in a complex, separable Hilbert space $\mathcal{H}$. As discussed in [90, Sect. VII.6], there are several (and in general, inequivalent) ways to reformulate such generalized spectral problems. For instance, if B is boundedly invertible, one may consider the spectral problem for the operators $B^{-1}A$ or AB^{-1}, and in some cases (e.g., if $B \geq \varepsilon I_{\mathcal{H}}$ for some $\varepsilon > 0$, a case also called a right-definite spectral problem) also that of $B^{-1/2}AB^{-1/2}$. Similarly, if A is boundedly invertible, the spectral problem for the linear pencil $A - zB$ can be reformulated in terms of the spectral problems for $A^{-1}B$ or BA^{-1}, and sometimes (e.g., if $A \geq \varepsilon I_{\mathcal{H}}$ for some $\varepsilon > 0$, a case also called a left-definite spectral problem)) in terms of that of $A^{-1/2}BA^{-1/2}$.

There exists an enormous body of literature for these kinds of generalized spectral problems and without any possibility of achieving completeness, we refer, for instance, to [2], [12], [55], [69], [73], [74], [75], [76], [100], [118], [133], and the extensive literature cited therein in the context of general boundary value problems. In the context of indefinite Sturm–Liouville-type boundary value problems we mention, for instance, [6], [8], [11], [13], [14], [15], [16], [18], [19], [20], [21], [23], [30], [31], [32], [37], [47], [52], [56], [85], [86], [88], [91], [92], [93], [101], [102], [119], [134], [137, Chs. 5, 11, 12], and again no attempt at a comprehensive account of the existing literature is possible due to the enormous volume of the latter.

The prime motivation behind our attempt to study certain left-definite eigenvalue problems is due to their natural occurrence in connection with the Camassa–Holm (CH) hierarchy. For a detailed treatment and an extensive list of references we refer to [60], [61, Ch. 5] and [62]. The first few equations of the CH hierarchy (cf., e.g., [61, Sect. 5.2] for a recursive approach to the CH hierarchy) explicitly read (with $u = u(x,t)$, $(x,t) \in \mathbb{R}^2$)

$$\begin{aligned}
\mathrm{CH}_0(u) &= 4u_{t_0} - u_{xxt_0} + u_{xxx} - 4u_x = 0, \\
\mathrm{CH}_1(u) &= 4u_{t_1} - u_{xxt_1} - 2uu_{xxx} - 4u_x u_{xx} + 24uu_x + c_1(u_{xxx} - 4u_x) = 0, \\
\mathrm{CH}_2(u) &= 4u_{t_2} - u_{xxt_2} + 2u^2 u_{xxx} - 8uu_x u_{xx} - 40u^2 u_x \\
&\quad + 2(u_{xxx} - 4u_x)\mathcal{G}\big(u_x^2 + 8u^2\big) - 8(4u - u_{xx})\mathcal{G}\big(u_x u_{xx} + 8uu_x\big) \\
&\quad + c_1(-2uu_{xxx} - 4u_x u_{xx} + 24uu_x) + c_2(u_{xxx} - 4u_x) = 0, \ \text{ etc.},
\end{aligned} \tag{1.1}$$

for appropriate constants c_ℓ, $\ell \in \mathbb{N}$. Here $\mathcal{G}$ is given by

$$\mathcal{G}\colon \begin{cases} L^\infty(\mathbb{R}; dx) \to L^\infty(\mathbb{R}; dx), \\ v \mapsto (\mathcal{G}v)(x) = \frac{1}{4}\int_{\mathbb{R}} dy\, e^{-2|x-y|} v(y), \quad x \in \mathbb{R}, \end{cases} \tag{1.2}$$

and one observes that $\mathcal{G}$ is the resolvent of minus the one-dimensional Laplacian at energy parameter equal to -4, that is,

$$\mathcal{G} = \bigg(-\frac{d^2}{dx^2} + 4\bigg)^{-1}. \tag{1.3}$$

The spectral problem underlying the CH hierarchy can then be cast in the form (with "prime" denoting d/dx),

$$\Phi'(z,x) = U(z,x)\Phi(z,x), \quad (z,x) \in \mathbb{C} \times \mathbb{R}, \tag{1.4}$$

where

$$\Phi(z,x) = \begin{pmatrix} \phi_1(z,x) \\ \phi_2(z,x) \end{pmatrix}, \quad U(z,x) = \begin{pmatrix} -1 & 1 \\ z[u_{xx}(x) - 4u(x)] & 1 \end{pmatrix}, \quad (z,x) \in \mathbb{C} \times \mathbb{R}. \tag{1.5}$$

Eliminating ϕ_2 in (1.4) then results in the scalar (weighted) spectral problem

$$-\phi''(z,x) + \phi(z,x) = z[u_{xx}(x) - 4u(x)]\phi(z,x), \quad (z,x) \in (\mathbb{C}\backslash\{0\}) \times \mathbb{R}. \tag{1.6}$$

In the specific context of the left-definite Camassa–Holm spectral problem we refer to [9], [10], [17], [19], [30], [31], [32], [33], [34] [35], [37], [51], [62], [87], [96], [98], [108], [109], [110], [111], and the literature cited therein.

Rather than directly studying (1.6) in this note, we will study some of its generalizations and hence focus on several spectral problems originating with the general Sturm–Liouville equation

$$-(p(x)\psi'(z,x))' + q(x)\psi(z,x) = zr(x)\psi(z,x), \quad (z,x) \in (\mathbb{C}\backslash\{0\}) \times \mathbb{R}, \tag{1.7}$$

under various hypotheses on the coefficients p, q, r to be described in more detail later on and with emphasis on the fact that r may change its sign. At this point we assume the following basic requirements on p, q, r (but we emphasize that later on we will consider vastly more general situations where q and r are permitted to lie in certain classes of distributions):

Hypothesis 1.1.

(i) *Suppose that $p > 0$ a.e. on $\mathbb{R}$, $p^{-1} \in L^1_{\text{loc}}(\mathbb{R}; dx)$, and that $q, r \in L^1_{\text{loc}}(\mathbb{R}; dx)$ are real-valued a.e. on $\mathbb{R}$. In addition, assume that $r \neq 0$ on a set of positive Lebesgue measure and that*

$$\pm \lim_{x \to \pm\infty} \int^x dx'\, p(x')^{-1/2} = \infty. \tag{1.8}$$

(ii) *Introducing the differential expression*

$$\tau = -\frac{d}{dx}p(x)\frac{d}{dx} + q(x), \quad x \in \mathbb{R}, \tag{1.9}$$

and the associated minimal operator $T_{\min}$ *in* $L^2(\mathbb{R};dx)$ *by*

$$\begin{aligned} &T_{\min}f = \tau f, \\ &f \in \mathrm{dom}(T_{\min}) = \big\{g \in L^2(\mathbb{R};dx) \,\big|\, g, (pg') \in AC_{\mathrm{loc}}(\mathbb{R}); \ \mathrm{supp}\,(g) \text{ compact}; \\ &\hspace{20em} \tau g \in L^2(\mathbb{R};dx)\big\}, \end{aligned} \tag{1.10}$$

we assume that for some $\varepsilon > 0$,

$$T_{\min} \geq \varepsilon I_{L^2(\mathbb{R};dx)}. \tag{1.11}$$

We note that our assumptions (1.8) and (1.11) imply that τ is in the limit point case at $+\infty$ and $-\infty$ (cf., e.g., [29], [59], [72], [122]). This permits one to introduce the maximally defined self-adjoint operator T in $L^2(\mathbb{R};dx)$ associated with τ by

$$\begin{aligned} &Tf = \tau f, \\ &f \in \mathrm{dom}(T) = \big\{g \in L^2(\mathbb{R};dx) \,\big|\, g, (pg') \in AC_{\mathrm{loc}}(\mathbb{R}); \ \tau g \in L^2(\mathbb{R};dx)\big\} \end{aligned} \tag{1.12}$$

(where $AC_{\mathrm{loc}}(\mathbb{R})$ denotes the set of locally absolutely continuous functions on $\mathbb{R}$). In particular, T is the closure of $T_{\min}$,

$$T = \overline{T_{\min}}, \tag{1.13}$$

and hence also

$$T \geq \varepsilon I_{L^2(\mathbb{R};dx)}. \tag{1.14}$$

Remark 1.2. By a result proven in Yafaev [138] (see also [68, pp. 110–115]), if $p = 1$ and $q \geq 0$ a.e. on $\mathbb{R}$, (1.14) holds for some $\varepsilon > 0$ if and only if there exists $c_0 > 0$ such that for some $a > 0$,

$$\int_x^{x+a} dx'\, q(x') \geq c_0, \quad x \in \mathbb{R}. \tag{1.15}$$

If p is bounded below by some $\varepsilon_0 > 0$ (which we may choose smaller than one), one has

$$\int_{\mathbb{R}} dx \big[p(x)|u'(x)|^2 + q(x)|u(x)|^2\big] \geq \varepsilon_0 \int_{\mathbb{R}} dx \big[|u'(x)|^2 + \varepsilon_0^{-1} q(x)|u(x)|^2\big]. \tag{1.16}$$

Hence (1.15) is then still sufficient for (1.14) to hold.

We also note that Theorem 3 in [138] shows that $q \geq 0$ is not necessary for (1.14) to hold. In fact, if $q_2 \geq 0$, but $\int_a^{a+1} dx\, q_2(x) \leq c$ for all $a \in \mathbb{R}$, one finds

$$-(c + 4c^2) \int_{\mathbb{R}} dx\, |u(x)|^2 \leq \int_{\mathbb{R}} dx \big[|u'(x)|^2 - q_2(x)|u(x)|^2\big]. \tag{1.17}$$

Hence if $p = 1$ and $q = \varepsilon + c + 4c^2 - q_2$ one obtains (1.14) even though q may assume negative values.

Given these preparations, we now associate the weighted eigenvalue equation (1.7) with a standard self-adjoint spectral problem of the form

$$\begin{aligned} &\underline{T^{-1/2}rT^{-1/2}}\chi = \zeta\chi, \\ &\chi(\zeta,x) = \big(T^{1/2}\psi(z,\cdot)\big)(x), \quad \zeta = 1/z \in \mathbb{C}\backslash\{0\},\ x \in \mathbb{R}, \end{aligned} \tag{1.18}$$

for the integral operator $\underline{T^{-1/2}rT^{-1/2}}$ in $L^2(\mathbb{R};dx)$, subject to certain additional conditions on p, q, r. We use the particular notation $\underline{T^{-1/2}rT^{-1/2}}$ to underscore the particular care that needs to be taken with interpreting this expression as a bounded, self-adjoint operator in $L^2(\mathbb{R};dx)$ (pertinent details can be found in (2.37) and, especially, in (3.98)). It is important to note that in contrast to a number of papers that find it necessary to use different Hilbert spaces in connection with a left-definite spectral problem (in some cases the weight r is replaced by $|r|$, in other situations the new Hilbert space is coefficient-dependent), our treatment works with one and the same underlying Hilbert space $L^2(\mathbb{R};dx)$.

We emphasize that rewriting (1.7) in the form (1.18) is not new. In particular, in the context of the CH spectral problem (1.6) this has briefly been used, for instance, in [36] (in the periodic case), in [33] (in the context of the CH scattering problem), in [62] (in connection with real-valued algebro-geometric CH solutions), and in [109] (in connection with CH flows and Fredholm determinants). However, apart from the approach discussed in [18], [19], [20], most investigations associated with the CH spectral problem (1.6) appear to focus primarily on certain Liouville–Green transformations which transform (1.6) into a Schrödinger equation for some effective potential coefficient (see, e.g., [30], [31], [32]). This requires additional assumptions on the coefficients which in general can be avoided in the context of (1.18). Indeed, the change of variables

$$\mathbb{R} \ni x \mapsto t = \int_0^x dx'\, p(x')^{-1}, \tag{1.19}$$

turns the equation $-(pu')' + qu = zru$ on $\mathbb{R}$ into

$$\begin{aligned} &-v'' + Qv = zRv \ \text{ on } \bigg(-\int_{-\infty}^0 dx'\, p(x')^{-1}, \int_0^{+\infty} dx'\, p(x')^{-1}\bigg), \\ &v(t) = u(x(t)), \quad Q(t) = p(x(t))q(x(t)), \quad R(t) = p(x(t))r(x(t)). \end{aligned} \tag{1.20}$$

However, assuming for instance, $\pm\int^{\pm\infty} dx\, p(x) = \infty$, the change of variables is only unitary between the spaces $L^2(\mathbb{R};dx)$ and $L^2(\mathbb{R};dx/p(x))$ and hence necessitates a change in the underlying measure.

The primary aim of this note is to sketch a few instances in which the integral operator approach in (1.18) naturally, and in a straightforward manner, leads to much more general spectral results and hence is preferable to the Liouville–Green approach. In particular, we are interested in generalized situations, where the coefficients q and r lie in certain classes of distributions. To the best of our knowledge, this level of generality is new in this context.

In Section 2 we analyze basic spectral theory of $\underline{T^{-1/2}rT^{-1/2}}$ in $L^2(\mathbb{R}; dx)$ assuming Hypothesis 1.1 and appropriate additional assumptions on p, q, r. The more general case where q and r lie in certain classes of distributions is treated in detail in Section 3. There we heavily rely on supersymmetric methods and Miura transformations. This approach exploits the intimate relationship between spectral theory for Schrödinger operators factorized into first-order differential operators and that of an associated Dirac-type operator. Section 4 is devoted to applications in the special case where q and r are periodic (for simplicity we take $p = 1$). We permit q to lie in a class of distributions and r to be a signed measure, which underscores the novelty of our approach. Three appendices provide ample background results: Appendix A is devoted to basic facts on relative boundedness and compactness of operators and forms; the supersymmetric formalism relating Schrödinger and Dirac-type operators is presented in Appendix B, and details on sesquilinear forms and their associated operators are provided in Appendix C.

Finally, we briefly summarize some of the notation used in this paper: Let $\mathcal{H}$ be a separable complex Hilbert space, $(\cdot,\cdot)_{\mathcal{H}}$ the scalar product in $\mathcal{H}$ (linear in the second factor), and $I_{\mathcal{H}}$ the identity operator in $\mathcal{H}$. Next, let T be a linear operator mapping (a subspace of) a Banach space into another, with $\operatorname{dom}(T)$, $\operatorname{ran}(T)$, and $\ker(T)$ denoting the domain, range, and kernel (i.e., null space) of T. The closure of a closable operator S is denoted by $\overline{S}$.

The spectrum, essential spectrum, point spectrum, discrete spectrum, absolutely continuous spectrum, and resolvent set of a closed linear operator in $\mathcal{H}$ will be denoted by $\sigma(\cdot)$, $\sigma_{\rm ess}(\cdot)$, $\sigma_{\rm p}(\cdot)$, $\sigma_{\rm d}(\cdot)$, $\sigma_{\rm ac}(\cdot)$, and $\rho(\cdot)$, respectively.

The Banach spaces of bounded and compact linear operators in $\mathcal{H}$ are denoted by $\mathcal{B}(\mathcal{H})$ and $\mathcal{B}_\infty(\mathcal{H})$, respectively. Similarly, the Schatten–von Neumann (trace) ideals will subsequently be denoted by $\mathcal{B}_s(\mathcal{H})$, $s \in (0,\infty)$. The analogous notation $\mathcal{B}(\mathcal{X}_1, \mathcal{X}_2)$, $\mathcal{B}_\infty(\mathcal{X}_1, \mathcal{X}_2)$, etc., will be used for bounded and compact operators between two Banach spaces $\mathcal{X}_1$ and $\mathcal{X}_2$. Moreover, $\mathcal{X}_1 \hookrightarrow \mathcal{X}_2$ denotes the continuous embedding of the Banach space $\mathcal{X}_1$ into the Banach space $\mathcal{X}_2$. Throughout this manuscript we use the convention that if X denotes a Banach space, X^* denotes the *adjoint space* of continuous conjugate linear functionals on X, also known as the *conjugate dual* of X.

In the bulk of this note, $\mathcal{H}$ will typically represent the space $L^2(\mathbb{R}; dx)$. Operators of multiplication by a function $V \in L^1_{\rm loc}(\mathbb{R}; dx)$ in $L^2(\mathbb{R}; dx)$ will by a slight abuse of notation again be denoted by V (rather than the frequently used, but more cumbersome, notation M_V) and unless otherwise stated, will always assumed to be maximally defined in $L^2(\mathbb{R}; dx)$ (i.e., $\operatorname{dom}(V) = \big\{f \in L^2(\mathbb{R}; dx) \,\big|\, Vf \in L^2(\mathbb{R}; dx)\big\}$). Moreover, in subsequent sections, the identity operator $I_{L^2(\mathbb{R};dx)}$ in $L^2(\mathbb{R}; dx)$ will simply be denoted by I for brevity.

The symbol $\mathcal{D}(\mathbb{R})$ denotes the space of test functions $C_0^\infty(\mathbb{R})$ with its usual (inductive limit) topology. The corresponding space of continuous linear functionals on $\mathcal{D}(\mathbb{R})$ is denoted by $\mathcal{D}'(\mathbb{R})$ (i.e., $\mathcal{D}'(\mathbb{R}) = C_0^\infty(\mathbb{R})'$).

2. General spectral theory of $\underline{T^{-1/2}rT^{-1/2}}$

In this section we derive some general spectral properties of $\underline{T^{-1/2}rT^{-1/2}}$ which reproduce some known results that were originally derived in the CH context of (1.6), but now we prove them under considerably more general conditions on the coefficients p, q, r, and generally, with great ease. In this section p, q, r will satisfy Hypothesis 1.1 and appropriate additional assumptions. (The case where q, r lie in certain classes of distributions will be treated in Section 3.)

For a quick summary of the notions of relatively bounded and compact operators and forms frequently used in this section, we refer to Appendix A.

Before analyzing the operator $\underline{T^{-1/2}rT^{-1/2}}$ we recall three useful results:

We denote by T_0 (minus) the usual Laplacian in $L^2(\mathbb{R}; dx)$ defined by

$$\begin{aligned} &T_0 f = -f'', \\ &f \in \mathrm{dom}(T_0) = \{g \in L^2(\mathbb{R}; dx) \,|\, g, g' \in AC_{\mathrm{loc}}(\mathbb{R});\ g'' \in L^2(\mathbb{R}; dx)\} = H^2(\mathbb{R}), \end{aligned} \tag{2.1}$$

where $H^m(\mathbb{R})$, $m \in \mathbb{N}$, abbreviate the usual Sobolev spaces of functions whose distributional derivatives up to order m lie in $L^2(\mathbb{R}; dx)$.

In the following it is useful to introduce the spaces of locally uniformly L^p-integrable functions on $\mathbb{R}$,

$$L^p_{\mathrm{loc\,unif}}(\mathbb{R}; dx) = \left\{f \in L^p_{\mathrm{loc}}(\mathbb{R}; dx) \,\middle|\, \sup_{a\in\mathbb{R}} \left(\int_a^{a+1} dx\, |f(x)|^p\right) < \infty\right\}, \tag{2.2}$$

$p \in [1, \infty)$. Equivalently, let

$$\eta \in C_0^\infty(\mathbb{R}), \quad 0 \le \eta \le 1, \quad \eta|_{B(0;1)} = 1, \tag{2.3}$$

with $B(x; r) \subset \mathbb{R}$ the open ball centered at $x_0 \in \mathbb{R}$ and radius $r > 0$, then

$$L^p_{\mathrm{loc\,unif}}(\mathbb{R}; dx) = \left\{f \in L^p_{\mathrm{loc}}(\mathbb{R}; dx) \,\middle|\, \sup_{a\in\mathbb{R}} \|\eta(\cdot - a) f\|_{L^p(\mathbb{R};dx)} < \infty\right\}, \quad p \in [1, \infty). \tag{2.4}$$

We refer to Appendix A for basic notions in connection with relatively bounded linear operators.

Theorem 2.1 ([125, Theorem 2.7.1], [127, p. 35]). *Let $V, w \in L^2_{\mathrm{loc}}(\mathbb{R}; dx)$. Then the following conditions* (i)–(iv) *are equivalent:*

(i) $$\mathrm{dom}(w) \supseteq \mathrm{dom}\left(T_0^{1/2}\right) = H^1(\mathbb{R}). \tag{2.5}$$

(ii) $$w \in L^2_{\mathrm{loc\,unif}}(\mathbb{R}; dx). \tag{2.6}$$

(iii) *For some $C > 0$,*

$$\begin{aligned} \|wf\|^2_{L^2(\mathbb{R};dx)} \le C\Big[\|T_0^{1/2} f\|^2_{L^2(\mathbb{R};dx)} + \|f\|^2_{L^2(\mathbb{R};dx)}\Big], \\ f \in \mathrm{dom}\left(T_0^{1/2}\right) = H^1(\mathbb{R}). \end{aligned} \tag{2.7}$$

(iv) *For all $\varepsilon > 0$, there exists $C_\varepsilon > 0$ such that:*

$$\|wf\|^2_{L^2(\mathbb{R};dx)} \leq \varepsilon \big\|T_0^{1/2} f\big\|^2_{L^2(\mathbb{R};dx)} + C_\varepsilon \|f\|^2_{L^2(\mathbb{R};dx)}, \quad f \in \operatorname{dom}\big(T_0^{1/2}\big) = H^1(\mathbb{R}). \tag{2.8}$$

Moreover, also the following conditions (v)–(viii) *are equivalent:*

(v)
$$\operatorname{dom}(V) \supseteq \operatorname{dom}(T_0) = H^2(\mathbb{R}). \tag{2.9}$$

(vi)
$$V \in L^2_{\text{loc unif}}(\mathbb{R}; dx). \tag{2.10}$$

(vii) *For some $C > 0$,*

$$\|Vf\|^2_{L^2(\mathbb{R};dx)} \leq C\Big[\|T_0 f\|^2_{L^2(\mathbb{R};dx)} + \|f\|^2_{L^2(\mathbb{R};dx)}\Big], \quad f \in \operatorname{dom}(T_0) = H^2(\mathbb{R}). \tag{2.11}$$

(viii) *For all $\varepsilon > 0$, there exists $C_\varepsilon > 0$ such that:*

$$\|Vf\|^2_{L^2(\mathbb{R};dx)} \leq \varepsilon \|T_0 f\|^2_{L^2(\mathbb{R};dx)} + C_\varepsilon \|f\|^2_{L^2(\mathbb{R};dx)}, \quad f \in \operatorname{dom}(T_0) = H^2(\mathbb{R}). \tag{2.12}$$

In fact, it is possible to replace $T_0^{1/2}$ by any polynomial $P_m\big(T_0^{1/2}\big)$ of degree $m \in \mathbb{N}$ in connection with items (i)–(iv).

We emphasize the remarkable fact that according to items (iii), (iv) and (vii), (viii), relative form and operator boundedness is actually equivalent to *infinitesimal* form and operator boundedness in Theorem 2.1.

For completeness, we briefly sketch some of the principal ideas underlying items (i)–(iv) in Theorem 2.1, particularly, focusing on item (ii): That item (i) implies item (iii) is of course a consequence of the closed graph theorem. Exploiting continuity of $f \in H^1(\mathbb{R})$, yields for arbitrary $\varepsilon > 0$,

$$\begin{aligned} |f(x)|^2 - |f(x')|^2 &= \int_{x'}^{x} dy\, \big[\overline{f(y}f'(y) + \overline{f'(y)}f(y)\big] \\ &\leq \varepsilon \int_{\mathcal{I}} dy\, |f'(y|^2 + \varepsilon^{-1} \int_{\mathcal{I}} dy\, |f(y)|^2, \quad f \in H^1(\mathbb{R}),\ x, x' \in \mathcal{I}, \end{aligned} \tag{2.13}$$

with $\mathcal{I} \subset \mathbb{R}$ an arbitrary interval of length one. The use of the mean value theorem for integrals then permits one to choose $x' \in \mathcal{I}$ such that

$$|f(x')|^2 = \int_{\mathcal{I}} dy\, |f(y)|^2 \tag{2.14}$$

implying

$$|f(x)|^2 \leq \varepsilon \int_{\mathcal{I}} dx'\, |f'(x')|^2 + (1 + \varepsilon^{-1}) \int_{\mathcal{I}} dx'\, |f(x')|^2, \quad f \in H^1(\mathbb{R}),\ x \in \mathcal{I}, \tag{2.15}$$

and hence after summing over all intervals $\mathcal{I}$ of length one, and using boundedness of $f \in H^1(\mathbb{R})$,

$$|f(x)|^2 \leq \|f\|^2_{L^\infty(\mathbb{R};dx)} \leq \varepsilon \|f'\|^2_{L^2(\mathbb{R};dx)} + (1 + \varepsilon^{-1}) \|f\|^2_{L^2(\mathbb{R};dx)}, \tag{2.16}$$

$f \in H^1(\mathbb{R})$, $x \in \mathbb{R}$. Multiplying (2.15) by $|w(x)|^2$ and integrating with respect to x over $\mathcal{I}$ yields

$$\int_{\mathcal{I}} dx\,|w(x)|^2|f(x)|^2 \le \varepsilon C_0 \int_{\mathcal{I}} dx'\,|f'(x')|^2 + (1+\varepsilon^{-1})C_0 \int_{\mathcal{I}} dx'\,|f(x')|^2, \quad (2.17)$$

and summing again over all intervals $\mathcal{I}$ of length one implies

$$\|wf\|^2_{L^2(\mathbb{R};dx)} \le \varepsilon C_0 \|f'\|^2_{L^2(\mathbb{R})} + \big(1+\varepsilon^{-1}\big)C_0\|f\|^2_{L^2(\mathbb{R};dx)}, \quad f \in H^1(\mathbb{R}), \quad (2.18)$$

where

$$C_0 := \sup_{a\in\mathbb{R}} \bigg(\int_a^{a+1} dx\,|w(x)|^2\bigg) < \infty, \quad (2.19)$$

illustrating the sufficiency part of condition $w \in L^2_{\text{loc unif}}(\mathbb{R};dx)$ in item (ii) for item (iv) to hold.

Next, consider $\psi(x) = e^{1-x^2}$, $\psi_a(x) = \psi(x-a)$, $x, a \in \mathbb{R}$. Then

$$\begin{aligned}\int_a^{a+1} dx\,|w(x)|^2 &\le \int_{\mathbb{R}} dx\,\big[|w(x)||\psi_a(x)|\big]^2 &(2.20)\\ &\le C\Big[\big\|T_0^{1/2}\psi_a\big\|^2_{L^2(\mathbb{R};dx)} + \|\psi_a\|^2_{L^2(\mathbb{R};dx)}\Big] &(2.21)\\ &\le \widehat{C}\big[\|\psi'\|^2_{L^2(\mathbb{R};dx)} + \|\psi\|^2_{L^2(\mathbb{R};dx)}\big] = \widetilde{C}, &(2.22)\end{aligned}$$

with $\widetilde{C}$ independent of a, illustrates necessity of the condition $w \in L^2_{\text{loc unif}}(\mathbb{R};dx)$ in item (ii) for item (iii) to hold.

Given $\varepsilon > 0$, there exists $\eta(\varepsilon) > 0$, such that the obvious inequality

$$\begin{aligned}&\|f'\|^2_{L^2(\mathbb{R};dx)} \le \varepsilon\big\|T_0^{m/2}\big\|^2_{L^2(\mathbb{R};dx)} + \eta(\varepsilon)\|f\|^2_{L^2(\mathbb{R};dx)},\\ &\qquad f \in \operatorname{dom}\big(T_0^{m/2}\big),\ m \in \mathbb{N},\ m \ge 2,\end{aligned} \quad (2.23)$$

holds. It suffices applying the Fourier transform and using $|p| \le \varepsilon|p|^m + \eta(\varepsilon)$, $m \in \mathbb{N}$, $m \ge 2$ to extend this to polynomials in $T_0^{1/2}$. This illustrates the sufficiency of the condition $V \in L^2_{\text{loc unif}}(\mathbb{R};dx)$ in item (vi) for item (viii) to hold.

We note that items (i)–(iv) in Theorem 2.1 are mentioned in [127, p. 35] without proof, but the crucial hint that $f \in H^1(\mathbb{R})$ implies that $f \in AC_{\text{loc}}(\mathbb{R}) \cap L^\infty(\mathbb{R};dx)$, is made there. We also remark that Theorem 2.7.1 in [125] is primarily concerned with items (v)–(viii) in Theorem 2.1. Nevertheless, its method of proof also yields the results (2.1)–(2.8), in particular, it contains the fundamental inequality (2.18).

Next, we also recall the following result (we refer to Appendix A for details on the notion of relative compactness for linear operators):

Theorem 2.2 ([125, Theorem 3.7.5], [126, Sects. 15.7, 15.9]). *Let* $w \in L^2_{\text{loc}}(\mathbb{R};dx)$. *Then the following conditions* (i)–(iii) *are equivalent:*

(i) $\qquad w$ *is* $T_0^{1/2}$*-compact.* (2.24)

(ii) $\qquad w$ *is* T_0*-compact.* (2.25)

(iii)
$$\lim_{|a|\to\infty}\left(\int_a^{a+1} dx\,|w(x)|^2\right)=0. \tag{2.26}$$

In fact, it is possible to replace $T_0^{1/2}$ *by any polynomial* $P_m\big(T_0^{1/2}\big)$ *of degree* $m\in\mathbb{N}$ *in item* (i).

We note that $w\in L^2_{\text{loc}}(\mathbb{R};dx)$ together with condition (2.26) imply that $w\in L^2_{\text{loc unif}}(\mathbb{R};dx)$ (cf. [126, p. 378]).

It is interesting to observe that the if and only if characterizations (2.1)–(2.8) for relative (resp., infinitesimal) form boundedness mentioned by Simon [127, p. 35], and those in (2.24)–(2.26) for relative (form) compactness by Schechter in the first edition of [126, Sects. 15.7, 15.9], were both independently published in 1971.

In the context of Theorems 2.1 and 2.2 we also refer to [5] for interesting results on necessary and sufficient conditions on relative boundedness and relative compactness for perturbations of Sturm–Liouville operators by lower-order differential expressions on a half-line (in addition, see [26], [77]).

We will also use the following result on trace ideals. To fix our notation, we denote by $f(X)$ the operator of multiplication by the measurable function f on $\mathbb{R}$, and similarly, we denote by $g(P)$ the operator defined by the spectral theorem for a measurable function g (equivalently, the operator of multiplication by the measurable function g in Fourier space $L^2(\mathbb{R};dp)$), where P denotes the self-adjoint (momentum) operator defined by

$$Pf=-if',\quad \operatorname{dom}(P)=H^1(\mathbb{R}). \tag{2.27}$$

Theorem 2.3 ([128, Theorem 4.1]). *Let* $f\in L^s(\mathbb{R};dx)$, $g\in L^s(\mathbb{R};dx))$, $s\in[2,\infty)$. *Then*

$$f(X)g(P)\in\mathcal{B}_s\big(L^2(\mathbb{R};dx)\big) \tag{2.28}$$

and

$$\|f(X)g(P)\|_{\mathcal{B}_s(L^2(\mathbb{R};dx))}\leq(2\pi)^{-1/s}\|f\|_{L^s(\mathbb{R};dx)}\|g\|_{L^s(\mathbb{R};dx)}. \tag{2.29}$$

If $s=2$, f *and* g *are both nonzero on a set of positive Lebesgue measure, and* $f(X)g(P)\in\mathcal{B}_2\big(L^2(\mathbb{R};dx)\big)$, *then*

$$f,g\in L^2(\mathbb{R};dx). \tag{2.30}$$

Given these preparations, we introduce the following convenient assumption:

Hypothesis 2.4. *In addition to the assumptions in Hypothesis* 1.1 *suppose that the form domain of* T *is given by*

$$\operatorname{dom}\big(T^{1/2}\big)=\operatorname{dom}\big(T_0^{1/2}\big)=H^1(\mathbb{R}). \tag{2.31}$$

Assuming for some positive constants c and C that

$$0<c\leq p\leq C\ \text{ a.e. on } \mathbb{R}, \tag{2.32}$$

an application of Theorem 2.1 (i), (ii) shows that (2.31) holds if $q\in L^1_{\text{loc}}(\mathbb{R};dx)$ satisfies

$$q\in L^1_{\text{loc unif}}(\mathbb{R};dx). \tag{2.33}$$

Indeed, since by the comment following Hypothesis 1.1, T is essentially self-adjoint on $\mathrm{dom}(T_{\min})$, $T \geq \varepsilon I$ for some $\varepsilon > 0$, and $\mathrm{dom}\big(T^{1/2}\big) = H^1(\mathbb{R})$, the sesquilinear form $\mathfrak{Q}_T$ associated with T is of the form

$$\begin{aligned} &\mathfrak{Q}_T(f,g) = \int_{\mathbb{R}} dx\, p(x)\overline{f'(x)}g'(x) + \int_{\mathbb{R}} dx\, q(x)\overline{f(x)}g(x), \\ &f,g \in \mathrm{dom}(\mathfrak{Q}_T) = \mathrm{dom}\big(T^{1/2}\big) = \mathrm{dom}\big(T_0^{1/2}\big) = H^1(\mathbb{R}). \end{aligned} \tag{2.34}$$

Hence, by Theorem 2.1 (i), (ii), this is equivalent to (2.33) keeping in mind that q is such that (1.11) holds.

Our first result then reads as follows:

Theorem 2.5. *Assume Hypothesis* 2.4.

(i) *Then*

$$|r|^{1/2}T^{-1/2} \in \mathcal{B}\big(L^2(\mathbb{R}; dx)\big) \tag{2.35}$$

if and only if

$$r \in L^1_{\mathrm{loc\,unif}}(\mathbb{R}; dx). \tag{2.36}$$

In particular, if (2.36) *holds, introducing*

$$\underline{T^{-1/2}rT^{-1/2}} = \big[|r|^{1/2}T^{-1/2}\big]^* \,\mathrm{sgn}(r)\big[|r|^{1/2}T^{-1/2}\big], \tag{2.37}$$

one concludes that

$$\underline{T^{-1/2}rT^{-1/2}} \in \mathcal{B}\big(L^2(\mathbb{R}; dx)\big). \tag{2.38}$$

(ii) *Let $r_0 \in \mathbb{R}$. Then*

$$|r - r_0|^{1/2}T^{-1/2} \in \mathcal{B}_\infty\big(L^2(\mathbb{R}; dx)\big) \tag{2.39}$$

if and only if

$$\lim_{|a|\to\infty} \bigg(\int_a^{a+1} dx\, |r(x) - r_0|\bigg) = 0. \tag{2.40}$$

In particular, if (2.40) *holds, introducing*

$$\begin{aligned} &\underline{T^{-1/2}(r - r_0)T^{-1/2}} \\ &\quad = \big[|r - r_0|^{1/2}T^{-1/2}\big]^* \,\mathrm{sgn}(r - r_0)\big[|r - r_0|^{1/2}T^{-1/2}\big], \end{aligned} \tag{2.41}$$

one concludes that

$$\underline{T^{-1/2}(r - r_0)T^{-1/2}} \in \mathcal{B}_\infty\big(L^2(\mathbb{R}; dx)\big). \tag{2.42}$$

(iii) *Let $r_0 \in \mathbb{R}$. Then*

$$|r - r_0|^{1/2}T^{-1/2} \in \mathcal{B}_2\big(L^2(\mathbb{R}; dx)\big) \tag{2.43}$$

if and only if

$$\int_{\mathbb{R}} dx\, |r(x) - r_0| < \infty. \tag{2.44}$$

In particular, if (2.44) *holds, then*

$$\underline{T^{-1/2}(r-r_0)T^{-1/2}} \in \mathcal{B}_1\big(L^2(\mathbb{R};dx)\big). \tag{2.45}$$

Proof. (i) By hypothesis (2.31) and the closed graph theorem one concludes that

$$\big[(T_0+I)^{1/2}T^{-1/2}\big] \in \mathcal{B}\big(L^2(\mathbb{R};dx)\big). \tag{2.46}$$

$\big($and analogously, $\big[(T_0+I)^{1/2}T^{-1/2}\big]^{-1} = T^{1/2}(T_0+I)^{-1/2} \in \mathcal{B}\big(L^2(\mathbb{R};dx)\big)\big)$. The equivalence of (2.35) and (2.36) then follows from (2.1) and (2.6) and the fact that

$$|r|^{1/2}T^{-1/2} = \big[|r|^{1/2}(T_0+I)^{-1/2}\big]\big[(T_0+I)^{1/2}T^{-1/2}\big]. \tag{2.47}$$

The inclusion (2.38) immediately follows from (2.35) and (2.37).

(ii) The equivalence of (2.39) and (2.40) follows from (2.24) and (2.26). The inclusion (2.42) then follows from (2.41), (2.46), and (2.47) with r replaced by $r-r_0$.

(iii) The equivalence of (2.43) and (2.44) follows from (2.29) and (2.30), employing again (2.46) and the fact that $(|p|^2+1)^{-1/2} \in L^2(\mathbb{R};dp)$. The relation (2.45) once more follows from (2.41), (2.46), and (2.47) with r replaced by $r-r_0$, and the fact that $S \in \mathcal{B}_1(\mathcal{H})$ if and only if $|S| \in \mathcal{B}_1(\mathcal{H})$ and hence if and only if $|S|^{1/2} \in \mathcal{B}_2(\mathcal{H})$. □

In the following we use the obvious notation for subsets of $\mathcal{M} \subset \mathbb{R}$ and constants $c \in \mathbb{R}$:

$$c\mathcal{M} = \{cx \in \mathbb{R} \,|\, x \in \mathcal{M}\}. \tag{2.48}$$

Corollary 2.6. *Assume Hypothesis* 2.4.

(i) *If* (2.40) *holds for some* $r_0 \in \mathbb{R}$, *then*

$$\sigma_{\rm ess}\big(\underline{T^{-1/2}rT^{-1/2}}\big) = \begin{cases} r_0\sigma_{\rm ess}\big(T^{-1}\big), & r_0 \in \mathbb{R}\backslash\{0\}, \\ \{0\}, & r_0 = 0. \end{cases} \tag{2.49}$$

(ii) *If* (2.44) *holds for some* $r_0 \in \mathbb{R}$, *then*

$$\sigma_{\rm ac}\big(\underline{T^{-1/2}rT^{-1/2}}\big) = \begin{cases} r_0\sigma_{\rm ac}\big(T^{-1}\big), & r_0 \in \mathbb{R}\backslash\{0\}, \\ \emptyset, & r_0 = 0. \end{cases} \tag{2.50}$$

Proof. For $r_0 \in \mathbb{R}\backslash\{0\}$ it suffices to use the decomposition

$$\underline{T^{-1/2}rT^{-1/2}} = \underline{T^{-1/2}[r_0+(r-r_0)]T^{-1/2}} = r_0T^{-1} + \underline{T^{-1/2}(r-r_0)T^{-1/2}} \tag{2.51}$$

and employ (2.42) together with Weyl's theorem (cf., e.g., [53, Sect. IX.2], [121, Sect. XIII.4], [135, Sect. 9.2]) to obtain (2.49), and combine (2.45) and the Kato–Rosenblum theorem (cf., e.g., [90, Sect. X.3], [120, Sect. XI.3], [135, Sect. 11.1]) to obtain (2.50).

In the case $r_0 = 0$ relation (2.49) holds since $\underline{T^{-1/2}rT^{-1/2}} \in \mathcal{B}_\infty\big(L^2(\mathbb{R};dx)\big)$ and $L^2(\mathbb{R};dx)$ is infinite dimensional. By the same argument one obtains (2.50) for $r_0 = 0$. □

In connection with (2.49) we also recall that by the spectral mapping theorem for self-adjoint operators A in $\mathcal{H}$,

$$0 \neq z \in \sigma_{\rm ess}\big((A - z_0 I_{\mathcal{H}})^{-1}\big), \ z_0 \in \rho(A), \ \text{ if and only if } \ z^{-1} + z_0 \in \sigma_{\rm ess}(A) \quad (2.52)$$

(cf., e.g., [121, Sect. XIII.4]). Finally, we mention that there exists a large body of results on determining essential and absolutely continuous spectra for Sturm–Liouville-type operators T associated with the differential expressions of the type $\tau = -\frac{d}{dx}p(x)\frac{d}{dx} + q(x)$, $x \in \mathbb{R}$. We refer, for instance, to [45, XIII.7], [115, Chs. 2, 4], [116, Sect. 24], and the literature cited therein.

Remark 2.7. While it is well known that for T densely defined and closed in $\mathcal{H}$,

$$\begin{aligned} &T \text{ is bounded (resp., compact, Hilbert–Schmidt)} \\ &\quad \text{if and only if } T^*T \text{ is bounded (resp., compact, trace class)}, \end{aligned} \quad (2.53)$$

the following example, communicated to us by G. Teschl [130], shows that if S is bounded and self-adjoint in $\mathcal{H}$ with spectrum $\sigma(S) = \{-1, 1\}$ then

$$T \text{ bounded is } \textbf{not} \text{ equivalent to } \overline{T^*ST} \text{ bounded} \quad (2.54)$$

assuming T^*ST to be densely defined in $\mathcal{H}$ (and hence closable in $\mathcal{H}$, since T^*ST is symmetric). Indeed, considering

$$T = \begin{pmatrix} A & 0 \\ 0 & A^{-1} \end{pmatrix}, \quad A = A^*, \quad A \geq I_{\mathcal{H}}, \quad S = \begin{pmatrix} 0 & I_{\mathcal{H}} \\ I_{\mathcal{H}} & 0 \end{pmatrix}, \quad (2.55)$$

then

$$\overline{T^*ST} = S, \quad (2.56)$$

and hence $\overline{T^*ST}$ is bounded, but T is unbounded if A is chosen to be unbounded.

Thus one cannot assert on abstract grounds that

$$\underline{T^{-1/2} r T^{-1/2}} = \big[|r|^{1/2} T^{-1/2}\big]^* \operatorname{sgn}(r) |r|^{1/2} T^{-1/2} \quad (2.57)$$

is bounded if and only if $|r|^{1/2}T^{-1/2}$ is. In fact, this is utterly wrong as we shall discuss in the following Section 3. Indeed, focusing directly on $|r|^{1/2}T^{-1/2}$ instead of $\underline{T^{-1/2}rT^{-1/2}}$ ignores crucial oscillations of r that permit one to considerably enlarge the class of admissible weights r. In particular, thus far we relied on estimates of the type

$$\big\| |q|^{1/2} f \big\|_{L^2(\mathbb{R};dx)}^2 \leq C \Big[\big\| T_0^{1/2} f \big\|_{L^2(\mathbb{R};dx)}^2 + \|f\|_{L^2(\mathbb{R};dx)}^2 \Big], \quad f \in H^1(\mathbb{R}), \quad (2.58)$$

equivalently,

$$\int_{\mathbb{R}} dx\, |q(x)| |f(x)|^2 \leq \big\| [T_0 + I]^{1/2} f \big\|_{L^2(\mathbb{R};dx)}^2, \quad f \in H^1(\mathbb{R}). \quad (2.59)$$

Consequently, we ignored all oscillations of q (and hence, r). Instead, we should focus on estimating

$$\bigg| \int_{\mathbb{R}} dx\, q(x) |f(x)|^2 \bigg| \leq \big\| [T_0 + I]^{1/2} f \big\|_{L^2(\mathbb{R};dx)}^2, \quad f \in H^1(\mathbb{R}), \quad (2.60)$$

and this will be the focus of the next Section 3.

3. Distributional coefficients

In this section we extend our previous considerations where $q, r \in L^1_{\text{loc unif}}(\mathbb{R}; dx)$, to the case where q and r are permitted to lie in a certain class of distributions. The extension to distributional coefficients will be facilitated by employing supersymmetric methods and an underlying Miura transformation. This approach permits one to relate spectral theory for Schrödinger operators factorized into a product of first-order differential operators with that of an associated Dirac-type operator.

We start with some background (cf., e.g., [70, Chs. 4–6], [103, Chs. 2, 3, 11], [112, Ch. 3]) and fix our notation in connection with Sobolev spaces. Introducing

$$L^2_s(\mathbb{R}) = L^2\Big(\mathbb{R}; \big(1 + |p|^2\big)^s dp\Big), \quad s \in \mathbb{R}, \tag{3.1}$$

and identifying,

$$L^2_0(\mathbb{R}) = L^2(\mathbb{R}; dp) = \big(L^2(\mathbb{R}; dp)\big)^* = \big(L^2_0(\mathbb{R})\big)^*, \tag{3.2}$$

one gets the chain of Hilbert spaces with respect to the pivot space $L^2_0(\mathbb{R}) = L^2(\mathbb{R}; dp)$,

$$L^2_s(\mathbb{R}) \subset L^2(\mathbb{R}; dp) \subset L^2_{-s}(\mathbb{R}) = \big(L^2_s(\mathbb{R})\big)^*, \quad s > 0. \tag{3.3}$$

Next, we introduce the maximally defined operator G_0 of multiplication by the function $\big(1 + |\cdot|^2\big)^{1/2}$ in $L^2(\mathbb{R}; dp)$,

$$\begin{aligned} &(G_0 f)(p) = \big(1 + |p|^2\big)^{1/2} f(p), \\ &f \in \operatorname{dom}(G_0) = \Big\{g \in L^2(\mathbb{R}; dp) \,\Big|\, \big(1 + |\cdot|^2\big)^{1/2} g \in L^2(\mathbb{R}; dp)\Big\}. \end{aligned} \tag{3.4}$$

The operator G_0 extends to an operator defined on the entire scale $L^2_s(\mathbb{R})$, $s \in \mathbb{R}$, denoted by $\widetilde{G}_0$, such that

$$\widetilde{G}_0 : L^2_s(\mathbb{R}) \to L^2_{s-1}(\mathbb{R}), \quad \big(\widetilde{G}_0\big)^{-1} : L^2_s(\mathbb{R}) \to L^2_{s+1}(\mathbb{R}), \text{ bijectively}, \quad s \in \mathbb{R}. \tag{3.5}$$

In particular, while

$$I : L^2(\mathbb{R}; dp) \to \big(L^2(\mathbb{R}; dp)\big)^* = L^2(\mathbb{R}; dp) \tag{3.6}$$

represents the standard identification operator between $L^2_0(\mathbb{R}) = L^2(\mathbb{R}; dp)$ and its adjoint space, $\big(L^2(\mathbb{R}; dp)\big)^* = \big(L^2_0(\mathbb{R})\big)^*$, via Riesz's lemma, we emphasize that we will not identify $\big(L^2_s(\mathbb{R})\big)^*$ with $L^2_s(\mathbb{R})$ when $s > 0$. In fact, it is the operator $\widetilde{G}_0^2$ that provides a unitary map

$$\widetilde{G}_0^2 : L^2_s(\mathbb{R}) \to L^2_{s-2}(\mathbb{R}), \quad s \in \mathbb{R}. \tag{3.7}$$

In particular,

$$\widetilde{G}_0^2 : L^2_1(\mathbb{R}) \to L^2_{-1}(\mathbb{R}) = \big(L^2_1(\mathbb{R})\big)^* \text{ is a unitary map}, \tag{3.8}$$

and we refer to (C.40) for an abstract analog of this fact.

Denoting the Fourier transform on $L^2(\mathbb{R}; dp)$ by $\mathcal{F}$, and then extended to the entire scale $L^2_s(\mathbb{R})$, $s \in \mathbb{R}$, more generally, to $\mathcal{S}'(\mathbb{R})$ by $\widetilde{\mathcal{F}}$ (with $\widetilde{\mathcal{F}} : \mathcal{S}'(\mathbb{R}) \to \mathcal{S}'(\mathbb{R})$ a homeomorphism), one obtains the scale of Sobolev spaces via

$$H^s(\mathbb{R}) = \widetilde{\mathcal{F}} L^2_s(\mathbb{R}), \quad s \in \mathbb{R}, \quad L^2(\mathbb{R}; dx) = \mathcal{F} L^2(\mathbb{R}; dp), \tag{3.9}$$

and hence,

$$\mathcal{F} G_0 \mathcal{F}^{-1} = (T_0 + I)^{1/2} : H^1(\mathbb{R}) \to L^2(\mathbb{R}; dx), \text{ bijectively}, \tag{3.10}$$

$$\widetilde{\mathcal{F}} \widetilde{G}_0 \widetilde{\mathcal{F}}^{-1} = \big(\widetilde{T}_0 + \widetilde{I}\big)^{1/2} : H^s(\mathbb{R}) \to H^{s-1}(\mathbb{R}), \text{ bijectively}, \quad s \in \mathbb{R}, \tag{3.11}$$

$$\widetilde{\mathcal{F}} \big(\widetilde{G}_0\big)^{-1} \widetilde{\mathcal{F}}^{-1} = \big(\widetilde{T}_0 + \widetilde{I}\big)^{-1/2} : H^s(\mathbb{R}) \to H^{s+1}(\mathbb{R}), \text{ bijectively}, \quad s \in \mathbb{R}. \tag{3.12}$$

We recall that T_0 was defined as

$$T_0 = -d^2/dx^2, \quad \operatorname{dom}(T_0) = H^2(\mathbb{R}), \tag{3.13}$$

in (2.1), but now the extension $\widetilde{T}_0$ of T_0 is defined on the entire Sobolev scale according to (3.11),

$$\big(\widetilde{T}_0 + \widetilde{I}\big) : H^s(\mathbb{R}) \to H^{s-2}(\mathbb{R}) \text{ is a unitary map}, \quad s \in \mathbb{R}, \tag{3.14}$$

and the special case $s = 1$ again corresponds to (C.26), (C.40),

$$\big(\widetilde{T}_0 + \widetilde{I}\big) : H^1(\mathbb{R}) \to H^{-1}(\mathbb{R}) = \big(H^1(\mathbb{R})\big)^* \text{ is a unitary map}. \tag{3.15}$$

In addition, we note that

$$H^0(\mathbb{R}) = L^2(\mathbb{R}; dx), \quad \big(H^s(\mathbb{R})\big)^* = H^{-s}(\mathbb{R}), \quad s \in \mathbb{R}, \tag{3.16}$$

$$\mathcal{S}(\mathbb{R}) \subset H^s(\mathbb{R}) \subset H^{s'}(\mathbb{R}) \subset L^2(\mathbb{R}; dx) \subset H^{-s'}(\mathbb{R}) \subset H^{-s}(\mathbb{R}) \subset \mathcal{S}'(\mathbb{R}), \quad s > s' > 0. \tag{3.17}$$

Moreover, we recall that $H^s(\mathbb{R})$ is conveniently and alternatively introduced as the completion of $C_0^\infty(\mathbb{R})$ with respect to the norm $\|\cdot\|_s$,

$$H^s(\mathbb{R}) = \overline{C_0^\infty(\mathbb{R})}^{\|\cdot\|_s}, \quad s \in \mathbb{R}, \tag{3.18}$$

where for $\psi \in C_0^\infty(\mathbb{R})$ and $s \in \mathbb{R}$,

$$\|\psi\|_s = \bigg(\int_{\mathbb{R}} d\xi \, \big(1+|\xi|^{2s}\big) |\widehat{\psi}(\xi)|^2\bigg)^{1/2}, \quad \widehat{\psi}(\xi) = (2\pi)^{-1/2} \int_{\mathbb{R}} dx \, e^{-i\xi x} \psi(x). \tag{3.19}$$

Equivalently,

$$H^s(\mathbb{R}) = \bigg\{ u \in \mathcal{S}'(\mathbb{R}) \, \bigg| \, \|u\|^2_{H^s(\mathbb{R})} = \int_{\mathbb{R}^n} d\xi \, \big(1 + |\xi|^{2s}\big) |\widehat{u}(\xi)|^2 < \infty \bigg\}, \quad s \in \mathbb{R}. \tag{3.20}$$

Similarly,

$$\begin{aligned} H^s_{\mathrm{loc}}(\mathbb{R}) &= \big\{ u \in \mathcal{D}'(\mathbb{R}) \, \big| \, \|\psi \, u\|_{H^s(\mathbb{R})} < \infty \text{ for all } \psi \in C_0^\infty(\mathbb{R}) \big\} \\ &= \big\{ u \in \mathcal{D}'(\mathbb{R}) \, \big| \, \|\eta(\cdot - a) \, u\|_{H^s(\mathbb{R})} < \infty \text{ for all } a \in \mathbb{R} \big\}, \quad s \in \mathbb{R} \end{aligned} \tag{3.21}$$

(cf. [70, p. 140]), and

$$H^s_{\text{loc unif}}(\mathbb{R}) = \Big\{u \in H^s_{\text{loc}}(\mathbb{R}) \,\Big|\, \sup_{a\in\mathbb{R}} \|\eta(\cdot - a)u\|_{H^s(\mathbb{R})} < \infty\Big\}, \quad s \in \mathbb{R}, \tag{3.22}$$

with η defined in (2.3).

Moreover, as proven in [43, Sect. 2] (cf. also [83], [113], [114]) elements $q \in H^{-1}_{\text{loc}}(\mathbb{R}) \subset \mathcal{D}'(\mathbb{R})$ can be represented by

$$q = q_2' \text{ for some } q_2 \in L^2_{\text{loc}}(\mathbb{R}; dx). \tag{3.23}$$

Similarly, if $q \in H^{s-1}(\mathbb{R})$ for some $s \geq 0$, [83, Lemma 2.1] proves the representation

$$q = v_\infty + v_s' \text{ for some } v_\infty \in H^\infty(\mathbb{R}),\ v_s \in H^s(\mathbb{R}), \tag{3.24}$$

where

$$H^\infty(\mathbb{R}) = \bigcap_{t\geq 0} H^t(\mathbb{R}) \subset C^\infty(\mathbb{R}). \tag{3.25}$$

In particular, if $q \in H^{-1}(\mathbb{R})$ one has the representation

$$q = v_\infty + q_2' \text{ for some } v_\infty \in H^\infty(\mathbb{R}),\ q_2 \in L^2(\mathbb{R}; dx). \tag{3.26}$$

Next, for $q \in H^{-1}_{\text{loc unif}}(\mathbb{R})$, [78, Theorem 2.1] proves the representation

$$q = q_1 + q_2' \text{ for some } q_j \in L^j_{\text{loc unif}}(\mathbb{R}; dx),\ j = 1, 2. \tag{3.27}$$

The decomposition $q = q_1 + q_2'$ in (3.27) is nonunique. In fact, also the representation

$$q = q_\infty + q_2' \text{ for some } q_\infty \in L^\infty(\mathbb{R}; dx),\ q_2 \in L^2_{\text{loc unif}}(\mathbb{R}; dx) \tag{3.28}$$

is proved in [78, Theorem 2.1]. Finally, if $q \in H^{-1}_{\text{loc}}(\mathbb{R})$ is periodic with period $\omega > 0$, [78, Remark 2.3] (see also [43, Proposition 1]) provides the representation

$$q = c + q_2' \text{ for some } c \in \mathbb{C},\ q_2 \in L^2_{\text{loc unif}}(\mathbb{R}; dx),\ q_2 \text{ periodic with period } \omega > 0. \tag{3.29}$$

Next, we turn to sequilinear forms $\mathfrak{Q}_q$ generated by a distribution $q \in \mathcal{D}'(\mathbb{R})$ as follows: For $f, g \in C_0^\infty(\mathbb{R})$, $\overline{f}$ (the complex conjugate of f) is a multiplier for q, that is, $\overline{f}q = q\overline{f} \in \mathcal{D}'(\mathbb{R})$ and hence the distributional pairing

$${}_{\mathcal{D}'(\mathbb{R})}\langle qf, g\rangle_{\mathcal{D}(\mathbb{R})} = (fq)(g) = q(\overline{f}g) = \mathfrak{Q}_q(f, g), \quad f, g \in C_0^\infty(\mathbb{R}), \tag{3.30}$$

is well defined and thus determines a sesquilinear form $\mathfrak{Q}_q(\cdot,\cdot)$ defined on $\mathcal{D}(\mathbb{R}) = C_0^\infty(\mathbb{R})$. The distribution $q \in \mathcal{D}'(\mathbb{R})$ is called a *multiplier* from $H^1(\mathbb{R})$ to $H^{-1}(\mathbb{R})$ if (3.30) continuously extends from $C_0^\infty(\mathbb{R})$ to $H^1(\mathbb{R})$, that is, for some $C > 0$,

$$|\mathfrak{Q}_q(f, g)| \leq C\|f\|_{H^1(\mathbb{R})}\|g\|_{H^1(\mathbb{R})}, \quad f, g \in C_0^\infty(\mathbb{R}), \tag{3.31}$$

and hence one defines this extension $\widetilde{\mathfrak{Q}}_q$ via

$$\begin{gathered}\widetilde{\mathfrak{Q}}_q(f, g) = \lim_{n\to\infty} \mathfrak{Q}_q(f_n, g_n), \quad f, g \in H^1(\mathbb{R}),\ f_n, g_n \in C_0^\infty(\mathbb{R}), \\ \text{assuming } \lim_{n\to\infty} \|f - f_n\|_{H^1(\mathbb{R})} = 0,\ \lim_{n\to\infty} \|g - g_n\|_{H^1(\mathbb{R})} = 0.\end{gathered} \tag{3.32}$$

(This extension is independent of the particular choices of sequences f_n, g_n and by polarization, (3.31) for $f = g$ suffices to yield the extension $\widetilde{\mathfrak{Q}}_q$ in (3.32).) The set of all multipliers from $H^1(\mathbb{R})$ to $H^{-1}(\mathbb{R})$ is usually denoted by $M\big(H^1(\mathbb{R}), H^{-1}(\mathbb{R})\big)$, equivalently, one could use the symbol $\mathcal{B}\big(H^1(\mathbb{R}), H^{-1}(\mathbb{R})\big)$, the bounded linear operators mapping $H^1(\mathbb{R})$ into $H^{-1}(\mathbb{R})$. Thus, for $q \in M\big(H^1(\mathbb{R}), H^{-1}(\mathbb{R})\big)$, the distributional pairing (3.30) extends to

$$_{H^{-1}(\mathbb{R})}\langle qf, g\rangle_{H^1(\mathbb{R})} = \widetilde{\mathfrak{Q}}_q(f, g), \quad f, g \in H^1(\mathbb{R}). \tag{3.33}$$

Theorem 3.1 ([7], [103, Sects. 2.5, 11.4], [104], [106], [117]). *Assume that* $q \in \mathcal{D}'(\mathbb{R})$ *generates the sesquilinear form* $\mathfrak{Q}_q$ *as in* (3.30). *Then the following conditions* (i)–(iii) *are equivalent:*

(i) *q is form bounded with respect to T_0, that is, for some $C > 0$,*

$$|\mathfrak{Q}_q(f, f)| \leq C\|f\|^2_{H^1(\mathbb{R})} = C\big[\|f'\|^2_{L^2(\mathbb{R};dx)} + \|f\|^2_{L^2(\mathbb{R};dx)}\big], \quad f \in C_0^\infty(\mathbb{R}), \tag{3.34}$$

equivalently,

$$q \in M\big(H^1(\mathbb{R}), H^{-1}(\mathbb{R})\big). \tag{3.35}$$

(ii) *q is infinitesimally form bounded with respect to T_0, that is, for all $\varepsilon > 0$, there exists $C_\varepsilon > 0$, such that,*

$$|\mathfrak{Q}_q(f, f)| \leq \varepsilon\|f'\|^2_{L^2(\mathbb{R};dx)} + C_\varepsilon\|f\|^2_{L^2(\mathbb{R};dx)}, \quad f \in H^1(\mathbb{R}). \tag{3.36}$$

(iii) *q is of the form*

$$q = q_1 + q_2', \quad \text{where } q_j \in L^j_{\text{loc unif}}(\mathbb{R}; dx), \ j = 1, 2. \tag{3.37}$$

Equivalently (cf. (3.22), (3.27)*),*

$$q \in H^{-1}_{\text{loc unif}}(\mathbb{R}). \tag{3.38}$$

Of course, if (3.34) (equivalently, (3.36)) holds, it extends to $\widetilde{\mathfrak{Q}}_q$ and all $f \in H^1(\mathbb{R})$.

Theorem 3.2 ([103, Sect. 11.4], [104], [106]). *Assume that* $q \in \mathcal{D}'(\mathbb{R})$. *Then the following conditions* (i) *and* (ii) *are equivalent:*

(i) *q is form compact with respect to T_0, that is, the map*

$$q\colon H^1(\mathbb{R}) \to H^{-1}(\mathbb{R}) \text{ is compact.} \tag{3.39}$$

(ii) *q is of the form*

$$q = q_1 + q_2', \quad \text{where } q_j \in L^j_{\text{loc unif}}(\mathbb{R}; dx), \ j = 1, 2, \tag{3.40}$$

and

$$\lim_{|a|\to\infty}\bigg(\int_a^{a+1} dx\, |q_1(x)|\bigg) = 0, \ \lim_{|a|\to\infty}\bigg(\int_a^{a+1} dx\, |q_2(x)|^2\bigg) = 0. \tag{3.41}$$

We should emphasize that the references [7], [103, Sects. 2.5, 11.4], [106], [117] in connection with Theorems 3.1 and 3.2, primarily focus on the multi-dimensional situation. In particular, the methods employed in Maz'ya and Verbitsky [104], [105], [106], [107], and Maz'ya and Shaposhnikova [103] rely on Bessel capacity methods. A considerably simplified approach to this circle of ideas, based on the existence of positive solutions of the underlying zero-energy Schrödinger operator (more generally, an equation of the type $-\text{div}(A\nabla u)+qu = 0$ in arbitrary open sets $\Omega \in \mathbb{R}^n$, with A satisfying an ellipticity condition) appeared in [81]. The special one-dimensional case is explicitly treated in [7], [104], and [117].

Remark 3.3. If $q \in \mathcal{D}'(\mathbb{R})$ is real valued and one of the conditions (i)–(iii) in Theorem 3.1 is satisfied, then the form sum

$$\mathfrak{Q}_T(f,g) = \mathfrak{Q}_{T_0}(f,g) + q(\overline{f}g), \quad f,g \in \text{dom}(\mathfrak{Q}_T) = H^1(\mathbb{R}), \tag{3.42}$$

defines a closed, densely defined, symmetric sesquilinear form $\mathfrak{Q}_T$ in $L^2(\mathbb{R}; dx)$, bounded from below. The self-adjoint operator T in $L^2(\mathbb{R}; dx)$, bounded from below, and uniquely associated to the form $\mathfrak{Q}_T$ then can be described as follows,

$$\begin{aligned} &Tf = \tau f, \quad \tau f = -(f' - q_2 f)' - q_2(f' - q_2 f) + (q_1 - q_2^2)f, \qquad (3.43)\\ &f \in \text{dom}(T) = \big\{g \in L^2(\mathbb{R}; dx) \,\big|\, g, (g' - q_2 g) \in AC_{\text{loc}}(\mathbb{R}),\, \tau g \in L^2(\mathbb{R}; dx)\big\}. \end{aligned}$$

In particular, the differential expression τ formally corresponds to a Schrödinger operator with distributional potential $q \in H^{-1}_{\text{loc unif}}(\mathbb{R})$,

$$\tau = -(d^2/dx^2) + q(x), \quad q = q_1 + q_2', \quad q_j \in L^j_{\text{loc unif}}(\mathbb{R}; dx),\ j = 1,2. \tag{3.44}$$

This is a consequence of the direct methods established in [7], [78]–[80], [83], [123], [124], [136], and of the Weyl–Titchmarsh theory approach to Schrödinger operators with distributional potentials developed in [49] (see also [48], [50], and the detailed list of references therein). In particular, since τ is assumed to be bounded from below, τ is in the limit point case at $\pm\infty$, rendering the maximally defined operator T in (3.43) to be self-adjoint (see also [4] and [49]). We will provide further details on $\text{dom}(T)$ in Remark 3.8.

Next, we turn to an elementary alternative approach to this circle of ideas in the real-valued context, based on the concept of Miura transformations (cf. [24], [39], [48], [57], [58], [66], [67], [83], [84], [132, Ch. 5], and the extensive literature cited therein)

$$\begin{cases} L^2_{\text{loc}}(\mathbb{R}; dx) \to H^{-1}_{\text{loc}}(\mathbb{R}) \\ \phi \mapsto \phi^2 - \phi' \end{cases} \tag{3.45}$$

with associated self-adjoint Schrödinger operator $T_1 \geq 0$ in $L^2(\mathbb{R}; dx)$ given by

$$T_1 = A^*A, \tag{3.46}$$

with A the closed operator defined in in $L^2(\mathbb{R}; dx)$ by

$$\begin{aligned} &Af = \alpha f, \quad \alpha f = f' + \phi f, \\ &f \in \text{dom}(A) = \big\{g \in L^2(\mathbb{R}; dx) \,\big|\, g \in AC_{\text{loc}}(\mathbb{R}),\, \alpha g \in L^2(\mathbb{R}; dx)\big\}, \end{aligned} \tag{3.47}$$

implying,

$$\begin{aligned} &A^* f = \alpha^+ f, \quad \alpha^+ f = -f' + \phi f, \\ &f \in \operatorname{dom}(A^*) = \big\{ g \in L^2(\mathbb{R}; dx) \,\big|\, g \in AC_{\mathrm{loc}}(\mathbb{R}),\, \alpha^+ g \in L^2(\mathbb{R}; dx) \big\}. \end{aligned} \tag{3.48}$$

Closedness of A and the fact that A^* is given by (3.48) was proved in [83] (the extension to $\phi \in L^1_{\mathrm{loc}}(\mathbb{R}; dx)$, ϕ real-valued, was treated in [48]). In addition, it was proved in [83] that

$$C_0^\infty(\mathbb{R}) \text{ is an operator core for } A \text{ and } A^*. \tag{3.49}$$

Thus, T_1 acts as,

$$\begin{aligned} &T_1 f = \tau_1 f, \quad \tau_1 f = \alpha^+ \alpha f = -(f' + \phi f)' + \phi(f' + \phi f), \\ &f \in \operatorname{dom}(T_1) = \big\{ g \in L^2(\mathbb{R}; dx) \,\big|\, g, \alpha g \in AC_{\mathrm{loc}}(\mathbb{R}), \tau_1 g \in L^2(\mathbb{R}; dx) \big\}. \end{aligned} \tag{3.50}$$

In particular, τ_1 is formally of the type,

$$\tau_1 = -(d^2/dx^2) + V_1(x), \quad V_1 = \phi^2 - \phi', \quad \phi \in L^2_{\mathrm{loc}}(\mathbb{R}; dx), \tag{3.51}$$

displaying the Riccati equation connection between V_1 and ϕ in connection with Miura's transformation (3.45).

Theorem 3.4 ([83]). *Assume that $q \in H^{-1}_{\mathrm{loc}}(\mathbb{R})$ is real valued. Then the following conditions* (i)–(iii) *are equivalent:*

(i) *$q = \phi^2 - \phi'$ for some real-valued $\phi \in L^2_{\mathrm{loc}}(\mathbb{R}; dx)$.*
(ii) *$(-d^2/dx^2) + q \geq 0$ in the sense of distributions, that is,*

$$\begin{aligned} (f', f')_{L^2(\mathbb{R};dx)} + q(\overline{f} f) = {}_{H^{-1}(\mathbb{R})}\langle (-f'' + qf), f \rangle_{H^1(\mathbb{R})} \geq 0 \\ \text{for all } f \in C_0^\infty(\mathbb{R}). \end{aligned} \tag{3.52}$$

(iii) *$[(-d^2/dx^2) + q]\psi = 0$ has a positive solution $0 < \psi \in H^1_{\mathrm{loc}}(\mathbb{R})$.*

We note that multi-dimensional extensions this circle of ideas are studied in great depth in [81].

Theorem 3.5 ([83]). *Assume that $q \in H^{s-1}(\mathbb{R})$, $s \geq 0$, is real valued. Then the following conditions* (i) *and* (ii) *are equivalent:*

(i) *$q = \phi^2 - \phi'$ for some real-valued $\phi \in H^s(\mathbb{R})$.*
(ii) *$(-d^2/dx^2) + q \geq 0$ in the sense of distributions* (*cf.* (3.52)) *and $q = q_1 + q_2'$ for some $q_j \in L^j(\mathbb{R}; dx)$, $j = 1, 2$.*

The following appears to be a new result:

Theorem 3.6. *Assume that $q \in H^{-1}_{\mathrm{loc\,unif}}(\mathbb{R})$ is real valued. Then the following conditions* (i)–(iii) *are equivalent:*

(i) *$q = \phi^2 - \phi'$ for some real-valued $\phi \in L^2_{\mathrm{loc\,unif}}(\mathbb{R}; dx)$.*
(ii) *$(-d^2/dx^2) + q \geq 0$ in the sense of distributions* (*cf.* (3.52)).
(iii) *$[(-d^2/dx^2) + q]\psi = 0$ has a positive solution $0 < \psi \in H^1_{\mathrm{loc}}(\mathbb{R})$.*

Proof. We will show that (ii) $\Longrightarrow$ (iii) $\Longrightarrow$ (i) $\Longrightarrow$ (ii).

Given item (ii), that is, $q \in H^{-1}_{\text{loc unif}}(\mathbb{R})$ is real valued and $(-d^2/dx^2)+q \geq 0$, one concludes the existence of $0 < \psi_0 \in H^1_{\text{loc}}(\mathbb{R})$ such that $-\psi_0'' + q\psi_0 = 0$ by Theorem 3.4 (iii). Thus, item (iii) follows.

Introducing

$$\phi_0 = -\psi_0'/\psi_0, \tag{3.53}$$

one infers that

$$\phi_0 \in L^2_{\text{loc}}(\mathbb{R}; dx) \text{ is real valued and } q = \phi_0^2 - \phi_0'. \tag{3.54}$$

Next, introducing A_0 and A_0^* as in (3.47) and (3.48), with α replaced by $\alpha_0 = (d/dx)+\phi_0$ (and analogously for α^+), we now introduce the sesquilinear form $\dot{\mathfrak{Q}}_{T_1}$ and its closure, $\mathfrak{Q}_{T_1}$, by

$$\begin{aligned} \dot{\mathfrak{Q}}_{T_1}(f,g) &= (A_0 f, A_0 g)_{L^2(\mathbb{R};dx)}, \quad f,g \in \operatorname{dom}\big(\dot{\mathfrak{Q}}_{T_1}\big) = C_0^\infty(\mathbb{R}), \\ \mathfrak{Q}_{T_1}(f,g) &= (A_0 f, A_0 g)_{L^2(\mathbb{R};dx)}, \quad f,g \in \operatorname{dom}(\mathfrak{Q}_{T_1}) = \operatorname{dom}(A_0), \end{aligned} \tag{3.55}$$

with $0 \leq T_1 = A_0^* A_0$ the uniquely associated self-adjoint operator.

Since by hypothesis $q \in H^{-1}_{\text{loc unif}}(\mathbb{R})$, (2.46) implies (cf. [78]) that q can be written as

$$q = q_1 + q_2' \text{ for some } q_j \in L^j_{\text{loc unif}}(\mathbb{R}; dx),\ j = 1,2, \tag{3.56}$$

and hence, we also introduce the sesquilinear form $\dot{\widehat{\mathfrak{Q}}}$ and its closure, $\widehat{\mathfrak{Q}}$ (cf. [78] for details),

$$\begin{aligned} \dot{\widehat{\mathfrak{Q}}}(f,g) &= (f',g')_{L^2(\mathbb{R};dx)} - (f',q_2 g)_{L^2(\mathbb{R};dx)} - (q_2 f, g')_{L^2(\mathbb{R};dx)} \\ &\quad + \big(|q_1|^{1/2} f, \operatorname{sgn}(q_1)|q_1|^{1/2} g\big)_{L^2(\mathbb{R};dx)}, \quad f,g \in \operatorname{dom}\big(\dot{\widehat{\mathfrak{Q}}}\big) = C_0^\infty(\mathbb{R}), \end{aligned} \tag{3.57}$$

$$\begin{aligned} \widehat{\mathfrak{Q}}(f,g) &= (f',g')_{L^2(\mathbb{R};dx)} - (f',q_2 g)_{L^2(\mathbb{R};dx)} - (q_2 f, g')_{L^2(\mathbb{R};dx)} \\ &\quad + \big(|q_1|^{1/2} f, \operatorname{sgn}(q_1)|q_1|^{1/2} g\big)_{L^2(\mathbb{R};dx)}, \quad f,g \in \operatorname{dom}\big(\widehat{\mathfrak{Q}}\big) = H^1(\mathbb{R}). \end{aligned} \tag{3.58}$$

Since

$$\mathfrak{Q}_{T_1}(f,g) = \widehat{\mathfrak{Q}}(f,g) = (f',g')_{L^2(\mathbb{R};dx)} + q(\overline{f}g), \quad f,g \in C_0^\infty(\mathbb{R}), \tag{3.59}$$

and $C_0^\infty(\mathbb{R})$ is a form core for $\mathfrak{Q}_{T_1}$ (cf. (3.49)) and $\widehat{\mathfrak{Q}}$, one concludes that $\mathfrak{Q}_{T_1} = \widehat{\mathfrak{Q}}$ and hence

$$\operatorname{dom}(\mathfrak{Q}_{T_1}) = \operatorname{dom}(A_0) = \operatorname{dom}\big(\widehat{\mathfrak{Q}}\big) = H^1(\mathbb{R}). \tag{3.60}$$

A comparison of (3.47) (with α replaced by α_0) and (3.60) implies that $\phi_0 g \in L^2(\mathbb{R}; dx)$ for $g \in \operatorname{dom}(A_0) = H^1(\mathbb{R})$, and hence,

$$\operatorname{dom}(\phi_0) \supseteq H^1(\mathbb{R}). \tag{3.61}$$

An application of Theorem 2.1 (i), (ii) then finally yields

$$\phi_0 \in L^2_{\text{loc unif}}(\mathbb{R}; dx), \tag{3.62}$$

which together with (3.54) implies item (i).

Finally, given $\phi \in L^2_{\mathrm{loc\,unif}}(\mathbb{R}; dx)$, ϕ real-valued, such that $q = \phi^2 - \phi'$, one computes, with $\alpha = (d/dx) + \phi$,

$$\begin{aligned} 0 \leq \|\alpha f\|^2_{L^2(\mathbb{R};dx)} = \|f'\|^2_{L^2(\mathbb{R};dx)} + q\big(|f|^2\big) = {}_{H^{-1}(\mathbb{R})}\langle(-f'' + qf), f\rangle_{H^1(\mathbb{R})}, \\ f \in C_0^\infty(\mathbb{R}), \end{aligned} \tag{3.63}$$

and hence item (i) implies item (ii). □

Thus, Theorem 3.6 further illustrates the results by Bak and Shkalikov [7] and Maz'ya and Verbitsky [104], [105], [106] (specialized to the one-dimensional situation) recorded in Theorem 3.1 in the particular case where q is real valued.

In connection with Theorem 3.6 (i), we also recall the following useful result:

Lemma 3.7 ([80]). *Assume that $q \in H^{-1}_{\mathrm{loc\,unif}}(\mathbb{R})$ is real valued and of the form $q = \phi^2 - \phi'$ for some real-valued $\phi \in L^2_{\mathrm{loc}}(\mathbb{R}; dx)$. Then, actually,*

$$\phi \in L^2_{\mathrm{loc\,unif}}(\mathbb{R}; dx). \tag{3.64}$$

Remark 3.8. Combining (3.42)–(3.44), (3.50), (3.51), (3.55), and (3.61) (identifying ϕ and ϕ_0 as well as T and T_1) then yields the following apparent improvement over the domain characterizations (3.43), (3.50),

$$\begin{aligned} &T_1 f = \tau_1 f, \quad \tau f = -(f' + \phi f)' + \phi(f' + \phi f), \\ &f \in \mathrm{dom}(T_1) = \big\{g \in L^2(\mathbb{R}; dx) \,\big|\, g, \alpha g \in AC_{\mathrm{loc}}(\mathbb{R}), \\ &\qquad\qquad g', \phi g \in L^2(\mathbb{R}; dx),\ \tau_1 g \in L^2(\mathbb{R}; dx)\big\}, \end{aligned} \tag{3.65}$$

with (3.51) staying in place. In fact, (3.50) and (3.65) are, of course, equivalent; the former represents a minimal characterization of $\mathrm{dom}(T_1)$.

Remark 3.9. Given $q = \phi^2 - \phi'$, $\phi \in L^2_{\mathrm{loc\,unif}}(\mathbb{R}; dx)$ as in Theorems 3.4–3.6, the question of uniqueness of ϕ for prescribed $q \in H^{-1}_{\mathrm{loc}}(\mathbb{R})$ arises naturally. This has been settled in [83] and so we briefly summarize some pertinent facts. Since $\phi = -\psi'/\psi$ for some $0 < \psi \in H^1_{\mathrm{loc}}(\mathbb{R})$, uniqueness of ϕ is equivalent to uniqueness of $\psi > 0$ satisfying $[(-d^2/dx^2) + q]\psi = 0$. Thus, suppose $0 < \psi_0 \in H^1_{\mathrm{loc}}(\mathbb{R})$ is a solution of $[(-d^2/dx^2) + q]\psi = 0$. Then, the general, real-valued solution of $[(-d^2/dx^2) + q]\psi = 0$ is of the type

$$\psi(x) = C_1\psi_0(x) + C_2\psi_0(x)\int_0^x dx'\,\psi_0(x')^{-2}, \quad x \in \mathbb{R},\ C_j \in \mathbb{R},\ j = 1, 2. \tag{3.66}$$

Next, introducing

$$c_\pm = \pm \lim_{x\to\pm\infty} \int_0^x dx'\,\psi_0(x')^{-2} \in (0, +\infty], \tag{3.67}$$

and defining $c_\pm^{-1} = 0$ if $c_\pm = +\infty$, all positive solutions $0 < \psi$ on $\mathbb{R}$ of $[(-d^2/dx^2) + q]\psi = 0$ are given by

$$\psi(x) = \psi_0(x)\bigg[1 + c\int_0^x dx'\,\psi_0(x')^{-2}\bigg], \quad c \in \big[-c_+^{-1}, c_-^{-1}\big]. \tag{3.68}$$

Consequently,

$$
\begin{aligned}
&0 < \psi_0 \in H^1_{\mathrm{loc}}(\mathbb{R}) \ \text{ is the unique solution of } \ [(-d^2/dx^2) + q]\psi = 0 \\
&\qquad \text{if and only if } \ \pm \lim_{x\to\pm\infty} \int_0^x dx' \, \psi_0(x')^{-2} = \infty.
\end{aligned} \tag{3.69}
$$

On the other hand, if at least one of $\pm \lim_{x\pm\infty} \int_0^x dx' \, \psi_0(x')^{-2} < \infty$, $[(-d^2/dx^2)+q]\psi = 0$ has a one (real) parameter family of positive solutions on $\mathbb{R}$ lying in $H^1_{\mathrm{loc}}(\mathbb{R})$ given by (3.68). Without going into further details, we note that Weyl–Titchmarsh solutions $\psi_\pm(\lambda,\cdot)$ corresponding to T in (3.43) for energies $\lambda < \inf(\sigma(T))$, are actually constant multiples of Hartman's principal solutions $T\widehat{\psi}_\pm(\lambda,\cdot) = \lambda\widehat{\psi}_\pm(\lambda,\cdot)$, that is, those that satisfy $\pm \int^{\pm\infty} dx' \, \big[\widehat{\psi}_\pm(\lambda,x')\big]^{-2} = \infty$.

Theorem 3.10. *Assume that $q \in H^{-1}_{\mathrm{loc}}(\mathbb{R})$ is real valued and suppose in addition that $(-d^2/dx^2)+q \geq 0$ in the sense of distributions* (*cf.* (3.52))*. Then the following conditions* (i)–(iv) *are equivalent:*

(i) *q is form compact with respect to T_0, that is, the map*

$$q\colon H^1(\mathbb{R}) \to H^{-1}(\mathbb{R}) \ \text{ is compact.} \tag{3.70}$$

(ii) *q is of the form $q = \phi^2 - \phi'$, where $\phi \in L^2_{\mathrm{loc\,unif}}(\mathbb{R}; dx)$ is real valued and*

$$\lim_{|a|\to\infty} \bigg(\int_a^{a+1} dx \, \phi(x)^2\bigg) = 0. \tag{3.71}$$

(iii) *The operator of multiplication by ϕ is $T_0^{1/2}$-compact.*

(iv) *The operator of multiplication by ϕ is $P_m\big(T_0^{1/2}\big)$-compact, where P_m is a polynomial of degree $m \in \mathbb{N}$.*

Proof. By Theorem 3.4, $(-d^2/dx^2) + q \geq 0$ in the sense of distributions implies that q is of the form $q = \phi^2 - \phi'$ for some real-valued $\phi \in L^2_{\mathrm{loc}}(\mathbb{R})$. By Lemma 3.7, one actually concludes that $\phi \in L^2_{\mathrm{loc\,unif}}(\mathbb{R})$. The equivalence of items (i) and (ii) then follows from Theorem 3.2 since upon identifying $q_1 = \phi^2$, $q_2 = \phi$, the two limiting relations in (3.41) are equivalent to (3.71). Equivalence of condition (3.71) and item (iii) is guaranteed by Theorem 2.2. □

At this point it is worth recalling a few additional details of the supersymmetric formalism started in (3.45)–(3.51), whose abstract roots can be found in Appendix B: Assuming $\phi \in L^2_{\mathrm{loc\,unif}}(\mathbb{R}; dx)$ to be real valued (we note, however, that this supersymmetric formalism extends to the far more general situation where $\phi \in L^1_{\mathrm{loc}}(\mathbb{R}; dx)$ is real valued, in fact, it extends to the situation where ϕ is matrix valued, see [48] for a detailed treatment of these matters), one has

$$A = (d/dx) + \phi, \quad A^* = -(d/dx) + \phi, \quad \mathrm{dom}(A) = \mathrm{dom}(A^*) = H^1(\mathbb{R}), \tag{3.72}$$

$$T_1 = A^*A = -(d^2/dx^2) + V_1, \quad V_1 = \phi^2 - \phi', \tag{3.73}$$

$$T_2 = AA^* = -(d^2/dx^2) + V_2, \quad V_2 = \phi^2 + \phi', \tag{3.74}$$

$$D = \begin{pmatrix} 0 & A^* \\ A & 0 \end{pmatrix} \text{ in } L^2(\mathbb{R}; dx) \oplus L^2(\mathbb{R}; dx), \tag{3.75}$$

$$D^2 = \begin{pmatrix} A^*A & 0 \\ 0 & AA^* \end{pmatrix} = T_1 \oplus T_2 \text{ in } L^2(\mathbb{R}; dx) \oplus L^2(\mathbb{R}; dx). \tag{3.76}$$

As a consequence, one can show (cf. [48]) the Weyl–Titchmarsh solutions, $\psi_\pm$, for D, T_1, T_2 satisfy

$$\psi_{D,1,\pm}(\zeta, x) = \psi_{T_1,\pm}(z, x), \quad z = \zeta^2, \ \zeta \in \mathbb{C}\backslash\mathbb{R}, \tag{3.77}$$

$$\psi_{T_2,\pm}(z, x) = c_1(z)(A\psi_{T_1,\pm})(z, x), \tag{3.78}$$

with $c_1(z)$ a normalization constant. Similarly, after interchanging the role of T_1 and T_2,

$$\psi_{D,2,\pm}(\zeta, x) = \psi_{T_2,\pm}(z, x), \quad z = \zeta^2, \ \zeta \in \mathbb{C}\backslash\mathbb{R}, \tag{3.79}$$

$$\psi_{T_1,\pm}(z, x) = c_2(z)(A^*\psi_{T_2,\pm})(z, x), \tag{3.80}$$

again with $c_2(z)$ a normalization constant. Here,

$$\Psi_{D,\pm}(\zeta, x) = \begin{pmatrix} \psi_{D,1,\pm}(\zeta, x) \\ \psi_{D,2,\pm}(\zeta, x) \end{pmatrix} \tag{3.81}$$

are the Weyl–Titchmarsh solutions for $D = \left(\begin{smallmatrix} 0 & A^* \\ A & 0 \end{smallmatrix}\right)$.

The (generalized, or renormalized) Weyl–Titchmarsh m-functions for D, T_1, T_2 satisfy:

$$m_{D,\pm}(\zeta, x_0) = \frac{1}{\zeta}\widehat{m}_{T_1,\pm}(z, x_0) = \frac{-\zeta}{\widehat{m}_{T_2,\pm}(z, x_0)}, \tag{3.82}$$

where x_0 is a fixed reference point (typically, $x_0 = 0$), and

$$\widehat{m}_{T_1,\pm}(z, x_0) = \frac{\psi^{[1,1]}_{T_1,\pm}(z, x_0)}{\psi_{T_1,\pm}(z, x_0)} = \frac{(A\psi_{T_1,\pm})(z, x_0)}{\psi_{T_1,\pm}(z, x_0)}, \tag{3.83}$$

$$\widehat{m}_{T_2,\pm}(z, x_0) = \frac{\psi^{[1,2]}_{T_2,\pm}(z, x_0)}{\psi_{T_2,\pm}(z, x_0)} = \frac{(-A^*\psi_{T_2,\pm})(z, x_0)}{\psi_{T_2,\pm}(z, x_0)}. \tag{3.84}$$

Here, $y^{[1,1]} = Ay = [y' + \phi y]$ is the quasi-derivative corresponding to T_1 and $y^{[1,2]} = -A^*y = [y' - \phi y]$ is the quasi-derivative corresponding to T_2.

Thus, spectral properties of D instantly translate into spectral properties of T_j, $j = 1, 2$, and vice versa (the latter with the exception of the zero spectral parameter). In particular, $\phi \in L^2_{\text{loc unif}}(\mathbb{R}; dx) \subset L^2_{\text{loc}}(\mathbb{R}; dx)$ in D is entirely "standard" (in fact, even $\phi \in L^1_{\text{loc}}(\mathbb{R}; dx)$ in D is entirely standard, see, e.g., [28] and the extensive literature cited therein), while the potentials $V_j = (-1)^j\phi' + \phi^2$, $j = 1, 2$, involve the distributional coefficient $\phi' \in H^{-1}_{\text{loc unif}}(\mathbb{R})$. (We also note that while in this paper the Dirac operator D only involves the $L^2_{\text{loc}}(\mathbb{R}; dx)$-coefficient ϕ, Dirac-type operators with distributional potentials have been studied in the literature, see, for instance [3, App. J] and [27].) In particular, spectral results for the "standard" one-dimensional Dirac-type operator D imply corresponding spectral results for Schrödinger operators bounded from below, with (real-valued) distributional potentials. Some applications of this spectral correspondence between D and T_j, $j = 1, 2$, to inverse spectral theory, local Borg–Marchenko uniqueness results, etc., were treated in [48]. In Section 4 we will apply this spectral correspondence to derive some Floquet theoretic results in connection with the Schrödinger operators T_j and hence for the distributional potentials $[\phi^2 + (-1)^j \phi'] \in H^{-1}_{\text{loc unif}}(\mathbb{R})$, $j = 1, 2$.

Remark 3.11. For simplicity we restricted ourselves to the special case $p = 1$ in Theorems 3.4–3.6 and Remarks 3.8 and 3.9. However, assuming

$$0 < p, p^{-1} \in L^\infty(\mathbb{R}; dx), \quad 0 < r, r^{-1} \in L^\infty(\mathbb{R}; dx), \tag{3.85}$$

the observations thus far in this section extend to the case where

$$\begin{aligned} \tau_1 f = \alpha^+ \alpha f &= -f'' + [\phi^2 - \phi'] f \\ &= -(f' + \phi f)' + \phi(f' + \phi f) \end{aligned} \tag{3.86}$$

is replaced by

$$\begin{aligned} \tau_1 f = \beta^+ \beta f &= r^{-1} \Big[- (pf')' + \big[p\phi^2 - (p\phi)'\big] f \Big] \\ &= r^{-1} \Big[- [p(f' + \phi f)]' + \phi [p(f' + \phi f)] \Big], \end{aligned} \tag{3.87}$$

where

$$\begin{aligned} \beta f &= (pr)^{-1/2} [p(f' + \phi f)], \\ \beta^+ f &= -(pr)^{-1} \Big\{ p \Big[\big[(pr)^{1/2} f\big]' - \phi \big[(pr)^{1/2} f\big] \Big] \Big\}. \end{aligned} \tag{3.88}$$

Remark 3.12. We only dwelled on

$$\operatorname{dom}\big(|T|^{1/2}\big) = H^1(\mathbb{R}) \tag{3.89}$$

to derive a number of *if and only if* results. For practitioners in this field, the sufficient conditions on q, r in terms of the $L^j_{\text{loc unif}}(\mathbb{R}; dx)$, $j = 1, 2$, and boundedness conditions on $0 < p, p^{-1}$, yielding form boundedness (i.e., self-adjointness) results, relative compactness, and trace class results, all work as long as one ensures

$$\operatorname{dom}\big(|T|^{1/2}\big) \subseteq H^1(\mathbb{R}). \tag{3.90}$$

This permits larger classes of coefficients p, q, r for which one can prove these types of self-adjointness and spectral results.

Before returning to our principal object, the Birman–Schwinger-type operator $\underline{T^{-1/2} r T^{-1/2}}$, but now in the context of distributional coefficients q and r, we briefly examine the well-known example of point interactions:

Example 3.13 (Delta distributions).

$$q_1(x) = 0, \quad q_2(x) = \begin{cases} 1, & x > x_0, \\ 0, & x < x_0, \end{cases} \quad \text{then} \quad q = q_2' = \delta_{x_0}, \quad x_0 \in \mathbb{R}. \tag{3.91}$$

Introducing the operator

$$A_{\alpha,x_0} = \frac{d}{dx} - \frac{\alpha}{2}\begin{cases} 1, & x > x_0, \\ -1, & x < x_0, \end{cases} \quad \text{that is, } \phi(x) = \frac{\alpha}{2}\operatorname{sgn}(x - x_0), \quad \alpha, x_0 \in \mathbb{R},$$
$$\operatorname{dom}(A_{\alpha,x_0}) = H^1(\mathbb{R}), \tag{3.92}$$

in $L^2(\mathbb{R}; dx)$, one infers that

$$A^*_{\alpha,x_0} A_{\alpha,x_0} = -\Delta_{\alpha,x_0} + (\alpha^2/4)I. \tag{3.93}$$

Here $-\Delta_{\alpha,x_0} = -d^2/dx^2 + \alpha\,\delta_{x_0}$ in $L^2(\mathbb{R}; dx)$ represents the self-adjoint realization of the one-dimensional point interaction (cf. [3, Ch. I.3]), that is, the Schrödinger operator with a delta function potential of strength (coupling constant) α centered at $x_0 \in \mathbb{R}$.

This extends to sums of delta distributions supported on a discrete set (Kronig–Penney model, etc.).

Next we apply this distributional approach to the Birman–Schwinger-type operator $\underline{\underline{T^{-1/2} r T^{-1/2}}}$. We outline the basic ideas in the following three steps:

Step 1. Assume $p, p^{-1} \in L^\infty(\mathbb{R}; dx)$, $p > 0$ a.e. on $\mathbb{R}$.

Step 2. Suppose $q = q_1 + q_2'$, where $q_j \in L^j_{\text{loc unif}}(\mathbb{R}; dx)$, $j = 1, 2$, are real valued. This uniquely defines a self-adjoint operator T in $L^2(\mathbb{R}; dx)$, bounded from below, $T \geq cI$ for some $c \in \mathbb{R}$, as the form sum $T = -(d/dx)p(d/dx) + q$ of $-(d/dx)p(d/dx)$ and the distribution $q = q_1 + q_2' \in \mathcal{D}'(\mathbb{R})$. Then

$$\operatorname{dom}\big(|T|^{1/2}\big) = H^1(\mathbb{R}). \tag{3.94}$$

If in addition, $\lim_{|a|\to\infty}\Big(\int_a^{a+1} dx\,|q_1(x) - c_1|\Big) = 0$ for some constant $c_1 \in \mathbb{R}$, and $\lim_{|a|\to\infty}\Big(\int_a^{a+1} dx\,|q_2(x)|^2\Big) = 0$, one again obtains results on essential spectra.

Step 3. Suppose without loss of generality, that $T \geq cI$, $c > 0$, and introduce $r = r_1 + r_2'$, $r_j \in L^j_{\text{loc unif}}(\mathbb{R}; dx)$ real-valued, $j = 1, 2$. This uniquely defines a bounded self-adjoint operator $\underline{\underline{T^{-1/2} r T^{-1/2}}}$ in $L^2(\mathbb{R}; dx)$ as described next: First write

$$\begin{aligned} &\underline{\underline{T^{-1/2} r T^{-1/2}}} \\ &\quad = \big[(T_0 + I)^{1/2} T^{-1/2}\big]^* \big[(T_0 + I)^{-1/2} r (T_0 + I)^{-1/2}\big] \big[(T_0 + I)^{1/2} T^{-1/2}\big]. \end{aligned} \tag{3.95}$$

Next, one interprets $(T_0 + I)^{-1/2} r (T_0 + I)^{-1/2}$ as follows: Employing T_0 and its extension, $\widetilde{T}_0$, to the entire Sobolev scale $H^s(\mathbb{R})$ in (3.9)–(3.15), in particular, we will employ the mapping properties, $\big(\widetilde{T}_0 + I\big)^{-1/2} : H^s(\mathbb{R}) \to H^{s+1}(\mathbb{R})$, $s \in \mathbb{R}$. Thus, using

$$\big[(T_0 + I)^{1/2} T^{-1/2}\big], \big[(T_0 + I)^{1/2} T^{-1/2}\big]^* \in \mathcal{B}\big(L^2(\mathbb{R}; dx)\big), \tag{3.96}$$

and

$$\Big[\underbrace{\big(\widetilde{T}_0+I\big)^{-1/2}}_{\in\mathcal{B}(H^{-1}(\mathbb{R}),L^2(\mathbb{R};dx))}\underbrace{r}_{\in\mathcal{B}(H^1(\mathbb{R}),H^{-1}(\mathbb{R}))}\underbrace{\big(\widetilde{T}_0+I\big)^{-1/2}}_{\in\mathcal{B}(L^2(\mathbb{R};dx),H^1(\mathbb{R}))}\Big]\in\mathcal{B}\big(L^2(\mathbb{R};dx)\big), \tag{3.97}$$

finally yields

$$\begin{aligned}\underline{\underline{T^{-1/2}rT^{-1/2}}} &= \big[(T_0+I)^{1/2}T^{-1/2}\big]^*\Big[\big(\widetilde{T}_0+I\big)^{-1/2}r\big(\widetilde{T}_0+I\big)^{-1/2}\Big]\\ &\quad\times\big[(T_0+I)^{1/2}T^{-1/2}\big]\in\mathcal{B}\big(L^2(\mathbb{R};dx)\big).\end{aligned} \tag{3.98}$$

Hence, our reformulated left-definite generalized eigenvalue problem becomes again a standard self-adjoint spectral problem in $L^2(\mathbb{R};dx)$,

$$\underline{\underline{T^{-1/2}rT^{-1/2}}}\,\chi=\frac{1}{z}\chi,\quad z\in\mathbb{C}\backslash\{0\}, \tag{3.99}$$

associated with the bounded, self-adjoint operator $\underline{\underline{T^{-1/2}rT^{-1/2}}}$ in $L^2(\mathbb{R};dx)$, yet this time we permit distributional coefficients satisfying

$$p,p^{-1}\in L^\infty(\mathbb{R};dx),\ p>0 \text{ a.e. on } \mathbb{R}, \tag{3.100}$$

$$q=q_1+q_2',\ q_j\in L^j_{\text{loc unif}}(\mathbb{R};dx) \text{ real-valued},\ j=1,2, \tag{3.101}$$

$$r=r_1+r_2',\ r_j\in L^j_{\text{loc unif}}(\mathbb{R};dx) \text{ real-valued},\ j=1,2, \tag{3.102}$$

with T defined as the self-adjoint, lower-semibounded operator uniquely associated with the lower-bounded, closed sesquilinear form $\mathfrak{Q}_T$ in $L^2(\mathbb{R};dx)$ given by (cf. (3.30))

$$\begin{aligned}\mathfrak{Q}_T(f,g) &= \big(p^{1/2}f',p^{1/2}g'\big)_{L^2(\mathbb{R};dx)}+q(\overline{f}g) &(3.103)\\ &= \big(p^{1/2}f',p^{1/2}g'\big)_{L^2(\mathbb{R};dx)}-(f',q_2g)_{L^2(\mathbb{R};dx)}-(q_2f,g')_{L^2(\mathbb{R};dx)}\\ &\quad+\big(|q_1|^{1/2}f,\operatorname{sgn}(q_1)|q_1|^{1/2}g\big)_{L^2(\mathbb{R};dx)}, &(3.104)\\ &= \big(p^{-1/2}(pf'-q_2f),p^{-1/2}(pg'-q_2g)\big)_{L^2(\mathbb{R};dx)} &(3.105)\\ &\quad+\big(|q_1|^{1/2}f,\operatorname{sgn}(q_1)|q_1|^{1/2}g\big)_{L^2(\mathbb{R};dx)}-\big(p^{-1/2}q_2f,p^{-1/2}q_2g\big)_{L^2(\mathbb{R};dx)},\\ &\qquad f,g\in\operatorname{dom}(\mathfrak{Q}_T)=H^1(\mathbb{R}).\end{aligned}$$

In particular, T corresponds to the differential expression $\tau=-(d/dx)p(d/dx)+q(x)$, $x\in\mathbb{R}$, and hence is explicitly given by

$$\begin{aligned}&Tf=\tau f,\quad \tau f=-(pf'-q_2f)'-p^{-1}q_2(pf'-q_2f)+\big(q_1-p^{-1}q_2^2\big)f,\\ &f\in\operatorname{dom}(T)=\big\{g\in L^2(\mathbb{R};dx)\,\big|\,g,(pg'-q_2g)\in AC_{\text{loc}}(\mathbb{R}),\ \tau g\in L^2(\mathbb{R};dx)\big\}.\\ &\qquad=\big\{g\in L^2(\mathbb{R};dx)\,\big|\,g,(pg'-q_2g)\in AC_{\text{loc}}(\mathbb{R}),\ \tau g\in L^2(\mathbb{R};dx)\\ &\qquad\qquad(pg'-q_2g)\in L^2(\mathbb{R};dx)\big\}.\end{aligned} \tag{3.106}$$

Without loss of generality we assume $T\geq cI$ for some $c>0$.

4. The case of periodic coefficients

In this section we apply some of the results collected in Sections 2 and 3 to the special, yet important, case where all coefficients are periodic with a fixed period. For simplicity, we will choose $p = 1$ throughout, but we emphasize that including the nonconstant, periodic coefficient p can be done in a standard manner as discussed in Remark 3.11. It is not our aim to present a thorough treatment of Floquet theory, rather, we intend to illustrate some of the scope underlying the approach developed in this paper.

One recalls that $q \in H^{-1}_{\text{loc}}(\mathbb{R})$ is called *periodic with period* $\omega > 0$ if

$$ {}_{H^{-1}(\mathbb{R})}\langle q, f(\cdot - \omega)\rangle_{H^1(\mathbb{R})} q = {}_{H^{-1}(\mathbb{R})}\langle q, f\rangle_{H^1(\mathbb{R})}, \quad f \in H^1(\mathbb{R}). \tag{4.1} $$

By (3.29), if $q \in H^{-1}_{\text{loc}}(\mathbb{R})$ is periodic, it can be written as $q = q_1 + q_2'$, where q_1 is a constant and $q_2 \in L^2_{\text{loc unif}}(\mathbb{R}; dx)$ is periodic with period ω. The analogous statement applies, of course, to the coefficient r in the differential equation (1.7), assuming (3.102) to hold. Introducing the abbreviations $Q = q - zr$, $Q_1 = q_1 - zr_1$, and $Q_2 = q_2 - zr_2$ and the quasi-derivative $y^{[1]} = y' - Q_2 y$ we may now write (1.7) as

$$ \tau y = -(y^{[1]})' - Q_2 y^{[1]} + (Q_1 - Q_2^2) y = 0, \tag{4.2} $$

or, equivalently, as the first-order system

$$ \begin{pmatrix} y \\ y^{[1]} \end{pmatrix}' = \begin{pmatrix} Q_2 & 1 \\ Q_1 - Q_2^2 & -Q_2 \end{pmatrix} \begin{pmatrix} y \\ y^{[1]} \end{pmatrix}. \tag{4.3} $$

Existence and uniqueness for the corresponding initial value problem as well as the constancy of the modified Wronskian,

$$ W(f, g)(x) = f(x) g^{[1]}(x) - f^{[1]}(x) g(x), \tag{4.4} $$

were established in [49]. As a consequence, the monodromy map

$$ M(z) : y \mapsto y(\cdot + \omega) \tag{4.5} $$

maps the two-dimensional space of solutions of equation (4.2) onto itself and has determinant 1 (as usual this is seen most easily by introducing a standard basis u_1, u_2 defined by the initial values $u_1(c) = u_2^{[1]}(c) = 1$ and $u_1^{[1]}(c) = u_2(c) = 0$). The trace of $M(z)$, given by $u_1(c+\omega) + u_2^{[1]}(c+\omega)$, is real which implies that the eigenvalues $\rho(z)$ and $1/\rho(z)$ of $M(z)$ (the Floquet multipliers) are either both real, or else, are complex conjugates of each other, in which case they both lie on the unit circle. The proof of Theorem 2.7 in [49] may also be adapted to show that, for each fixed point x, the functions $u_1(x)$, $u_2(x)$, $u_1^{[1]}(x)$, and $u_2^{[1]}(x)$ are entire functions of growth order $1/2$ with respect to z. In particular, $\operatorname{tr}_{\mathbb{C}^2}(M(\cdot))$ is an entire function of growth order $1/2$.

We start by focusing on the operator T as discussed in (3.42)–(3.44).

Throughout this section we make the following assumptions:

Hypothesis 4.1. *Assume that* $q \in H^{-1}_{\rm loc}(\mathbb{R})$ *is real valued and periodic with period* $\omega > 0$ *(and hence, actually,* $q \in H^{-1}_{\rm loc\, unif}(\mathbb{R})$*). Define* T *in* $L^2(\mathbb{R}; dx)$ *according to* (3.42)*–*(3.44) *and suppose that* $T \geq 0$.

Lemma 4.2. *Assume Hypothesis* 4.1*. Then there exists* $\phi_0 \in L^2_{\rm loc\, unif}(\mathbb{R}; dx)$*, real-valued and periodic of period* $\omega > 0$*, such that* $q = \phi_0^2 - \phi_0'$.

Proof. It suffices to note that (as in the standard case where $q \in L^1_{\rm loc}(\mathbb{R})$ is real valued and periodic with period $\omega > 0$) the Weyl–Titchmarsh solutions $\psi_{T,\pm}(z, \cdot\,)$ satisfy

$$\psi_{T,\pm}(z,x) > 0, \quad z < 0, \ x \in \mathbb{R}, \tag{4.6}$$

which extends by continuity to $z = 0$, that is,

$$\psi_{T,\pm}(0,x) > 0, \quad x \in \mathbb{R}, \tag{4.7}$$

although, $\psi_{T,\pm}(0,\cdot)$ may no longer lie in L^2 near $\pm\infty$ and hence cease to be a Weyl–Titchmarsh solution. (By oscillation theory, cf. [49], a zero of $\psi_{T,\pm}(0,\cdot)$ would contradict $T \geq 0$.) Using the Floquet property of $\psi_{T,\pm}(z, \cdot\,)$, $\phi_\pm$ defined by

$$\phi_\pm(x) = \psi'_{T,\pm}(0,x)/\psi_{T,\pm}(0,x), \quad x \in \mathbb{R}, \tag{4.8}$$

satisfies

$$\phi_\pm \in L^2_{\rm loc}(\mathbb{R}), \ \ \phi_\pm(\cdot) \text{ is periodic with period } \omega > 0, \tag{4.9}$$

in particular,

$$\phi_\pm \in L^2_{\rm loc\, unif}(\mathbb{R}) \text{ and } q = \phi_\pm^2 - \phi_\pm'. \tag{4.10}$$

(If $\inf(\sigma(T)) = 0$, one has $\psi_{T,+}(0,x) = \psi_{T,-}(0,x)$ and hence $\phi_+ = \phi_-$.) □

Given Hypothesis 4.1, Lemma 4.2 guarantees the existence of a real-valued, ω-periodic $\phi \in L^2_{\rm loc\, unif}(\mathbb{R}; dx)$ such that $q = \phi^2 - \phi'$ and hence we can identify the operator T in $L^2(\mathbb{R}; dx)$ with $T_1 = A^*A$ in (3.46) (resp., (3.72)), where A and A^* defined as in (3.47) and (3.48) (resp., (3.72)). In addition, we define the periodic Dirac-type operator D in $L^2(\mathbb{R}; dx) \oplus L^2(\mathbb{R}; dx)$ by (3.75).

Since $\phi \in L^2([0,\omega]; dx)$, for any $\varepsilon > 0$ and all $g \in H^1((0,\omega))$, one has

$$\begin{aligned} \|\phi g\|^2_{L^2([0,\omega];dx)} &\leq \varepsilon \|g'\|^2_{L^2([0,\omega];dx)} \\ &\quad + \|\phi\|^2_{L^2([0,\omega];dx)} \big[\omega^{-1} + \|\phi\|^2_{L^2([0,\omega];dx)} \varepsilon^{-1}\big] \|g\|^2_{L^2([0,\omega];dx)} \end{aligned} \tag{4.11}$$

(cf. [125, p. 19–20, 37]). Utilizing (4.11), one can introduce the reduced Dirac-type operator D_θ in $L^2([0,\omega]; dx)$, $\theta \in [0, 2\pi]$, by

$$D_\theta = \begin{pmatrix} 0 & A_\theta^* \\ A_\theta & 0 \end{pmatrix} \text{ in } L^2([0,\omega]; dx) \oplus L^2([0,\omega]; dx), \tag{4.12}$$

where

$$A_\theta = (d/dx) + \phi, \quad \operatorname{dom}(A_\theta) = \big\{g \in H^1((0,\omega)) \,\big|\, g(\omega) = e^{i\theta} g(0)\big\}, \tag{4.13}$$

$$A_\theta^* = -(d/dx) + \phi, \quad \operatorname{dom}(A_\theta^*) = \big\{g \in H^1((0,\omega)) \,\big|\, g(\omega) = e^{i\theta} g(0)\big\}, \tag{4.14}$$

and A_θ (and hence A_θ^*) is closed in $L^2([0,\omega]; dx)$, implying self-adjointness of D_θ.

Employing the identity (3.76), $D^2 = T_1 \oplus T_2$, and analogously for D_θ^2,

$$D_\theta^2 = \begin{pmatrix} A_\theta^* A_\theta & 0 \\ 0 & A_\theta A_\theta^* \end{pmatrix} = T_{1,\theta} \oplus T_{2,\theta} \text{ in } L^2([0,\omega]; dx) \oplus L^2([0,\omega]; dx), \quad (4.15)$$

$$T_{1,\theta} = A_\theta^* A_\theta, \quad T_{2,\theta} = A_\theta A_\theta^* \text{ in } L^2([0,\omega]; dx), \quad (4.16)$$

and applying the standard direct integral formalism combined with Floquet theory to D, D_θ (cf., [22, App. to Ch. 10], [46], [121, Sect. XIII.16]), where

$$L^2(\mathbb{R}; dx) \simeq \frac{1}{2\pi} \int_{[0,2\pi]}^{\oplus} d\theta \, L^2([0,\omega]; dx), \quad (4.17)$$

then yields the following result (with $\simeq$ abbreviating unitary equivalence):

Theorem 4.3. *Assume Hypothesis* 4.1. *Then the periodic Dirac operator D (cf.* (3.75)) *satisfies*

$$D \simeq \frac{1}{2\pi} \int_{[0,2\pi]}^{\oplus} d\theta \, D_\theta, \quad (4.18)$$

with respect to the direct integral decomposition (4.17), *and $\sigma_p(D) = \sigma_{sc}(D) = \emptyset$. Moreover, $\sigma(D)$ is purely absolutely continuous of uniform spectral multiplicity equal to two, and $\sigma(D)$ consists of a union of compact intervals accumulating at $+\infty$ and $-\infty$.*

In addition, the spectra of T_j (cf. (3.73), (3.74)) *satisfy $\sigma_p(T_j) = \sigma_{sc}(T_j) = \emptyset$, in fact, $\sigma(T_j)$ is purely absolutely continuous of uniform spectral multiplicity equal to two, and $\sigma(T_j)$ consists of a union of compact intervals accumulating at $+\infty$, $j = 1, 2$.*

We note in passing that the spectral properties of T_j, $j = 1, 2$, alternatively, also follow from the m-function relations (3.83), (3.84). In fact, applying the results in [48], one can extend Theorem 4.3 to the case where $\phi \in L^1_{\text{loc}}(\mathbb{R}; dx)$ is real valued and periodic of period $\omega > 0$, but we will not pursue this any further in this paper.

The supersymmetric approach linking (periodic, quasi-periodic, finite-gap, etc.) Schrödinger and Dirac-type operators has been applied repeatedly in the literature, see, for instance, [40], [57], [58], [66], [67], [94], and the extensive literature cited therein. In addition, we note that spectral theory (gap and eigenvalue asymptotics, etc.) for Schrödinger operators with periodic distributional potentials has been thoroughly investigated in [41], [42], [43], [44], [78], [79], [82], [89], [95], [97], [113], [114].

We now investigate the eigenvalues associated with the differential equation (1.7) and quasi-periodic boundary conditions utilizing the operator $\underline{T^{-1/2} r T^{-1/2}}$ in $L^2([0,\omega]; dx)$ when r is a measure. More precisely, let $R : [0,\omega] \to \mathbb{R}$ be a left-continuous real-valued function of bounded variation and μ_R the associated signed measure. We associate with R the following map

$$r : H^1((0,\omega)) \to H^{-1}((0,\omega)) \quad (4.19)$$

via the Lebesgue–Stieltjes integral,

$$ {}_{H^{-1}(\mathbb{R})}\langle rf, g\rangle_{H^1(\mathbb{R})} = \int_0^\omega d\mu_R(x)\,\overline{f(x)}g(x), \quad f,g \in H^1((0,\omega)). \tag{4.20}$$

One notes that the map r defined in terms of (4.19), (4.20) is bounded.

We also write $R = R_+ - R_-$ where $R_\pm$ are both left-continuous and nondecreasing and thus give rise to positive finite measures on $[0,\omega]$.

Thus,

$$K \in \mathcal{B}\big(L^2([0,\omega];dx), H^1((0,\omega))\big) \text{ implies } K^*rK \in \mathcal{B}\big(L^2([0,\omega];dx)\big). \tag{4.21}$$

Similarly,

$$K \in \mathcal{B}_\infty\big(L^2([0,\omega];dx), H^1((0,\omega))\big) \text{ implies } K^*rK \in \mathcal{B}_\infty\big(L^2([0,\omega];dx)\big). \tag{4.22}$$

Lemma 4.4. *Suppose $K \in \mathcal{B}_\infty\big((L^2([0,\omega];dx), H^1((0,\omega))\big)$ is compact and that $C_0^\infty((0,\omega)) \subset \operatorname{ran}(K)$. In addition, assume that R is a real-valued function of bounded variation on $[0,\omega]$ and define r as in* (4.19), (4.20). *Then K^*rK has infinitely many positive* (*resp., negative*) *eigenvalues unless R_+* (*resp., R_-*) *is a pure jump function with only finitely many jumps* (*if any*).

Proof. Without loss of generality we may assume that $\omega = 1$ and we may also restrict attention to R_+ only. Accordingly, suppose that the measure associated with R_+ has a continuous part or that R_+ has infinitely many jumps, but, that by way of contradiction, K^*rK has only finitely many (say, $N \geq 0$) positive eigenvalues. We will show below that there is a positive number ℓ and $N+1$ sets $\Omega_1, \dots, \Omega_{N+1}$, which have a distance of at least ℓ from each other and from the endpoints of $[0,1]$, for which $\int_{\Omega_j} d\mu_R > 0$.

For any ε, with $0 < \varepsilon < \ell/2$, let J_ε be the Friedrichs mollifier as introduced, for instance, in [1, Sect. 2.28]. Applying [1, Theorem 2.29], the functions

$$g_{j,\varepsilon} = J_\varepsilon * \chi_{\Omega_j}, \quad j = 1, \dots, N+1, \tag{4.23}$$

satisfy the following properties:

(i) $g_{j,\varepsilon} \in C_0^\infty((0,1)) \subset \operatorname{ran}(K)$,
(ii) $g_{j,\varepsilon}$ are zero at points which are further than ε away from Ω_j,
(iii) $\lim_{\varepsilon\downarrow 0} \|g_{j,\varepsilon} - \chi_{\Omega_j}\|_{L^2([0,1];dx)} = 0$,
(iv) $|g_{j,\varepsilon}(x)| \leq 1$.

Property (i) implies that there are functions $f_{j,\varepsilon} \in L^2([0,1];dx)$ such that $g_{j,\varepsilon} = Kf_{j,\varepsilon}$ since $C_0^\infty((0,1)) \subset \operatorname{ran}(K)$. By property (iii), $g_{j,\varepsilon} \to \chi_{\Omega_j}$ pointwise a.e. on $(0,1)$ as $\varepsilon \downarrow 0$, and hence the dominated convergence theorem implies that

$$ {}_{H^{-1}(\mathbb{R})}\langle rKf_{j,\varepsilon}, Kf_{j,\varepsilon}\rangle_{H^1(\mathbb{R})} = \int_{[0,1]} d\mu_R(x)\,|g_{j,\varepsilon}(x)|^2 \underset{\varepsilon\downarrow 0}{\longrightarrow} \int_{\Omega_j} d\mu_R(x) > 0. \tag{4.24}$$

Hence we may fix $\varepsilon > 0$ in such a way that

$$\int_{[0,1]} d\mu_R(x)\,|(Kf_{j,\varepsilon})(x)|^2 > 0, \quad j = 1, \dots, N+1. \tag{4.25}$$

Next, by property (ii) mentioned above, the supports of the $g_{j,\varepsilon}$ are pairwise disjoint, implying

$$\left|\sum_{j=1}^{N+1} c_j g_{j,\varepsilon}\right|^2 = \sum_{j+1}^{N+1} |c_j|^2 |g_{j,\varepsilon}|^2 \tag{4.26}$$

for any choice of $c_j \in \mathbb{C}$, $j = 1, \dots, N+1$.

Assume now that $f = \sum_{j=1}^{N+1} c_j f_{j,\varepsilon}$, where at least one of the coefficients $c_j \neq 0$. Then equations (4.25) and (4.26) imply

$$\begin{aligned}(f, K^*rKf)_{L^2([0,1];dx)} &= \int_{[0,1]} d\mu_R(x)\, |(Kf)(x)|^2 \\ &= \sum_{j=1}^{N+1} |c_j|^2 \int_{[0,1]} d\mu_R(x)\, |(Kf_{j,\varepsilon})(x)|^2 > 0.\end{aligned} \tag{4.27}$$

We will now prove that for some choices of the coefficients c_j, the expression $(f, K^*rKf)_{L^2([0,1];dx)}$ cannot be positive so that one arrives at a contradiction to (4.27), proving that there must be infinitely many positive eigenvalues. To do so, we denote the nonzero eigenvalues and eigenfunctions of the compact, self-adjoint operator K^*rK by λ_k and φ_k, respectively. More specifically, assume that the positive eigenvalues have labels $k = 1, \dots, N$, while the labels of the non-positive eigenvalues are chosen from the non-positive integers. The spectral theorem, applied to K^*rK, yields

$$\begin{aligned}0 < (f, K^*rKf)_{L^2([0,1];dx)} &= \sum_{k=-\infty}^{N} \lambda_k |(\varphi_k, f)_{L^2([0,1];dx)}|^2 \\ &\leq \sum_{k=1}^{N} \lambda_k |(\varphi_k, f)_{L^2([0,1];dx)}|^2\end{aligned} \tag{4.28}$$

for any $f \in L^2([0,1]; dx)$. If $N = 0$, this is the desired contradiction. If $N \geq 1$, the inequality (4.28) shows that no non-zero element of $L^2([0,1]; dx)$ can be orthogonal to all the eigenfunctions associated with positive eigenvalues. However, the underdetermined system

$$\sum_{j=1}^{N+1} c_j (\varphi_k, f_{j,\varepsilon})_{L^2([0,1];dx)} = (\varphi_k, f)_{L^2([0,1];dx)} = 0, \quad k = 1, \dots, N, \tag{4.29}$$

has nontrivial solutions $(c_1, \dots, c_N)$ proving that $f = \sum_{j=1}^{N+1} c_j f_{j,\varepsilon}$ is orthogonal to all the eigenfunctions associated with positive eigenvalues so that we again arrive at a contradiction.

It remains to establish the existence of the sets Ω_j with the required properties. Recall that, by Lebesgue's decomposition theorem, $R = R_1 + R_2 + R_3$, where R_1 is absolutely continuous, R_2 is continuous but $R_2' = 0$ a.e. on $[0,1]$, and R_3 is a jump function and that these generate an absolutely continuous measure μ_1, a singular continuous measure μ_2, and a discrete measure μ_3 (i.e., one supported on a countable subset of $\mathbb{R}$), respectively. By Jordan's decomposition theorem, each of these measures may be split into its positive and negative part $\mu_{j,\pm}$, $j = 1,2,3$. We will denote the respective supports of these measures by $A_{j,\pm}$, $j = 1,2,3$. Note that $A_{j,+} \cap A_{j,-}$ is empty for each j by Hahn's decomposition theorem. We also define $R_{j,\pm}(x) = \mu_{j,\pm}([0,x])$.

First, we assume that the support $A_{1,+}$ of $\mu_{1,+}$ has positive Lebesgue measure. Since the supports of μ_2 and μ_3 have zero Lebesgue measure, they are subsets of a union of open intervals whose total length is arbitrarily small. Thus, we may find a set $\Omega \subset A_{1,+}$ of positive Lebesgue measure which avoids a neighborhood of the supports of μ_2 and μ_3 so that $\int_\Omega d\mu_R > 0$. Now define $M = \lceil (2N+3)/m(\Omega) \rceil$, with $m(\cdot)$ abbreviating Lebesgue measure and $\lceil x \rceil$ the smallest integer not smaller than x. Dividing the interval $[0,1]$ uniformly into M subintervals, each will have length not exceeding $\ell = m(\Omega)/(2N+3)$. Consequently, at least $2N+3$ of these intervals will intersect Ω in a set of positive Lebesgue measure and hence of positive μ_R-measure. $N+1$ of the latter ones will have a distance of at least ℓ from each other and from the endpoints of $[0,1]$. These intersections will be the sought after sets $\Omega_1, \dots, \Omega_{N+1}$.

Next assume $\mu_{1,+} = 0$ but $\mu_{2,+}([0,1]) = a_2 > 0$. Since $A_{3,-}$ is countable we have $\mu_{2,+}(A_{3,-}) = 0$. Also, of course, $\mu_{2,+}(A_{2,-}) = 0$. By the regularity of $\mu_{2,+}$ there is, for every positive ε, an open set W covering $A_{2,-} \cup A_{3,-}$ such that $\mu_{2,+}(W) < \varepsilon$. Set $\Omega = (0,1)\backslash\overline{W}$ and $\varepsilon = a_2/2$. Since $\overline{W} - W$ is countable we have $\mu_{2,+}(\Omega) = \mu_{2,+}((0,1)\backslash W) > a_2/2$. Since $R_{2,+}$ is uniformly continuous there is a $\delta > 0$ so that $R_{2,+}(y) - R_{2,+}(x) < a_2/(2(2N+3))$ as long as $0 < y - x < \delta$. Thus, splitting Ω in intervals of length at most δ, we have that at least $2N+3$ of these intervals have positive $\mu_{2,+}$-measure and $N+1$ of these have a positive distance from each other and from the endpoints of $[0,1]$. We denote these intervals by Ω'_1, $\dots$, Ω'_{N+1}. We now have $\mu_{2,+}(\Omega'_k) > 0$ but $\mu_{2,-}(\Omega'_k) = \mu_{3,-}(\Omega'_k) = 0$. However, it may still be the case that $\mu_{1,-}(\Omega'_k) > \mu_{2,+}(\Omega'_k)$. Regularity of $\mu_{1,-}$ allows us to find a set Ω_k such that $A_{2,+} \cap \Omega'_k \subset \Omega_k \subset \Omega'_k$ and $\mu_{1,-}(\Omega_k)$ are arbitrarily small. This way we may guarantee that $\mu(\Omega_k) > 0$ for $k = 1, \dots, N+1$.

Finally, assume that R_+ is a pure jump function, but with infinitely many jumps. Then we may choose pairwise disjoint intervals Ω_k about $N+1$ of the jump discontinuities of R_+ and we may choose them so small that their $\mu_{j,-}(\Omega_k)$ is smaller than the jump so that again $\mu(\Omega_k) > 0$ for $k = 1, \dots, N+1$. □

We emphasize that Lemma 4.4 applies, in particular, to the special case, where $d\mu_R(x) = r(x)dx$ is purely absolutely continuous on $\mathbb{R}$:

Corollary 4.5. *Suppose* $K \in \mathcal{B}_\infty(L^2([0,\omega]; dx))$ *is self-adjoint with* $\operatorname{ran}(K) \supseteq H^1((0,\omega))$. *Assume in addition that* $r \in L^1([0,\omega]; dx)$ *is real valued such that* $|r|^{1/2}K \in \mathcal{B}_\infty(L^2([0,\omega]; dx))$. *Then* $\underline{KrK} := [|r|^{1/2}K]^* \operatorname{sgn}(r)|r|^{1/2}K$ *has infinitely many positive (resp., negative) eigenvalues unless* $r_+ = 0$ *(resp.,* $r_- = 0$*) a.e. on* $(0,\omega)$.

Identifying T_θ in $L^2([0,\omega];dx)$ with $T_{1,\theta} = A_\theta^* A_\theta$ (in analogy to the identification of T in $L^2(\mathbb{R};dx)$ with $T_1 = A^*A$), recalling the construction of $\widetilde{T}$, $\widetilde{T}_\theta$ according to (C.26), (C.40), an application of Lemma 4.4, employing (C.44)–(C.49), then yields the following result:

Theorem 4.6. *Assume Hypothesis* 4.1, *suppose that* μ_R *is a signed measure, and let* r *be defined as in* (4.19), (4.20). *In addition, assume that* r *is periodic of period* $\omega > 0$.

(i) *Suppose that* $T_\theta \geq c_\theta I_{L^2([0,\omega];dx)}$ *for some* $c_\theta > 0$. *Then* $\big(\widetilde{T}_\theta\big)^{-1/2} r \big(\widetilde{T}_\theta\big)^{-1/2}$ *has infinitely many positive (resp., negative) eigenvalues unless* R_+ *(resp.,* R_-*) is a pure jump function with only finitely many jumps (if any).*

(ii) *Suppose that* $T \geq c I_{L^2(\mathbb{R};dx)}$ *for some* $c > 0$. *Then* $\sigma\Big(\big(\widetilde{T}\big)^{-1/2} r \big(\widetilde{T}\big)^{-1/2}\Big)$ *consists of a union of compact intervals accumulating at* 0 *unless* $R = 0$ *a.e. on* $(0,\omega)$. *In addition,*

$$-\psi'' + q\psi = zr\psi \tag{4.30}$$

has a conditional stability set (consisting of energies z *with at least one bounded solution on* $\mathbb{R}$*) composed of a sequence of intervals on* $(0,\infty)$ *tending to* $+\infty$ *and/or* $-\infty$, *unless* R_+ *and/or* R_- *is a pure jump function with only finitely many jumps (if any). Finally,*

$$\sigma_{\rm p}\Big(\big(\widetilde{T}\big)^{-1/2} r \big(\widetilde{T}\big)^{-1/2}\Big) = \emptyset. \tag{4.31}$$

Proof. Lemma 4.4, identifying K and $\big(\widetilde{T}_\theta\big)^{-1/2}$ (cf. (C.48) and our notational convention (C.49)) proves item (i).

As usual (see Eastham [46, Sect. 2.1] or Brown, Eastham, and Schmidt [25, Sect. 1.4]), the conditional stability set S of equation (4.30) is given by

$$S = \{\lambda \in \mathbb{R} \,|\, |\mathrm{tr}_{\mathbb{C}^2}(M(\lambda))| \leq 2\} \tag{4.32}$$

since, if $\lambda \in S$ and only then, the monodromy operator $M(\lambda)$ has at least one eigenvector associated with an eigenvalue of modulus 1. Since $\mathrm{tr}_{\mathbb{C}^2}(M(\cdot))$ is an analytic, hence, continuous function, the set $S^0 = \{\lambda \in \mathbb{R} \,|\, |\mathrm{tr}_{\mathbb{C}^2}(M(\lambda))| < 2\}$ is an open set and thus a union of open intervals. Moreover, $\{\lambda \in \mathbb{R} \,|\, \mathrm{tr}_{\mathbb{C}^2}(M(\lambda)) = 2\}$ (i.e., the set of periodic eigenvalues) and $\{\lambda \in \mathbb{R} \,|\, \mathrm{tr}_{\mathbb{C}^2}(M(\lambda)) = -2\}$ (i.e., the set of anti-periodic eigenvalues) are discrete sets without finite accumulation points. It follows that S is obtained as the union of the closures of each of the open intervals constituting S^0, equivalently, S is a union of closed intervals. One notes that the closure of several disjoint components of S^0 may form one closed interval in S.

Applying Lemma 4.4 to the case $K = \big(\widetilde{T}_\theta\big)^{-1/2}$ one obtains a countable number of eigenvalues $\zeta_n(\theta)$, $n \in \mathbb{Z}\backslash\{0\}$ which we may label so that $n\,\zeta_n(\theta) > 0$. These eigenvalues accumulate at zero (from either side). It is clear that equation (4.30) posed on the interval $[0,\omega]$ has a nontrivial solution satisfying the boundary conditions $\psi(\omega) = e^{i\theta}\psi(0)$ and $\psi^{[1]}(\omega) = e^{i\theta}\psi^{[1]}(0)$ precisely when $z = 1/\zeta_n(\theta)$ for some $n \in \mathbb{Z}\backslash\{0\}$. In particular, the endpoints of the conditional stability intervals, which correspond to the values $\theta = 0$ and $\theta = \pi$, tend to both, $+\infty$ and $-\infty$.

Finally, eigenfunctions $u \in L^2(\mathbb{R}; dx)$ of $(\widetilde{T})^{-1/2} r (\widetilde{T})^{-1/2}$ are related to solutions $y \in H^1(\mathbb{R})$ of (1.7) (with $p = 1$) via $y = (\widetilde{T})^{-1/2} u$. Since the basics of Floquet theory apply to (1.7) (cf. our comments at the beginning of this section and earlier in the current proof), the existence of Floquet multipliers $\rho(z)$ and $1/\rho(z)$ prevents (1.7) from having an $L^2(\mathbb{R}; dx)$ (let alone, $H^1(\mathbb{R})$) solution. Hence, the existence of an eigenfunction $u \in L^2(\mathbb{R}; dx)$ of $(\widetilde{T})^{-1/2} r (\widetilde{T})^{-1/2}$ would imply the contradiction $y \in H^1(\mathbb{R})$, implying (4.31). □

Theorem 4.6 considerably extends prior results by Constantin [30] (see also [31], [32]) on eigenvalue asymptotics for left-definite periodic Sturm–Liouville problems since no smoothness is assumed on q and r, in addition, q is permitted to be a distribution and r is extended from merely being a function to a measure. Moreover, it also extends results of Daho and Langer [37], Marletta and Zettl [102], and Philipp [119]: While these authors consider the nonsmooth setting, our result appears to be the first that permits periodic distributions, respectively, measures as coefficients.

Remark 4.7. In the special case where the measure $d\mu_R(x) = r(x)dx$ is purely absolutely continuous on $\mathbb{R}$, the fact that

$$\underline{T^{-1/2} r T^{-1/2}} \simeq \frac{1}{2\pi} \int_{[0,2\pi]}^{\oplus} \underline{T_\theta^{-1/2} r T_\theta^{-1/2}} \tag{4.33}$$

with respect to the decomposition (4.17), together with continuity of the eigenvalues of $\underline{T_\theta^{-1/2} r T_\theta^{-1/2}}$ with respect to θ, proves that $\sigma(\underline{T^{-1/2} r T^{-1/2}})$ consists of a union of compact intervals accumulating at 0 unless $r = 0$ a.e. on $(0, \omega)$.

Moreover, employing the methods in [65, Sect. 2], Theorem 4.6 (i) immediately extends to any choice of self-adjoint separated boundary conditions replacing the θ boundary conditions

$$g(\omega) = e^{i\theta} g(0), \quad g'(\omega) = e^{i\theta} g'(0), \quad \theta \in [0, 2\pi], \tag{4.34}$$

in $A_\theta^* A_\theta$ by separated ones of the type

$$\begin{aligned} &\sin(\alpha) g'(0) + \cos(\alpha) g(0) = 0, \\ &\sin(\beta) g'(\omega) + \cos(\beta) g(\omega) = 0, \quad \alpha, \beta \in [0, \pi]. \end{aligned} \tag{4.35}$$

We emphasize that the following Appendices A, B, and C do not contain new results. We offer them for the convenience of the reader with the goal of providing a fairly self-contained account, enhancing the readability of this manuscript.

Appendix A. Relative boundedness and compactness of operators and forms

In this appendix we briefly recall the notion of relatively bounded (resp., compact) and relatively form bounded (resp., form compact) perturbations of a self-adjoint operator A in some complex separable Hilbert space $\mathcal{H}$:

Definition A.1.

(i) Suppose that A is a self-adjoint operator in $\mathcal{H}$. A closed operator B in $\mathcal{H}$ is called *relatively bounded* (resp., *relatively compact*) *with respect to* A (in short, B is called A*-bounded* (resp., A*-compact*)), if

$$\begin{aligned}&\operatorname{dom}(B) \supseteq \operatorname{dom}(A) \text{ and}\\ &B(A - zI_{\mathcal{H}})^{-1} \in \mathcal{B}(\mathcal{H}) \text{ (resp., } \in \mathcal{B}_\infty(\mathcal{H})), \quad z \in \rho(A).\end{aligned} \tag{A.1}$$

(ii) Assume that A is self-adjoint and bounded from below (i.e., $A \geqslant cI_{\mathcal{H}}$ for some $c \in \mathbb{R}$). Then a densely defined and closed operator B in $\mathcal{H}$ is called *relatively form bounded* (resp., *relatively form compact*) *with respect to* A (in short, B is called A*-form bounded* (resp., A*-form compact*)), if

$$\begin{aligned}&\operatorname{dom}\left(|B|^{1/2}\right) \supseteq \operatorname{dom}\left(|A|^{1/2}\right) \text{ and}\\ &|B|^{1/2}((A + (1-c)I_{\mathcal{H}}))^{-1/2} \in \mathcal{B}(\mathcal{H}) \text{ (resp., } \in \mathcal{B}_\infty(\mathcal{H})).\end{aligned} \tag{A.2}$$

Remark A.2. (i) Using the polar decomposition of B (i.e., $B = U_B|B|$, with U_B a partial isometry), one observes that B is A-bounded (resp., A-compact) if and only if $|B|$ is A-bounded (resp., A-compact). Similarly, by (A.2), B is A-form bounded (resp., A-form compact), if and only if $|B|$ is.

(ii) Since$B$ is assumed to be closed (in fact, closability of B suffices) in Definition A.1 (i), the first condition $\operatorname{dom}(B) \supseteq \operatorname{dom}(A)$ in (A.1) already implies $B(A - zI_{\mathcal{H}})^{-1} \in \mathcal{B}(\mathcal{H})$, $z \in \rho(A)$, and hence the A-boundedness of B (cf. again [90, Remark IV.1.5], [135, Theorem 5.9]). By the same token, since $A^{1/2}$ and $|B|^{1/2}$ are closed, the requirement $\operatorname{dom}\left(|B|^{1/2}\right) \supseteq \operatorname{dom}\left(A^{1/2}\right)$ in Definition A.1 (ii), already implies that $|B|^{1/2}((A + (1-c)I_{\mathcal{H}}))^{-1/2} \in \mathcal{B}(\mathcal{H})$ (cf. [90, Remark IV.1.5], [135, Theorem 5.9]), and hence the first condition in (A.2) suffices in the relatively form bounded context.

(iii) In the special case where B is self-adjoint, condition (A.2) implies the existence of $\alpha \geq 0$ and $\beta \geq 0$, such that

$$\begin{aligned}\left|\left(|B|^{1/2}f, \operatorname{sgn}(B)|B|^{1/2}f\right)_{\mathcal{H}}\right| \leq \left\||B|^{1/2}f\right\|_{\mathcal{H}}^2 \leq \alpha\left\||A|^{1/2}f\right\|_{\mathcal{H}}^2 + \beta\|f\|_{\mathcal{H}}^2,&\\ f \in \operatorname{dom}\left(|A|^{1/2}\right).&\end{aligned} \tag{A.3}$$

(iv) In connection with relative boundedness, (A.1) can be replaced by the condition

$$\begin{aligned}&\operatorname{dom}(B) \supseteq \operatorname{dom}(A), \text{ and there exist numbers } a \geqslant 0,\, b \geqslant 0 \text{ such that}\\ &\|Bf\|_{\mathcal{H}} \leqslant a\|Af\|_{\mathcal{H}} + b\|f\|_{\mathcal{H}} \text{ for all } f \in \operatorname{dom}(A),\end{aligned} \tag{A.4}$$

or equivalently, by

$$\begin{aligned}&\operatorname{dom}(B) \supseteq \operatorname{dom}(A), \text{ and there exist numbers } \widetilde{a} \geqslant 0,\, \widetilde{b} \geqslant 0 \text{ such that}\\ &\|Bf\|_{\mathcal{H}}^2 \leqslant \widetilde{a}^2\|Af\|_{\mathcal{H}}^2 + \widetilde{b}^2\|f\|_{\mathcal{H}}^2 \text{ for all } f \in \operatorname{dom}(A).\end{aligned} \tag{A.5}$$

(v) If A is self-adjoint and bounded from below, the number α defined by

$$\alpha = \lim_{\mu\uparrow\infty} \left\|B(A+\mu I_{\mathcal{H}})^{-1}\right\|_{\mathcal{B}(\mathcal{H})} = \lim_{\mu\uparrow\infty} \left\||B|(A+\mu I_{\mathcal{H}})^{-1}\right\|_{\mathcal{B}(\mathcal{H})} \tag{A.6}$$

equals the greatest lower bound (i.e., the infimum) of the possible values for a in (A.4) (resp., for $\widetilde{a}$ in (A.5)). This number α is called the A-bound of B. Similarly, we call

$$\beta = \lim_{\mu\uparrow\infty} \left\||B|^{1/2}\big(|A|^{1/2}+\mu I_{\mathcal{H}}\big)^{-1}\right\|_{\mathcal{B}(\mathcal{H})} \tag{A.7}$$

the A-form bound of B (resp., $|B|$). If $\alpha = 0$ in (A.6) (resp., $\beta = 0$ in (A.7)) then B is called *infinitesimally bounded* (resp., *infinitesimally form bounded*) with respect to A.

We then have the following result:

Theorem A.3. *Assume that $A \geqslant 0$ is self-adjoint in $\mathcal{H}$.*

(i) *Let B be a closed, densely defined operator in $\mathcal{H}$ and suppose that* $\operatorname{dom}(B) \supseteq \operatorname{dom}(A)$. *Then B is A-bounded and hence* (A.4) *holds for some constants $a \geq 0$, $b \geq 0$. In addition, B is also A-form bounded,*

$$|B|^{1/2}(A+I_{\mathcal{H}})^{-1/2} \in \mathcal{B}(\mathcal{H}). \tag{A.8}$$

More specifically,

$$\left\||B|^{1/2}(A+I_{\mathcal{H}})^{-1/2}\right\|_{\mathcal{B}(\mathcal{H})} \leqslant (a+b)^{1/2}, \tag{A.9}$$

and hence, if B is A-bounded with A-bound α strictly less than one, $0 \leq \alpha < 1$ (cf. (A.6)*), then B is also A-form bounded with A-form bound β strictly less than one, $0 \leq \beta < 1$ (cf.* (A.7)*). In particular, if B is infinitesimally bounded with respect to A, then B is infinitesimally form bounded with respect to A.*

(ii) *Suppose that B is self-adjoint in $\mathcal{H}$, that* $\operatorname{dom}(B) \supseteq \operatorname{dom}(A)$, *and hence* (A.4) *holds for some constants $a \geq 0$, $b \geq 0$. Then*

$$\overline{(A+I_{\mathcal{H}})^{-1/2}B(A+I_{\mathcal{H}})^{-1/2}} \in \mathcal{B}(\mathcal{H}), \tag{A.10}$$

$$\left\|\overline{(A+I_{\mathcal{H}})^{-1/2}B(A+I_{\mathcal{H}})^{-1/2}}\right\|_{\mathcal{B}(\mathcal{H})} \leqslant (a+b). \tag{A.11}$$

We also recall the following result:

Theorem A.4. *Assume that $A \geqslant 0$ is self-adjoint in $\mathcal{H}$.*

(i) *Let B be a densely defined closed operator in $\mathcal{H}$ and suppose that* $\operatorname{dom}(B) \supseteq \operatorname{dom}(A)$. *In addition, assume that B is A-compact. Then B is also A-form compact,*

$$|B|^{1/2}(A+I_{\mathcal{H}})^{-1/2} \in \mathcal{B}_\infty(\mathcal{H}). \tag{A.12}$$

(ii) *Suppose that B is self-adjoint in $\mathcal{H}$ and that* $\operatorname{dom}(B) \supseteq \operatorname{dom}(A)$. *In addition, assume that B is A-compact. Then*

$$\overline{(A+I_{\mathcal{H}})^{-1/2}B(A+I_{\mathcal{H}})^{-1/2}} \in \mathcal{B}_\infty(\mathcal{H}). \tag{A.13}$$

For proofs of Theorems A.3 and A.4 under more general conditions on A and B, we refer to [63] and the detailed list of references therein.

Appendix B. Supersymmetric Dirac-type operators in a nutshell

In this appendix we briefly summarize some results on supersymmetric Dirac-type operators and commutation methods due to [39], [66], [131], and [132, Ch. 5] (see also [71]).

The standing assumption in this appendix will be the following.

Hypothesis B.1. *Let $\mathcal{H}_j$, $j = 1, 2$, be separable complex Hilbert spaces and*

$$A : \mathcal{H}_1 \supseteq \operatorname{dom}(A) \to \mathcal{H}_2 \tag{B.1}$$

be a densely defined, closed, linear operator.

We define the self-adjoint Dirac-type operator in $\mathcal{H}_1 \oplus \mathcal{H}_2$ by

$$Q = \begin{pmatrix} 0 & A^* \\ A & 0 \end{pmatrix}, \quad \operatorname{dom}(Q) = \operatorname{dom}(A) \oplus \operatorname{dom}(A^*). \tag{B.2}$$

Operators of the type Q play a role in supersymmetric quantum mechanics (see, e.g., the extensive list of references in [24]). Then,

$$Q^2 = \begin{pmatrix} A^*A & 0 \\ 0 & AA^* \end{pmatrix} \tag{B.3}$$

and for notational purposes we also introduce

$$H_1 = A^*A \text{ in } \mathcal{H}_1, \quad H_2 = AA^* \text{ in } \mathcal{H}_2. \tag{B.4}$$

In the following, we also need the polar decomposition of A and A^*, that is, the representations

$$A = V_A|A| = |A^*|V_A = V_A A^* V_A \text{ on } \operatorname{dom}(A) = \operatorname{dom}(|A|), \tag{B.5}$$
$$A^* = V_{A^*}|A^*| = |A|V_{A^*} = V_{A^*} A V_{A^*} \text{ on } \operatorname{dom}(A^*) = \operatorname{dom}(|A^*|), \tag{B.6}$$
$$|A| = V_{A^*}A = A^*V_A = V_{A^*}|A^*|V_A \text{ on } \operatorname{dom}(|A|), \tag{B.7}$$
$$|A^*| = V_A A^* = AV_{A^*} = V_A|A|V_{A^*} \text{ on } \operatorname{dom}(|A^*|), \tag{B.8}$$

where

$$|A| = (A^*A)^{1/2}, \quad |A^*| = (AA^*)^{1/2}, \quad V_{A^*} = (V_A)^*, \tag{B.9}$$
$$V_{A^*}V_A = P_{\overline{\operatorname{ran}(|A|)}} = P_{\overline{\operatorname{ran}(A^*)}}, \quad V_A V_{A^*} = P_{\overline{\operatorname{ran}(|A^*|)}} = P_{\overline{\operatorname{ran}(A)}}. \tag{B.10}$$

In particular, V_A is a partial isometry with initial set $\overline{\operatorname{ran}(|A|)}$ and final set $\overline{\operatorname{ran}(A)}$ and hence V_{A^*} is a partial isometry with initial set $\overline{\operatorname{ran}(|A^*|)}$ and final set $\overline{\operatorname{ran}(A^*)}$. In addition,

$$V_A = \begin{cases} \overline{A(A^*A)^{-1/2}} = \overline{(AA^*)^{-1/2}A} & \text{on } (\ker(A))^\perp, \\ 0 & \text{on } \ker(A). \end{cases} \tag{B.11}$$

Next, we collect some properties relating H_1 and H_2.

Theorem B.2 ([39]). *Assume Hypothesis* B.1 *and let ϕ be a bounded Borel measurable function on $\mathbb{R}$.*

(i) *One has*

$$\ker(A) = \ker(H_1) = (\operatorname{ran}(A^*))^{\perp}, \quad \ker(A^*) = \ker(H_2) = (\operatorname{ran}(A))^{\perp}, \tag{B.12}$$

$$V_A H_1^{n/2} = H_2^{n/2} V_A, \ n \in \mathbb{N}, \quad V_A \phi(H_1) = \phi(H_2) V_A. \tag{B.13}$$

(ii) *H_1 and H_2 are essentially isospectral, that is,*

$$\sigma(H_1)\backslash\{0\} = \sigma(H_2)\backslash\{0\}, \tag{B.14}$$

in fact,

$$A^*A[I_{\mathcal{H}_1} - P_{\ker(A)}] \text{ is unitarily equivalent to } AA^*[I_{\mathcal{H}_2} - P_{\ker(A^*)}]. \tag{B.15}$$

In addition,

$$\begin{aligned} &f \in \operatorname{dom}(H_1) \text{ and } H_1 f = \lambda^2 f, \ \lambda \neq 0, \\ &\quad \text{implies } Af \in \operatorname{dom}(H_2) \text{ and } H_2(Af) = \lambda^2 (Af), \end{aligned} \tag{B.16}$$

$$\begin{aligned} &g \in \operatorname{dom}(H_2) \text{ and } H_2\, g = \mu^2 g, \ \mu \neq 0, \\ &\quad \text{implies } A^*g \in \operatorname{dom}(H_1) \text{ and } H_1(A^*g) = \mu^2 (A^*g), \end{aligned} \tag{B.17}$$

with multiplicities of eigenvalues preserved.

(iii) *One has for $z \in \rho(H_1) \cap \rho(H_2)$,*

$$I_{\mathcal{H}_2} + z(H_2 - zI_{\mathcal{H}_2})^{-1} \supseteq A(H_1 - zI_{\mathcal{H}_1})^{-1}A^*, \tag{B.18}$$

$$I_{\mathcal{H}_1} + z(H_1 - zI_{\mathcal{H}_1})^{-1} \supseteq A^*(H_2 - zI_{\mathcal{H}_2})^{-1}A, \tag{B.19}$$

and

$$A^*\phi(H_2) \supseteq \phi(H_1)A^*, \quad A\phi(H_1) \supseteq \phi(H_2)A, \tag{B.20}$$

$$V_{A^*}\phi(H_2) \supseteq \phi(H_1)V_{A^*}, \quad V_A\phi(H_1) \supseteq \phi(H_2)V_A. \tag{B.21}$$

As noted by E. Nelson (unpublished), Theorem B.2 follows from the spectral theorem and the elementary identities,

$$Q = V_Q|Q| = |Q|V_Q, \tag{B.22}$$

$$\ker(Q) = \ker(|Q|) = \ker(Q^2) = (\operatorname{ran}(Q))^{\perp} = \ker(A) \oplus \ker(A^*), \tag{B.23}$$

$$\begin{aligned} &I_{\mathcal{H}_1\oplus\mathcal{H}_2} + z(Q^2 - zI_{\mathcal{H}_1\oplus\mathcal{H}_2})^{-1} \\ &\quad = Q^2(Q^2 - zI_{\mathcal{H}_1\oplus\mathcal{H}_2})^{-1} \supseteq Q(Q^2 - zI_{\mathcal{H}_1\oplus\mathcal{H}_2})^{-1}Q, \quad z \in \rho(Q^2), \end{aligned} \tag{B.24}$$

$$Q\phi(Q^2) \supseteq \phi(Q^2)Q, \tag{B.25}$$

where

$$V_Q = \begin{pmatrix} 0 & (V_A)^* \\ V_A & 0 \end{pmatrix} = \begin{pmatrix} 0 & V_{A^*} \\ V_A & 0 \end{pmatrix}. \tag{B.26}$$

In particular,

$$\ker(Q) = \ker(A) \oplus \ker(A^*), \quad P_{\ker(Q)} = \begin{pmatrix} P_{\ker(A)} & 0 \\ 0 & P_{\ker(A^*)} \end{pmatrix}, \tag{B.27}$$

and we also recall that

$$\mathfrak{S}_3 Q \mathfrak{S}_3 = -Q, \quad \mathfrak{S}_3 = \begin{pmatrix} I_{\mathcal{H}_1} & 0 \\ 0 & -I_{\mathcal{H}_2} \end{pmatrix}, \tag{B.28}$$

that is, Q and $-Q$ are unitarily equivalent. (For more details on Nelson's trick see also [129, Sect. 8.4], [132, Subsect. 5.2.3].) We also note that

$$\psi(|Q|) = \begin{pmatrix} \psi(|A|) & 0 \\ 0 & \psi(|A^*|) \end{pmatrix} \tag{B.29}$$

for Borel measurable functions ψ on $\mathbb{R}$, and

$$\overline{[Q|Q|^{-1}]} = \begin{pmatrix} 0 & (V_A)^* \\ V_A & 0 \end{pmatrix} = V_Q \text{ if } \ker(Q) = \{0\}. \tag{B.30}$$

Finally, we recall the following relationships between Q and H_j, $j = 1, 2$.

Theorem B.3 ([24], [131]). *Assume Hypothesis* B.1.

(i) *Introducing the unitary operator U on* $(\ker(Q))^{\perp}$ *by*

$$U = 2^{-1/2} \begin{pmatrix} I_{\mathcal{H}_1} & (V_A)^* \\ -V_A & I_{\mathcal{H}_2} \end{pmatrix} \text{ on } (\ker(Q))^{\perp}, \tag{B.31}$$

one infers that

$$UQU^{-1} = \begin{pmatrix} |A| & 0 \\ 0 & -|A^*| \end{pmatrix} \text{ on } (\ker(Q))^{\perp}. \tag{B.32}$$

(ii) *One has*

$$(Q - \zeta I_{\mathcal{H}_1 \oplus \mathcal{H}_2})^{-1} = \begin{pmatrix} \zeta(H_1 - \zeta^2 I_{\mathcal{H}_1})^{-1} & A^*(H_2 - \zeta^2 I_{\mathcal{H}_2})^{-1} \\ A(H_1 - \zeta^2 I_{\mathcal{H}_1})^{-1} & \zeta(H_2 - \zeta^2 I_{\mathcal{H}_2})^{-1} \end{pmatrix}, \quad \zeta^2 \in \rho(H_1) \cap \rho(H_2). \tag{B.33}$$

(iii) *In addition,*

$$\begin{pmatrix} f_1 \\ f_2 \end{pmatrix} \in \operatorname{dom}(Q) \text{ and } Q \begin{pmatrix} f_1 \\ f_2 \end{pmatrix} = \eta \begin{pmatrix} f_1 \\ f_2 \end{pmatrix}, \ \eta \neq 0, \quad \text{implies } f_j \in \operatorname{dom}(H_j) \text{ and } H_j f_j = \eta^2 f_j, \ j = 1, 2. \tag{B.34}$$

Conversely,

$$f \in \operatorname{dom}(H_1) \text{ and } H_1 f = \lambda^2 f, \ \lambda \neq 0, \quad \text{implies } \begin{pmatrix} f \\ \lambda^{-1} A f \end{pmatrix} \in \operatorname{dom}(Q) \text{ and } Q \begin{pmatrix} f \\ \lambda^{-1} A f \end{pmatrix} = \lambda \begin{pmatrix} f \\ \lambda^{-1} A f \end{pmatrix}. \tag{B.35}$$

Similarly,

$$g \in \operatorname{dom}(H_2) \text{ and } H_2 g = \mu^2 g, \ \mu \neq 0, \quad \text{implies } \begin{pmatrix} \mu^{-1} A^* g \\ g \end{pmatrix} \in \operatorname{dom}(Q) \text{ and } Q \begin{pmatrix} \mu^{-1} A^* g \\ g \end{pmatrix} = \mu \begin{pmatrix} \mu^{-1} A^* g \\ g \end{pmatrix}. \tag{B.36}$$

Appendix C. Sesquilinear forms and associated operators

In this appendix we describe a few basic facts on sesquilinear forms and linear operators associated with them following [64, Sect. 2]. Let $\mathcal{H}$ be a complex separable Hilbert space with scalar product $(\,\cdot\,,\,\cdot\,)_{\mathcal{H}}$ (antilinear in the first and linear in the second argument), $\mathcal{V}$ a reflexive Banach space continuously and densely embedded into $\mathcal{H}$. Then also $\mathcal{H}$ embeds continuously and densely into $\mathcal{V}^*$. That is,

$$\mathcal{V} \hookrightarrow \mathcal{H} \hookrightarrow \mathcal{V}^*. \tag{C.1}$$

Here the continuous embedding $\mathcal{H} \hookrightarrow \mathcal{V}^*$ is accomplished via the identification

$$\mathcal{H} \ni v \mapsto (\,\cdot\,, v)_{\mathcal{H}} \in \mathcal{V}^*, \tag{C.2}$$

and recall our convention in this manuscript that if X denotes a Banach space, X^* denotes the *adjoint space* of continuous conjugate linear functionals on X, also known as the *conjugate dual* of X.

In particular, if the sesquilinear form

$${}_{\mathcal{V}}\langle\,\cdot\,,\,\cdot\,\rangle_{\mathcal{V}^*} \colon \mathcal{V} \times \mathcal{V}^* \to \mathbb{C} \tag{C.3}$$

denotes the duality pairing between $\mathcal{V}$ and $\mathcal{V}^*$, then

$${}_{\mathcal{V}}\langle u, v\rangle_{\mathcal{V}^*} = (u, v)_{\mathcal{H}}, \quad u \in \mathcal{V},\ v \in \mathcal{H} \hookrightarrow \mathcal{V}^*, \tag{C.4}$$

that is, the $\mathcal{V}, \mathcal{V}^*$ pairing ${}_{\mathcal{V}}\langle\,\cdot\,,\,\cdot\,\rangle_{\mathcal{V}^*}$ is compatible with the scalar product $(\,\cdot\,,\,\cdot\,)_{\mathcal{H}}$ in $\mathcal{H}$.

Let $T \in \mathcal{B}(\mathcal{V}, \mathcal{V}^*)$. Since $\mathcal{V}$ is reflexive, $(\mathcal{V}^*)^* = \mathcal{V}$, one has

$$T \colon \mathcal{V} \to \mathcal{V}^*, \quad T^* \colon \mathcal{V} \to \mathcal{V}^* \tag{C.5}$$

and

$${}_{\mathcal{V}}\langle u, Tv\rangle_{\mathcal{V}^*} = {}_{\mathcal{V}^*}\langle T^*u, v\rangle_{(\mathcal{V}^*)^*} = {}_{\mathcal{V}^*}\langle T^*u, v\rangle_{\mathcal{V}} = \overline{{}_{\mathcal{V}}\langle v, T^*u\rangle_{\mathcal{V}^*}}. \tag{C.6}$$

Self-adjointness of T is then defined by $T = T^*$, that is,

$${}_{\mathcal{V}}\langle u, Tv\rangle_{\mathcal{V}^*} = {}_{\mathcal{V}^*}\langle Tu, v\rangle_{\mathcal{V}} = \overline{{}_{\mathcal{V}}\langle v, Tu\rangle_{\mathcal{V}^*}}, \quad u, v \in \mathcal{V}, \tag{C.7}$$

nonnegativity of T is defined by

$${}_{\mathcal{V}}\langle u, Tu\rangle_{\mathcal{V}^*} \geq 0, \quad u \in \mathcal{V}, \tag{C.8}$$

and *boundedness from below of T by $c_T \in \mathbb{R}$* is defined by

$${}_{\mathcal{V}}\langle u, Tu\rangle_{\mathcal{V}^*} \geq c_T \|u\|_{\mathcal{H}}^2, \quad u \in \mathcal{V}. \tag{C.9}$$

(By (C.4), this is equivalent to ${}_{\mathcal{V}}\langle u, Tu\rangle_{\mathcal{V}^*} \geq c_T\, {}_{\mathcal{V}}\langle u, u\rangle_{\mathcal{V}^*}$, $u \in \mathcal{V}$.)

Next, let the sesquilinear form $\mathfrak{a}(\,\cdot\,,\,\cdot\,) \colon \mathcal{V} \times \mathcal{V} \to \mathbb{C}$ (antilinear in the first and linear in the second argument) be *$\mathcal{V}$-bounded*, that is, there exists a $c_{\mathfrak{a}} > 0$ such that

$$|\mathfrak{a}(u, v)| \leqslant c_{\mathfrak{a}} \|u\|_{\mathcal{V}} \|v\|_{\mathcal{V}}, \quad u, v \in \mathcal{V}. \tag{C.10}$$

Then $\widetilde{A}$ defined by

$$\widetilde{A} \colon \begin{cases} \mathcal{V} \to \mathcal{V}^*, \\ v \mapsto \widetilde{A}v = \mathfrak{a}(\,\cdot\,, v), \end{cases} \tag{C.11}$$

satisfies

$$\widetilde{A} \in \mathcal{B}(\mathcal{V}, \mathcal{V}^*) \text{ and } {}_{\mathcal{V}}\langle u, \widetilde{A}v \rangle_{\mathcal{V}^*} = \mathfrak{a}(u, v), \quad u, v \in \mathcal{V}. \tag{C.12}$$

Assuming further that $\mathfrak{a}(\cdot\,,\cdot)$ is *symmetric*, that is,

$$\mathfrak{a}(u, v) = \overline{\mathfrak{a}(v, u)}, \quad u, v \in \mathcal{V}, \tag{C.13}$$

and that $\mathfrak{a}$ is $\mathcal{V}$-*coercive*, that is, there exists a constant $C_0 > 0$ such that

$$\mathfrak{a}(u, u) \geq C_0 \|u\|_{\mathcal{V}}^2, \quad u \in \mathcal{V}, \tag{C.14}$$

respectively, then,

$$\widetilde{A} \colon \mathcal{V} \to \mathcal{V}^* \text{ is bounded, self-adjoint, and boundedly invertible.} \tag{C.15}$$

Moreover, denoting by A the part of $\widetilde{A}$ in $\mathcal{H}$ defined by

$$\operatorname{dom}(A) = \big\{u \in \mathcal{V} \,|\, \widetilde{A}u \in \mathcal{H}\big\} \subseteq \mathcal{H}, \quad A = \widetilde{A}\big|_{\operatorname{dom}(A)} \colon \operatorname{dom}(A) \to \mathcal{H}, \tag{C.16}$$

then A is a (possibly unbounded) self-adjoint operator in $\mathcal{H}$ satisfying

$$A \geq C_0 I_{\mathcal{H}}, \tag{C.17}$$

$$\operatorname{dom}\big(A^{1/2}\big) = \mathcal{V}. \tag{C.18}$$

In particular,

$$A^{-1} \in \mathcal{B}(\mathcal{H}). \tag{C.19}$$

The facts (C.1)–(C.19) are a consequence of the Lax–Milgram theorem and the second representation theorem for symmetric sesquilinear forms. Details can be found, for instance, in [38, Sects. VI.3, VII.1], [53, Ch. IV], and [99].

Next, consider a symmetric form $\mathfrak{b}(\cdot\,,\cdot) \colon \mathcal{V} \times \mathcal{V} \to \mathbb{C}$ and assume that $\mathfrak{b}$ is *bounded from below by* $c_{\mathfrak{b}} \in \mathbb{R}$, that is,

$$\mathfrak{b}(u, u) \geq c_{\mathfrak{b}} \|u\|_{\mathcal{H}}^2, \quad u \in \mathcal{V}. \tag{C.20}$$

Introducing the scalar product $(\cdot\,,\cdot)_{\mathcal{V}_{\mathfrak{b}}} \colon \mathcal{V} \times \mathcal{V} \to \mathbb{C}$ (and the associated norm $\|\cdot\|_{\mathcal{V}_{\mathfrak{b}}}$) by

$$(u, v)_{\mathcal{V}_{\mathfrak{b}}} = \mathfrak{b}(u, v) + (1 - c_{\mathfrak{b}})(u, v)_{\mathcal{H}}, \quad u, v \in \mathcal{V}, \tag{C.21}$$

turns $\mathcal{V}$ into a pre-Hilbert space $(\mathcal{V}; (\cdot\,,\cdot)_{\mathcal{V}_{\mathfrak{b}}})$, which we denote by $\mathcal{V}_{\mathfrak{b}}$. The form $\mathfrak{b}$ is called *closed* in $\mathcal{H}$ if $\mathcal{V}_{\mathfrak{b}}$ is actually complete, and hence a Hilbert space. The form $\mathfrak{b}$ is called *closable* in $\mathcal{H}$ if it has a closed extension. If $\mathfrak{b}$ is closed in $\mathcal{H}$, then

$$|\mathfrak{b}(u, v) + (1 - c_{\mathfrak{b}})(u, v)_{\mathcal{H}}| \leqslant \|u\|_{\mathcal{V}_{\mathfrak{b}}} \|v\|_{\mathcal{V}_{\mathfrak{b}}}, \quad u, v \in \mathcal{V}, \tag{C.22}$$

and

$$|\mathfrak{b}(u, u) + (1 - c_{\mathfrak{b}})\|u\|_{\mathcal{H}}^2| = \|u\|_{\mathcal{V}_{\mathfrak{b}}}^2, \qquad u \in \mathcal{V}, \tag{C.23}$$

show that the form $\mathfrak{b}(\cdot\,,\cdot) + (1 - c_{\mathfrak{b}})(\cdot\,,\cdot)_{\mathcal{H}}$ is a symmetric, $\mathcal{V}$-bounded, and $\mathcal{V}$-coercive sesquilinear form. Hence, by (C.11) and (C.12), there exists a linear map

$$\widetilde{B}_{c_{\mathfrak{b}}} \colon \begin{cases} \mathcal{V}_{\mathfrak{b}} \to \mathcal{V}_{\mathfrak{b}}^*, \\ v \mapsto \widetilde{B}_{c_{\mathfrak{b}}} v = \mathfrak{b}(\cdot\,, v) + (1 - c_{\mathfrak{b}})(\cdot\,, v)_{\mathcal{H}}, \end{cases} \tag{C.24}$$

with

$$\begin{aligned}&\widetilde{B}_{c_{\mathfrak{b}}} \in \mathcal{B}(\mathcal{V}_{\mathfrak{b}}, \mathcal{V}_{\mathfrak{b}}^*) \text{ and} \\ &{}_{\mathcal{V}_{\mathfrak{b}}}\langle u, \widetilde{B}_{c_{\mathfrak{b}}} v\rangle_{\mathcal{V}_{\mathfrak{b}}^*} = \mathfrak{b}(u,v) + (1-c_{\mathfrak{b}})(u,v)_{\mathcal{H}}, \quad u,v \in \mathcal{V},\end{aligned} \tag{C.25}$$

in particular, $\widetilde{B}_{c_{\mathfrak{b}}}$ is bounded, self-adjoint, and boundedly invertible. Introducing the linear map

$$\widetilde{B} = \widetilde{B}_{c_{\mathfrak{b}}} + (c_{\mathfrak{b}} - 1)\widetilde{I} \colon \mathcal{V}_{\mathfrak{b}} \to \mathcal{V}_{\mathfrak{b}}^*, \tag{C.26}$$

where $\widetilde{I} \colon \mathcal{V}_{\mathfrak{b}} \hookrightarrow \mathcal{V}_{\mathfrak{b}}^*$ denotes the continuous inclusion (embedding) map of $\mathcal{V}_{\mathfrak{b}}$ into $\mathcal{V}_{\mathfrak{b}}^*$, $\widetilde{B}$ is bounded and self-adjoint, and one obtains a self-adjoint operator B in $\mathcal{H}$ by restricting $\widetilde{B}$ to $\mathcal{H}$,

$$\operatorname{dom}(B) = \big\{u \in \mathcal{V} \,\big|\, \widetilde{B}u \in \mathcal{H}\big\} \subseteq \mathcal{H}, \quad B = \widetilde{B}\big|_{\operatorname{dom}(B)} \colon \operatorname{dom}(B) \to \mathcal{H}, \tag{C.27}$$

satisfying the following properties:

$$B \geq c_{\mathfrak{b}} I_{\mathcal{H}}, \tag{C.28}$$

$$\operatorname{dom}\big(|B|^{1/2}\big) = \operatorname{dom}\big((B - c_{\mathfrak{b}} I_{\mathcal{H}})^{1/2}\big) = \mathcal{V}, \tag{C.29}$$

$$\mathfrak{b}(u,v) = \big(|B|^{1/2}u, U_B |B|^{1/2} v\big)_{\mathcal{H}} \tag{C.30}$$

$$= \big((B - c_{\mathfrak{b}} I_{\mathcal{H}})^{1/2}u, (B - c_{\mathfrak{b}} I_{\mathcal{H}})^{1/2}v\big)_{\mathcal{H}} + c_{\mathfrak{b}}(u,v)_{\mathcal{H}} \tag{C.31}$$

$$= {}_{\mathcal{V}_{\mathfrak{b}}}\langle u, \widetilde{B}v\rangle_{\mathcal{V}_{\mathfrak{b}}^*}, \quad u,v \in \mathcal{V}, \tag{C.32}$$

$$\mathfrak{b}(u,v) = (u, Bv)_{\mathcal{H}}, \quad u \in \mathcal{V},\ v \in \operatorname{dom}(B), \tag{C.33}$$

$$\begin{aligned}\operatorname{dom}(B) = \{v \in \mathcal{V} \,|\, &\text{there exists an } f_v \in \mathcal{H} \text{ such that} \\ &\mathfrak{b}(w,v) = (w, f_v)_{\mathcal{H}} \text{ for all } w \in \mathcal{V}\},\end{aligned} \tag{C.34}$$

$$\begin{aligned}&Bu = f_u, \quad u \in \operatorname{dom}(B), \\ &\operatorname{dom}(B) \text{ is dense in } \mathcal{H} \text{ and in } \mathcal{V}_{\mathfrak{b}}.\end{aligned} \tag{C.35}$$

Properties (C.34) and (C.35) uniquely determine B. Here U_B in (C.31) is the partial isometry in the polar decomposition of B, that is,

$$B = U_B|B|, \quad |B| = (B^*B)^{1/2} \geq 0. \tag{C.36}$$

The operator B is called the *operator associated with the form* $\mathfrak{b}$.

The norm in the Hilbert space $\mathcal{V}_{\mathfrak{b}}^*$ is given by

$$\|\ell\|_{\mathcal{V}_{\mathfrak{b}}^*} = \sup\{|{}_{\mathcal{V}_{\mathfrak{b}}}\langle u, \ell\rangle_{\mathcal{V}_{\mathfrak{b}}^*}| \,|\, \|u\|_{\mathcal{V}_{\mathfrak{b}}} \leqslant 1\}, \quad \ell \in \mathcal{V}_{\mathfrak{b}}^*, \tag{C.37}$$

with associated scalar product,

$$(\ell_1, \ell_2)_{\mathcal{V}_{\mathfrak{b}}^*} = {}_{\mathcal{V}_{\mathfrak{b}}}\big\langle \big(\widetilde{B} + (1 - c_{\mathfrak{b}})\widetilde{I}\big)^{-1}\ell_1, \ell_2\big\rangle_{\mathcal{V}_{\mathfrak{b}}^*}, \quad \ell_1, \ell_2 \in \mathcal{V}_{\mathfrak{b}}^*. \tag{C.38}$$

Since

$$\big\|\big(\widetilde{B} + (1 - c_{\mathfrak{b}})\widetilde{I}\big)v\big\|_{\mathcal{V}_{\mathfrak{b}}^*} = \|v\|_{\mathcal{V}_{\mathfrak{b}}}, \quad v \in \mathcal{V}, \tag{C.39}$$

the Riesz representation theorem yields

$$\big(\widetilde{B} + (1 - c_{\mathfrak{b}})\widetilde{I}\big) \in \mathcal{B}(\mathcal{V}_{\mathfrak{b}}, \mathcal{V}_{\mathfrak{b}}^*) \text{ and } \big(\widetilde{B} + (1 - c_{\mathfrak{b}})\widetilde{I}\big) \colon \mathcal{V}_{\mathfrak{b}} \to \mathcal{V}_{\mathfrak{b}}^* \text{ is unitary.} \tag{C.40}$$

In addition,

$$
\begin{aligned}
{}_{\mathcal{V}_{\mathfrak{b}}}\langle u, (\widetilde{B} + (1-c_{\mathfrak{b}})\widetilde{I})v\rangle_{\mathcal{V}_{\mathfrak{b}}^*} &= \big((B + (1-c_{\mathfrak{b}})I_{\mathcal{H}})^{1/2}u, (B + (1-c_{\mathfrak{b}})I_{\mathcal{H}})^{1/2}v\big)_{\mathcal{H}} \\
&= (u,v)_{\mathcal{V}_{\mathfrak{b}}}, \quad u,v \in \mathcal{V}_{\mathfrak{b}}.
\end{aligned} \tag{C.41}
$$

In particular,

$$
\big\|(B + (1-c_{\mathfrak{b}})I_{\mathcal{H}})^{1/2}u\big\|_{\mathcal{H}} = \|u\|_{\mathcal{V}_{\mathfrak{b}}}, \quad u \in \mathcal{V}_{\mathfrak{b}}, \tag{C.42}
$$

and hence

$$
(B + (1-c_{\mathfrak{b}})I_{\mathcal{H}})^{1/2} \in \mathcal{B}(\mathcal{V}_{\mathfrak{b}}, \mathcal{H}) \text{ and } (B + (1-c_{\mathfrak{b}})I_{\mathcal{H}})^{1/2} \colon \mathcal{V}_{\mathfrak{b}} \to \mathcal{H} \text{ is unitary.} \tag{C.43}
$$

The facts (C.20)–(C.43) comprise the second representation theorem of sesquilinear forms (cf. [53, Sect. IV.2], [54, Sects. 1.2–1.5], and [90, Sect. VI.2.6]).

We briefly supplement (C.20)–(C.43) with some considerations that hint at mapping properties of $\big(\widetilde{B}+(1-c_{\mathfrak{b}})\widetilde{I}\big)^{\pm 1/2}$ on a scale of spaces, which, for simplicity, we restrict to the triple of spaces $\mathcal{V}_{\mathfrak{b}}$, $\mathcal{H}$, and $\mathcal{V}_{\mathfrak{b}}^*$ in this appendix. We start by defining

$$
\big(\hat{B}_{c_{\mathfrak{b}}} + (1-c_{\mathfrak{b}})\hat{I}\big)^{1/2} \colon \begin{cases} \mathcal{V}_{\mathfrak{b}} \to \mathcal{H}, \\ v \mapsto (B + (1-c_{\mathfrak{b}})I_{\mathcal{H}})^{1/2}v, \end{cases} \tag{C.44}
$$

and similarly,

$$
\big(\check{B}_{c_{\mathfrak{b}}} + (1-c_{\mathfrak{b}})\check{I}\big)^{1/2} \colon \begin{cases} \mathcal{H} \to \mathcal{V}_{\mathfrak{b}}^*, \\ f \mapsto \mathfrak{b}\big(\,\cdot\,, (B + (1-c_{\mathfrak{b}})I_{\mathcal{H}})^{-1/2}f\big) \\ \qquad +(1-c_{\mathfrak{b}})\big(\,\cdot\,, (B + (1-c_{\mathfrak{b}})I_{\mathcal{H}})^{-1/2}f\big)_{\mathcal{H}}. \end{cases} \tag{C.45}
$$

Then both maps in (C.44) and (C.45) are bounded and boundedly invertible. In particular,

$$
\begin{aligned}
&\big(\hat{B}_{c_{\mathfrak{b}}} + (1-c_{\mathfrak{b}})\hat{I}\big)^{1/2} \in \mathcal{B}(\mathcal{V}_{\mathfrak{b}}, \mathcal{H}), \quad \big(\hat{B}_{c_{\mathfrak{b}}} + (1-c_{\mathfrak{b}})\hat{I}\big)^{-1/2} \in \mathcal{B}(\mathcal{H}, \mathcal{V}_{\mathfrak{b}}), \\
&\big(\check{B}_{c_{\mathfrak{b}}} + (1-c_{\mathfrak{b}})\check{I}\big)^{1/2} \in \mathcal{B}(\mathcal{H}, \mathcal{V}_{\mathfrak{b}}^*), \quad \big(\check{B}_{c_{\mathfrak{b}}} + (1-c_{\mathfrak{b}})\check{I}\big)^{-1/2} \in \mathcal{B}(\mathcal{V}_{\mathfrak{b}}^*, \mathcal{H}),
\end{aligned} \tag{C.46}
$$

and

$$
\begin{aligned}
&\big(\hat{B}_{c_{\mathfrak{b}}} + (1-c_{\mathfrak{b}})\hat{I}\big)^{1/2}\big(\check{B}_{c_{\mathfrak{b}}} + (1-c_{\mathfrak{b}})\check{I}\big)^{1/2} = \big(\widetilde{B} + (1-c_{\mathfrak{b}})\widetilde{I}\big) \in \mathcal{B}(\mathcal{V}_{\mathfrak{b}}, \mathcal{V}_{\mathfrak{b}}^*), \\
&\big(\check{B}_{c_{\mathfrak{b}}} + (1-c_{\mathfrak{b}})\check{I}\big)^{-1/2}\big(\hat{B}_{c_{\mathfrak{b}}} + (1-c_{\mathfrak{b}})\hat{I}\big)^{-1/2} = \big(\widetilde{B} + (1-c_{\mathfrak{b}})\widetilde{I}\big)^{-1} \in \mathcal{B}(\mathcal{V}_{\mathfrak{b}}^*, \mathcal{V}_{\mathfrak{b}}).
\end{aligned} \tag{C.47}
$$

Due to self-adjointness of $\widetilde{B}$ as a bounded map from $\mathcal{V}_{\mathfrak{b}}$ to $\mathcal{V}_{\mathfrak{b}}^*$ in the sense of (C.7), one finally obtains that

$$
\begin{aligned}
\Big(\big(\hat{B}_{c_{\mathfrak{b}}} + (1-c_{\mathfrak{b}})\hat{I}\big)^{\pm 1/2}\Big)^* &= \big(\check{B}_{c_{\mathfrak{b}}} + (1-c_{\mathfrak{b}})\check{I}\big)^{\pm 1/2}, \\
\Big(\big(\check{B}_{c_{\mathfrak{b}}} + (1-c_{\mathfrak{b}})\check{I}\big)^{\pm 1/2}\Big)^* &= \big(\hat{B}_{c_{\mathfrak{b}}} + (1-c_{\mathfrak{b}})\hat{I}\big)^{\pm 1/2}.
\end{aligned} \tag{C.48}
$$

Hence, we will follow standard practice in connection with chains of (Sobolev) spaces and refrain from painstakingly distinguishing the $\hat{}$- and $\check{}$-operations and simply resort to the notation

$$\big(\widetilde{B} + (1 - c_{\mathfrak{b}})\widetilde{I}\big)^{\pm 1/2} \tag{C.49}$$

for the operators in (C.46) in the bulk of this paper.

A special but important case of nonnegative closed forms is obtained as follows: Let $\mathcal{H}_j$, $j = 1, 2$, be complex separable Hilbert spaces, and $T\colon \operatorname{dom}(T) \to \mathcal{H}_2$, $\operatorname{dom}(T) \subseteq \mathcal{H}_1$, a densely defined operator. Consider the nonnegative form $\mathfrak{a}_T\colon \operatorname{dom}(T) \times \operatorname{dom}(T) \to \mathbb{C}$ defined by

$$\mathfrak{a}_T(u, v) = (Tu, Tv)_{\mathcal{H}_2}, \quad u, v \in \operatorname{dom}(T). \tag{C.50}$$

Then the form $\mathfrak{a}_T$ is closed (resp., closable) in $\mathcal{H}_1$ if and only if T is. If T is closed, the unique nonnegative self-adjoint operator associated with $\mathfrak{a}_T$ in $\mathcal{H}_1$, whose existence is guaranteed by the second representation theorem for forms, then equals $T^*T \geq 0$. In particular, one obtains in addition to (C.50),

$$\mathfrak{a}_T(u, v) = (|T|u, |T|v)_{\mathcal{H}_1}, \quad u, v \in \operatorname{dom}(T) = \operatorname{dom}(|T|). \tag{C.51}$$

Moreover, since

$$\begin{aligned} \mathfrak{b}(u, v) + (1 - c_{\mathfrak{b}})(u, v)_{\mathcal{H}} &= \big((B + (1 - c_{\mathfrak{b}})I_{\mathcal{H}})^{1/2}u, (B + (1 - c_{\mathfrak{b}})I_{\mathcal{H}})^{1/2}v\big)_{\mathcal{H}}, \\ &\qquad u, v \in \operatorname{dom}(b) = \operatorname{dom}\big(|B|^{1/2}\big) = \mathcal{V}, \end{aligned} \tag{C.52}$$

and $(B + (1 - c_{\mathfrak{b}})I_{\mathcal{H}})^{1/2}$ is self-adjoint (and hence closed) in $\mathcal{H}$, a symmetric, $\mathcal{V}$-bounded, and $\mathcal{V}$-coercive form is densely defined in $\mathcal{H} \times \mathcal{H}$ and closed in $\mathcal{H}$ (a fact we will be using in the proof of Theorem 2.3). We refer to [90, Sect. VI.2.4] and [135, Sect. 5.5] for details.

Next we recall that if $\mathfrak{a}_j$ are sesquilinear forms defined on $\operatorname{dom}(\mathfrak{a}_j)$, $j = 1, 2$, bounded from below and closed, then also

$$(\mathfrak{a}_1 + \mathfrak{a}_2)\colon \begin{cases} (\operatorname{dom}(\mathfrak{a}_1) \cap \operatorname{dom}(\mathfrak{a}_2)) \times (\operatorname{dom}(\mathfrak{a}_1) \cap \operatorname{dom}(\mathfrak{a}_2)) \to \mathbb{C}, \\ (u, v) \mapsto (\mathfrak{a}_1 + \mathfrak{a}_2)(u, v) = \mathfrak{a}_1(u, v) + \mathfrak{a}_2(u, v) \end{cases} \tag{C.53}$$

is bounded from below and closed (cf., e.g., [90, Sect. VI.1.6]).

Finally, we also recall the following perturbation theoretic fact: Suppose $\mathfrak{a}$ is a sesquilinear form defined on $\mathcal{V} \times \mathcal{V}$, bounded from below and closed, and let $\mathfrak{b}$ be a symmetric sesquilinear form bounded with respect to $\mathfrak{a}$ with bound less than one, that is, $\operatorname{dom}(\mathfrak{b}) \supseteq \mathcal{V} \times \mathcal{V}$, and that there exist $0 \leqslant \alpha < 1$ and $\beta \geqslant 0$ such that

$$|\mathfrak{b}(u, u)| \leqslant \alpha|\mathfrak{a}(u, u)| + \beta\|u\|_{\mathcal{H}}^2, \quad u \in \mathcal{V}. \tag{C.54}$$

Then

$$(\mathfrak{a} + \mathfrak{b})\colon \begin{cases} \mathcal{V} \times \mathcal{V} \to \mathbb{C}, \\ (u, v) \mapsto (\mathfrak{a} + \mathfrak{b})(u, v) = \mathfrak{a}(u, v) + \mathfrak{b}(u, v) \end{cases} \tag{C.55}$$

defines a sesquilinear form that is bounded from below and closed (cf., e.g., [90, Sect. VI.1.6]). In the special case where α can be chosen arbitrarily small, the form $\mathfrak{b}$ is called infinitesimally form bounded with respect to $\mathfrak{a}$.

Acknowledgment

We gratefully acknowledge valuable correspondence with Rostyslav Hryniv, Mark Malamud, Roger Nichols, Fritz Philipp, Barry Simon, Günter Stolz, and Gerald Teschl. In addition, we are indebted to Igor Verbitsky for helpful discussions. Finally, we sincerely thank the anonymous referee for numerous helpful comments improving the presentation of our results.

References

[1] R.A. Adams and J.J.F. Fournier, *Sobolev Spaces*, second edition, Academic Press, 2003.

[2] S. Alama, M. Avellaneda, P.A. Deift, and R. Hempel, *On the existence of eigenvalues of a divergence-form operator $A + \lambda B$ in a gap of $\sigma(A)$*, Asymptotic Anal. **8**, 311–344 (1994).

[3] S. Albeverio, F. Gesztesy, R. Høegh-Krohn, and H. Holden, *Solvable Models in Quantum Mechanics*, with an Appendix by P. Exner, 2nd ed., AMS Chelsea Publishing, Providence, RI, 2005.

[4] S. Albeverio, A. Kostenko, and M. Malamud, *Spectral theory of semibounded Sturm–Liouville operators with local interactions on a discrete set*, J. Math. Phys. **51**, 102102 (2010), 24 pp.

[5] T.G. Anderson and D.B. Hinton, *Relative boundedness and compactness theory for second-order differential operators*, J. Inequal. & Appl. **1**, 375–400 (1997).

[6] F.V. Atkinson and A.B. Mingarelli, *Asymptotics of the number of zeros and of the eigenvalues of general weighted Sturm–Liouville problems*, J. reine angew. Math. **375/376**, 380–393 (1987).

[7] J.-G. Bak and A.A. Shkalikov, *Multipliers in dual Sobolev spaces and Schrödinger operators with distribution potentials*, Math. Notes **71**, 587–594 (2002).

[8] R. Beals, *Indefinite Sturm–Liouville problems and half-range completeness*, J. Diff. Eq. **56**, 391–407 (1985).

[9] R. Beals, D.H. Sattinger, and J. Szmigielski, *Multipeakons and the classical moment problem*, Adv. Math. **154**, 229–257 (2000).

[10] R. Beals, D.H. Sattinger, and J. Szmigielski, *Periodic peakons and Calogero–Françoise flows*, J. Inst. Math. Jussieu **4**, No. 1, 1–27 (2005).

[11] J. Behrndt, *On the spectral theory of singular indefinite Sturm–Liouville operators*, J. Math. Anal. Appl. **334**, 1439–1449 (2007).

[12] J. Behrndt, *Spectral theory of elliptic differential operators with indefinite weights*, Proc. Roy. Soc. Edinburgh **143A**, 21–38 (2013).

[13] J. Behrndt, Q. Katatbeh, and C. Trunk, *Non-real eigenvalues of singular indefinite Sturm–Liouville operators*, Proc. Amer. Math. Soc. **137**, 3797–3806 (2009).

[14] J. Behrndt, R. Möws, and C. Trunk, *Eigenvalue estimates for singular left-definite Sturm–Liouville operators*, J. Spectral Th. **1**, 327–347 (2011).

[15] J. Behrndt and F. Philipp, *Spectral analysis of singular ordinary differential operators with indefinite weights*, J. Diff. Eq. **248**, 2015–2037 (2010).

[16] J. Behrndt and C. Trunk, *On the negative squares of indefinite Sturm–Liouville operators*, J. Diff. Eq. **238**, 491–519 (2007).

[17] C. Bennewitz, *On the spectral problem associated with the Camassa–Holm equation*, J. Nonlinear Math. Phys. **11**, 422–434 (2004).

[18] C. Bennewitz, B. M. Brown, and R. Weikard, *Inverse spectral and scattering theory for the half-line left-definite Sturm–Liouville problem*, SIAM J. Math. Anal. **40**, 2105–2131 (2009).

[19] C. Bennewitz, B.M. Brown, and R. Weikard, *A uniqueness result for one-dimensional inverse scattering*, Math. Nachr. **285**, 941–948 (2012).

[20] C. Bennewitz, B.M. Brown, and R. Weikard, *Scattering and inverse scattering for a left-definite Sturm–Liouville problem*, J. Diff. Eq. **253**, 2380–2419 (2012).

[21] C. Bennewitz and W.N. Everitt, *On second-order left-definite boundary value problems*, in *Ordinary Differential Equations and Operators,* (*Proceedings, Dundee,* 1982), W.N. Everitt and R.T. Lewis (eds.), Lecture Notes in Math., Vol. 1032, Springer, Berlin, 1983, pp. 31–67.

[22] A.M. Berthier, *Spectral Theory and Wave Operators for the Schrödinger Equation,* Research Notes in Mathematics, Vol. 71, Pitman, Boston, 1982.

[23] P. Binding and A. Fleige, *Conditions for an indefinite Sturm–Liouville Riesz basis property,* in *Recent Advances in Operator Theory in Hilbert and Krein Spaces,* J. Behrndt, K.-H. Förster, and C. Trunk (eds.), Operator Theory: Advances and Applications, Birkhäuser, Basel, Vol. 198, 2009, pp. 87–95.

[24] D. Bollé, F. Gesztesy, H. Grosse, W. Schweiger, and B. Simon, *Witten index, axial anomaly, and Krein's spectral shift function in supersymmetric quantum mechanics,* J. Math. Phys. **28**, 1512–1525 (1987).

[25] B.M. Brown, M.S.P. Eastham, and K.M. Schmidt, *Periodic Differential Operators,* Birkhäuser, 2013.

[26] R.C. Brown and D.B. Hinton, *Relative form boundedness and compactness for a second-order differential operator*, J. Comp. Appl. Math. **171**, 123–140 (2004).

[27] R. Carlone, M. Malamud, and A. Posilicano, *On the spectral theory of Gesztesy–Šeba realizations of 1-D Dirac operators with point interactions on a discrete set,* J. Diff. Eq. **254**, 3835–3902 (2013).

[28] S. Clark and F. Gesztesy, *Weyl–Titchmarsh M-function asymptotics and Borg-type theorems for Dirac operators,* Trans. Amer. Math. Soc. **354**, 3475–3534 (2002).

[29] S. Clark and F. Gesztesy, *On Povzner–Wienholtz-type self-adjointness results for matrix-valued Sturm–Liouville operators,* Proc. Roy. Soc. Edinburgh **133A**, 747–758 (2003).

[30] A. Constantin, *A general-weighted Sturm–Liouville problem,* Scuola Norm. Sup. **24**, 767–782 (1997).

[31] A. Constantin, *On the spectral problem for the periodic Camassa–Holm equation,* J. Math. Anal. Appl. **210**, 215–230 (1997).

[32] A. Constantin, *On the inverse spectral problem for the Camassa–Holm equation*, J. Funct. Anal. **155**, 352–363(1998).

[33] A. Constantin, *On the scattering problem for the Camassa–Holm equation*, Proc. Roy. Soc. London A **457**, 953–970 (2001).

[34] A. Constantin, V.S. Gerdjikov, and R.I. Rossen, *Inverse scattering transform for the Camassa–Holm equation*, Inverse Probl. **22**, 2197–2207 (2006).

[35] A. Constantin and J. Lenells, *On the inverse scattering approach to the Camassa–Holm equation*, J. Nonlinear Math. Phys. **10**, 252–255 (2003).

[36] A. Constantin and H.P. McKean, *A shallow water equation on the circle*, Commun. Pure Appl. Math. **52**, 949–982 (1999).

[37] K. Daho and H. Langer, *Sturm–Liouville operators with an indefinite weight function: The Periodic case*, Radcvi Mat. **2**, 165–188 (1986).

[38] R. Dautray and J.-L. Lions, *Mathematical Analysis and Numerical Methods for Science and Technology, Volume 2, Functional and Variational Methods*, Springer, Berlin, 2000.

[39] P.A. Deift, *Applications of a commutation formula*, Duke Math. J. **45**, 267–310 (1978).

[40] P. Djakov and B. Mityagin, *Multiplicities of the eigenvalues of periodic Dirac operators*, J. Diff. Eq. **210**, 178–216 (2005).

[41] P. Djakov and B. Mityagin, *Spectral gap asymptotics of one-dimensional Schrödinger operators with singular periodic potentials*, Integral Transforms Special Fcts. **20**, nos. 3-4, 265–273 (2009).

[42] P. Djakov and B. Mityagin, *Spectral gaps of Schrödinger operators with periodic singular potentials*, Dyn. PDE **6**, no. 2, 95–165 (2009).

[43] P. Djakov and B. Mityagin, *Fourier method for one-dimensional Schrödinger operators with singular periodic potentials*, in *Topics in Operator Theory, Vol. 2: Systems and Mathematical Physics*, J.A. Ball, V. Bolotnikov, J.W. Helton, L. Rodman, I.M. Spitkovsky (eds.), Operator Theory: Advances and Applications, Vol. 203, Birkhäuser, Basel, 2010, pp. 195–236.

[44] P. Djakov and B. Mityagin, *Criteria for existence of Riesz bases consisting of root functions of Hill and 1d Dirac operators*, J. Funct. Anal. **263**, 2300–2332 (2012).

[45] N. Dunford and J.T. Schwartz, *Linear Operators Part II: Spectral Theory*, Interscience, New York, 1988.

[46] M.S.P. Eastham, *The Spectral Theory of Periodic Differential Equations*, Scottish Academic Press, Edinburgh and London, 1973.

[47] J. Eckhardt, *Direct and inverse spectral theory of singular left-definite Sturm–Liouville operators*, J. Diff. Eq. **253**, 604–634 (2012).

[48] J. Eckhardt, F. Gesztesy, R. Nichols, and G. Teschl, *Supersymmetry and Schrödinger-type operators with distributional matrix-valued potentials*, arXiv:1206.4966, J. Spectral Theory, to appear.

[49] J. Eckhardt, F. Gesztesy, R. Nichols, and G. Teschl, *Weyl–Titchmarsh theory for Sturm–Liouville operators with distributional potentials*, Opuscula Math. **33**, 467–563 (2013).

[50] J. Eckhardt, F. Gesztesy, R. Nichols, and G. Teschl, *Inverse spectral theory for Sturm–Liouville operators with distributional coefficients*, J. London Math. Soc. (2) **88**, 801–828 (2013).

[51] J. Eckhardt and G. Teschl, *On the isospectral problem of the dispersionless Camassa–Holm equation*, Adv. Math. **235**, 469–495 (2013).

[52] J. Eckhardt and G. Teschl, *Sturm–Liouville operators with measure-valued coefficients*, J. Analyse Math. **120**, 151–224 (2013).

[53] D.E. Edmunds and W.D. Evans, *Spectral Theory and Differential Operators*, Clarendon Press, Oxford, 1989.

[54] W.G. Faris, *Self-Adjoint Operators*, Lecture Notes in Mathematics, Vol. **433**, Springer, Berlin, 1975.

[55] J. Fleckinger and M.L. Lapidus, *Eigenvalues of elliptic boundary value problems with an indefinite weight function*, Trans. Amer. Math. Soc. **295**, 305–324 (1986).

[56] G. Freiling, V. Rykhlov, and V. Yurko, *Spectral analysis for an indefinite singular Sturm–Liouville problem*, Appl. Anal. **81**, 1283–1305 (2002).

[57] F. Gesztesy, *On the modified Korteweg-de Vries equation*, in *Differential Equations with Applications in Biology, Physics, and Engineering*, J.A. Goldstein, F. Kappel, and W. Schappacher (eds.), Marcel Dekker, New York, 1991, pp. 139–183.

[58] F. Gesztesy, *Quasi-periodic, finite-gap solutions of the modified Korteweg-de Vries equation*, in *Ideas and Methods in Mathematical Analysis, Stochastics, and Applications*, S. Albeverio, J.E. Fenstad, H. Holden, and T. Lindstrøm (eds.), Vol. 1, Cambridge Univ. Press, Cambridge, 1992, pp. 428–471.

[59] F. Gesztesy, *A complete spectral characterizaton of the double commutation method*, J. Funct. Anal. **117**, 401–446 (1993).

[60] F. Gesztesy and H. Holden, *Algebro-geometric solutions of the Camassa–Holm hierarchy*, Rev. Mat. Iberoamericana **19**, 73–142 (2003).

[61] F. Gesztesy and H. Holden, *Soliton Equations and Their Algebro-Geometric Solutions. Vol. I:* $(1+1)$*-Dimensional Continuous Models*, Cambridge Studies in Advanced Mathematics, Vol. 79, Cambridge Univ. Press, 2003.

[62] F. Gesztesy and H. Holden, *Real-valued algebro-geometric solutions of the Camassa–Holm hierarchy*, Phil. Trans. Roy. Soc. A **366**, 1025–1054 (2008).

[63] F. Gesztesy, M. Malamud, M. Mitrea, and S. Naboko, *Generalized polar decompositions for closed operators in Hilbert spaces and some applications*, Integral Eq. Operator Th. **64**, 83–113 (2009).

[64] F. Gesztesy, M. Mitrea. *Nonlocal Robin Laplacians and some remarks on a paper by Filonov on eigenvalue inequalities.* J. Diff. Eq. **247**, 2871–2896 (2009).

[65] F. Gesztesy and R. Nichols, *Weak convergence of spectral shift functions for one-dimensional Schrödinger operators*, Math. Nachr. **285**, 1799–1838 (2012).

[66] F. Gesztesy, W. Schweiger, and B. Simon, *Commutation methods applied to the* $mKdV$*-equation*, Trans. Amer. Math. Soc. **324**, 465–525 (1991).

[67] F. Gesztesy and R. Svirsky, $(m)KdV$*-Solitons on the background of quasi-periodic finite-gap solutions*, Memoirs Amer. Math. Soc. **118** (563), 1–88 (1995).

[68] I.M. Glazman, *Direct Methods of Qualitative Spectral Analysis of Singular Differential Operators*, Israel Program for Scientific Translations, Jerusalem, 1965.

[69] T. Godoy, J.-P. Gossez, S. Paczka, *On the asymptotic behavior of the principal eigenvalues of some elliptic problems*, Ann. Mat. Pura Appl. **189**, 497–521 (2010).

[70] G. Grubb, *Distributions and Operators*, Graduate Texts in Mathematics, Vol. 252, Springer, New York, 2009.

[71] V. Hardt, A. Konstantinov, and R. Mennicken, *On the spectrum of the product of closed operators*, Math. Nachr. **215**, 91–102 (2000).

[72] P. Hartman *Differential equations with non-oscillatory eigenfunctions*, Duke Math. J. **15**, 697–709 (1948).

[73] R. Hempel, *A left-definite generalized eigenvalue problem for Schrödinger operators*, Habilitation, Dept. of Mathematics, University of Munich, Germany, 1987.

[74] P. Hess, *On the relative completeness of the generalized eigenvectors of elliptic eigenvalue problems with indefinite weight functions*, Math. Ann. **270**, 467–475 (1985).

[75] P. Hess, *On the asymptotic distribution of eigenvalues of some nonselfadjoint problems*, Bull. London Math. Soc. **18**, 181–184 (1986).

[76] P. Hess and T. Kato, *On some linear and nonlinear eigenvalue problems with an indefinite weight function*, Commun. Partial Diff. Eq. **5**, 999–1030 (1980).

[77] D. B. Hinton and S.C. Melescue, *Relative boundedness-compactness inequalities for a second order differential operator*, Math. Ineq. & Appls. **4**, 35–52 (2001).

[78] R.O. Hryniv and Ya.V. Mykytyuk, 1D *Schrödinger operators with periodic singular potentials*, Methods Funct. Anal. Topology **7**, no. 4, 31–42 (2001).

[79] R.O. Hryniv and Ya.V. Mykytyuk, 1D *Schrödinger operators with singular Gordon potentials*, Methods Funct. Anal. Topology **8**, no. 1, 36–48 (2002).

[80] R.O. Hryniv and Ya.V. Mykytyuk, *Self-adjointness of Schrödinger operators with singular potentials*, Meth. Funct. Anal. Topology **18**, 152–159 (2012).

[81] B.J. Jaye, V.G. Maz'ya, and I.E. Verbitsky, *Existence and regularity of positive solutions of elliptic equations of Schrödinger type*, J. Analyse Math. **118**, 577–621 (2012).

[82] T. Kappeler and C. Möhr, *Estimates for periodic and Dirichlet eigenvalues of the Schrödinger operator with singular potentials*, J. Funct. Anal. **186**, 62–91 (2001).

[83] T. Kappeler, P. Perry, M. Shubin, and P. Topalov, *The Miura map on the line*, Int. Math. Res. Notices, 2005, No. 50.

[84] T. Kappeler and P. Topalov, *Global fold structure of the Miura map on* $L^2(\mathbb{T})$, Int. Math. Res. Notices **2004**, No. 39, 2039–2068.

[85] I.M. Karabash, *A functional model, eigenvalues, and finite singular critical points for indefinite Sturm–Liouville operators*, in *Topics in Operator Theory. Volume 2. Systems and Mathematical Physics*, J.A. Ball, V. Bolotnikov, J.W. Helton, L. Rodman, I.M. Spitkovsky (eds.), Operator Theory: Advances and Applications, Vol. 203, Birkhäuser, Basel, 2010, pp. 247–287.

[86] I.M. Karabash and M.M. Malamud, *Indefinite Sturm–Liouville operators* $(\operatorname{sgn} x)\left(-\frac{d^2}{dx^2}+q(x)\right)$ *with finite-zone potentials*, Operators and Matrices **1**, 301–368 (2007).

[87] I.M. Karabash, A.S. Kostenko, and M.M. Malamud, *The similarity problem for J-nonnegative Sturm–Liouville operators*, J. Diff. Eq. **246**, 964–997 (2009).

[88] I. Karabash and C. Trunk, *Spectral properties of singular Sturm–Liouville operators with indefinite weight* $\operatorname{sgn} x$, Proc. Roy. Soc. Edinburgh **139A**, 483–503 (2009).

[89] M. Kato, *Estimates for the eigenvalues of Hill's operator with distributional coefficients*, Tokyo J. Math. **33**, 361–364 (2010).

[90] T. Kato, *Perturbation Theory for Linear Operators*, corr. printing of the 2nd ed., Springer, Berlin, 1980.

[91] Q. Kong, H. Wu, and A. Zettl, *Left-definite Sturm–Liouville problems*, J. Diff. Eq. **177**, 1–26 (2001).

[92] Q. Kong, H. Wu, and A. Zettl, *Singular left-definite Sturm–Liouville problems*, J. Diff. Eq. **206**, 1–29 (2004).

[93] Q. Kong, H. Wu, A. Zettl, and M. Möller, *Indefinite Sturm–Liouville problems*, Proc. Roy. Soc. Edinburgh **133A**, 639–652 (2003).

[94] E. Korotyaev, *Inverse problem for periodic "weighted" operators*, J. Funct. Anal. **170**, 188–218 (2000).

[95] E. Korotyaev, *Characterization of the spectrum of Schrödinger operators with periodic distributions*, Int. Math. Res. Notices **2003**, No. 37, 2019–2031.

[96] E. Korotyaev, *Inverse spectral problem for the periodic Camassa–Holm equation*, J. Nonlinear Math. Phys. **11**, 499–507 (2004).

[97] E. Korotyaev, *Sharp asymptotics of the quasimomentum*, Asymptot. Anal. **80**, 269–287 (2012).

[98] A. Kostenko, *The similarity problem for indefinite Sturm–Liouville operators with periodic coefficients*, Operators and Matrices **5**, 707–722 (2011).

[99] J.L. Lions, *Espaces d'interpolation et domaines de puissances fractionnaires d'opérateurs*, J. Math. Soc. Japan **14**, 233–241 (1962).

[100] B.V. Loginov and O.V. Makeeva, *The pseudoperturbation method in generalized eigenvalue problems*, Dokl. Math. **77**, 194–197 (2008).

[101] M. Marletta and A. Zettl, *Counting and computing eigenvalues of left-definite Sturm–Liouville problems*, J. Comp. Appl. Math. **148**, 65–75 (2002).

[102] M. Marletta and A. Zettl, *Floquet theory for left-definite Sturm–Liouville problems*, J. Math. Anal. Appl. **305**, 477–482 (2005).

[103] V.G. Maz'ya and T.O. Shaposhnikova, *Theory of Sobolev Multipliers*, Springer, Berlin, 2009.

[104] V.G. Maz'ya and I.E. Verbitsky, *Boundedness and compactness criteria for the one-dimensional Schrödinger operator*, in *Function Spaces, Interpolation Theory and Related Topics*, M. Cwikel, M. Engliš, A. Kufner, L.-E. Persson, G. Sparr (eds.), de Gruyter, Berlin, 2002, pp. 369–382.

[105] V.G. Maz'ya and I.E. Verbitsky, *The Schrödinger operator on the energy space: boundedness and compactness criteria*, Acta Math. **188**, 263–302 (2002).

[106] V.G. Maz'ya and I.E. Verbitsky, *Infinitesimal form boundedness and Trudinger's subordination for the Schrödinger operator*, Invent. Math. **162**, 81–136 (2005).

[107] V.G. Maz'ya and I.E. Verbitsky, *Form boundedness of the general second-order differential operator*, Commun. Pure Appl. Math. **59**, 1286–1329 (2006).

[108] H.P. McKean, *Addition for the acoustic equation*, Commun. Pure Appl. Math. **54**, 1271–1288 (2001).

[109] H.P. McKean, *Fredholm determinants and the Camassa–Holm hierarchy*, Commun. Pure Appl. Math. **56**, 638–680 (2003).

[110] H.P. McKean, *The Liouville correspondence between the Korteweg–de Vries and the Camassa–Holm hierarchies*, Commun. Pure Appl. Math. **56**, 998–1015 (2003).

[111] H. P. McKean, *Breakdown of the Camassa–Holm equation*, Commun. Pure Appl. Math. **57**, 416–418 (2004).

[112] W. McLean, *Strongly Elliptic Systems and Boundary Integral Equations*, Cambridge University Press, Cambridge, 2000.

[113] V.A. Mikhailets and V.M. Molyboga, *One-dimensional Schrödinger operators with singular periodic potentials*, Meth. Funct. Anal. Topology **14**, no. 2, 184–200 (2008).

[114] V.A. Mikhailets and V.M. Molyboga, *Spectral gaps of the one-dimensional Schrödinger operators with singular periodic potentials*, Meth. Funct. Anal. Topology **15**, no. 1, 31–40 (2009).

[115] E. Müller-Pfeifer, *Spectral Theory of Ordinary Differential Operators*, Ellis Horwood, Chichester, 1981.

[116] M.A. Naimark, *Linear Differential Operators, Part II*, Ungar, New York, 1968.

[117] M.I. Neiman-zade and A.A. Shkalikov, *Strongly elliptic operators with singular coefficients*, Russ. J. Math. Phys. **13**, 70–78 (2006).

[118] W.V. Petryshyn, *On the eigenvalue problem $Tu - \lambda Su = 0$ with unbounded and nonsymmetric operators T and S*, Phil. Trans. Roy. Soc. London A **262**, 413–458 (1968).

[119] F. Philipp, *Indefinite Sturm–Liouville operators with periodic coefficients*, Operators and Matrices **7**, 777–811 (2013).

[120] M. Reed and B. Simon, *Methods of Modern Mathematical Physics. III: Scattering Theory*, Academic Press, New York, 1979.

[121] M. Reed and B. Simon, *Methods of Modern Mathematical Physics. IV: Analysis of Operators*, Academic Press, New York, 1978.

[122] F. Rellich, *Halbbeschränkte gewöhnliche Differentialoperatoren zweiter Ordnung*, Math. Ann. **122**, 343–368 (1951).

[123] A.M Savchuk and A.A. Shkalikov, *Sturm–Liouville operators with singular potentials*, Math. Notes **66**, 741–753 (1999).

[124] A.M. Savchuk and A.A. Shkalikov, *Sturm–Liouville operators with distribution potentials*, Trans. Moscow Math. Soc. **2003**, 143–192.

[125] M. Schechter, *Operator Methods in Quantum Mechanics*, North–Holland, New York, 1981.

[126] M. Schechter, *Principles of Functional Analysis*, 2nd ed., Graduate Studies in Mathematics, Vol. 36, Amer. Math. Soc., Providence, RI, 2002.

[127] B. Simon, *Quantum Mechanics for Hamiltonians Defined as Quadratic Forms*, Princeton University Press, Princeton, NJ, 1971.

[128] B. Simon, *Trace Ideals and Their Applications*, 2nd ed., Mathematical Surveys and Monographs, Vol. 120, Amer. Math. Soc., Providence, RI, 2005.

[129] G. Teschl, *Mathematical Methods in Quantum Mechanics; With Applications to Schrödinger Operators*, Graduate Studies in Mathematics, Amer. Math. Soc., Vol. 99, RI, 2009.

[130] G. Teschl, private communication.

[131] B. Thaller, *Normal forms of an abstract Dirac operator and applications to scattering theory*, J. Math. Phys. **29**, 249–257 (1988).

[132] B. Thaller, *The Dirac Equation*, Springer, Berlin, 1992.

[133] C. Tretter, *Linear operator pencils $A - \lambda B$ with discrete spectrum*, Integral Eq. Operator Th. **37**, 357–373 (2000).

[134] H. Volkmer, *Sturm–Liouville problems with indefinite weights and Everitt's inequality*, Proc. Roy. Soc. Edinburgh **126A**, 1097–1112 (1996).

[135] J. Weidmann, *Linear Operators in Hilbert Spaces*, Graduate Texts in Mathematics, Vol. 68, Springer, New York, 1980.

[136] J. Weidmann, *Spectral Theory of Ordinary Differential Operators*, Lecture Notes in Math., Vol. 1258, Springer, Berlin, 1987.

[137] A. Zettl, *Sturm–Liouville Theory*, Mathematical Surveys and Monographs, Vol. 121, Amer. Math. Soc., Providence, RI, 2005.

[138] D.R. Yafaev, *On the spectrum of the perturbed polyharmonic operator*, in *Topics in Mathematical Physics, Vol. 5. Spectral Theory*, M.Sh. Birman (ed.), Consultants Bureau, New York, 1972, pp. 107–112.

Fritz Gesztesy
Department of Mathematics
University of Missouri
Columbia, MO 65211, USA
e-mail: gesztesyf@missouri.edu
URL: http://www.math.missouri.edu/personnel/faculty/gesztesyf.html

Rudi Weikard
Department of Mathematics
University of Alabama at Birmingham
Birmingham, AL 35294, USA
e-mail: rudi@math.uab.edu
URL: http://www.math.uab.edu/~rudi/

Operator Theory:
Advances and Applications, Vol. 240, 189–194

Remarks on Spaces of Compact Operators between Reflexive Banach Spaces

G. Godefroy

Abstract. We observe that if X and Y are two reflexive separable spaces such that the canonical map $J : X\hat{\otimes}Y \to X\check{\otimes}Y$ is injective, then every compact operator from X to Y^* is in the norm closure of finite rank operators, and every bounded operator $T \in L(X, Y^*)$ is uniform limit on compact sets of a sequence (R_n) of finite rank operators such that $\|R_n\| \leq \|T\|$. This would apply in particular to the case $X = Y$, i.e., to a reflexive Pisier space if such a space exists. We show that if $Z \subset L(X)$ is a subspace which strictly contains the space $K(X)$ of compact operators on a reflexive Banach space X, then $K(X)$ is not 1-complemented in Z, and it is locally 1-complemented in Z exactly when Z is contained in the closure of $K(X)$ with respect to the uniform convergence on compact subsets of X. Several consequences are spelled out.

Mathematics Subject Classification (2010). 46A32, 46B20.

Keywords. Spaces of compact operators, approximation properties, smooth norms.

1. Introduction

The purpose of this note is to apply some general results from geometry of Banach spaces (mainly, from duality and isometric theory) to the space of compact operators between reflexive Banach spaces on the real field. In Section 2, we consider pairs of reflexive Banach spaces such that the canonical map from the projective tensor product to the injective tensor product is one-to-one. We show that in this case, the corresponding spaces of operators behave "as if" X or Y has the approximation property, although it is not always so. Quite unexpectedly, James' techniques on norm-attaining linear functionals can be applied in this context (see [9] for an early use of such arguments). The results of Section 3 rely on the use of the Fréchet smoothness of the norm of $L(X)$ at certain operators of rank 1, in the spirit of [8] and [18]. The main result of this section is Lemma 3.2 which connects the metric position of an operator $T \in L(X)$ with respect to the space $K(X)$ and the uniform approximation of T on compact sets by compact operators.

2. Injectivity of the canonical map between tensor products

We start with the following lemma, whose roots go back to [15]. We refer to [4] for various applications of Petunin-Plichko's result.

Lemma 2.1. *Let X and Y be separable reflexive Banach spaces. Then $(X\hat{\otimes}Y)^* = L(X, Y^*)$ and the restriction map to $K(X, Y^*) \subset L(X, Y^*)$ defines a quotient map Q from $X\hat{\otimes}Y$ onto $K(X, Y^*)^*$.*

Proof. It is classical that $(X\hat{\otimes}Y)^* = L(X, Y^*)$ and this isometric identification can actually be used as definition of the projective tensor product. Since X is reflexive, any operator $T \in K(X, Y^*)$ attains its norm, and it easily follows that $K(X, Y^*)$ consists of norm-attaining elements of $(X\hat{\otimes}Y)^*$. The spaces $K(X, Y^*)$ and $X\hat{\otimes}Y$ are separable. The Lemma is therefore a special case of ([2], Lemma 2.5), which simply follows from the fact that the restriction to $K(X, Y^*)$ of the unit sphere of $X\hat{\otimes}Y$ is a separable James boundary (in the sense of [5]) of the unit ball of $K(X, Y^*)^*$. □

It is interesting to notice that this quotient map, which we obtain here with very general arguments relying ultimately on James' theorem [10] and Simons' inequality [17], can also be derived with an algebraic approach, adapted to the particular Banach spaces under consideration: indeed it is this same quotient map Q which appears in the commutative diagram of [8] (see the proof of Proposition 1.1 there), relying on the previous work [3].

We now prove the following:

Theorem 2.2. *If X and Y are separable reflexive spaces such that the canonical map $J : X\hat{\otimes}Y \to X\check{\otimes}Y$ is injective, then every compact operator from X to Y^* is in the norm closure of finite rank operators, and every bounded operator $T \in L(X, Y^*)$ is uniform limit on compact sets of a sequence (R_n) of finite rank operators such that $\|R_n\| \leq \|T\|$.*

Proof. Since the map J is one-to-one, the simple tensors $x^* \otimes y^*$ separate the space $X\hat{\otimes}Y$ and thus their linear span $X^* \otimes Y^* \subset K(X, Y^*)$ separate $X\hat{\otimes}Y$. It follows now from Lemma 2.1 that Q is a canonical isometry from $X\hat{\otimes}Y$ onto $K(X, Y^*)^*$, and that the space of compact operators from X to Y^* is the norm-closure of the space of finite rank operators, in other words that $K(X, Y^*) = X^*\check{\otimes}Y^*$. Now we can dualize the equation $K(X, Y^*)^* = X\hat{\otimes}Y$ and get $K(X, Y^*)^{**} = (X\hat{\otimes}Y)^* = L(X, Y^*)$.

If X and Y are reflexive, the space $K(X, Y^*)^{**}$ consists of all operators $T \in L(X, Y^*)$ which are uniform limits on compact sets of compact operators ([8], Cor. 1.2) and the control of the norm follows from ([8], Theorem 1.5). Finally we can replace compact operators by finite rank ones since $K(X, Y^*) = X^*\check{\otimes}Y^*$. □

The simplest case where Theorem 2.2 applies is when X or Y has the approximation property (see [12], page 3). In this case, Theorem 2.2 is well known and it goes back to Grothendieck's fundamental works. When $Y = X^*$, the map J is one-to-one (if and) only if X has the approximation property (see [12], Theorem 0.3) and this is a special case of Theorem 2.2. However, the assumptions of Theorem 2.2 are also satisfied when X and Y have type 2 (see [13], Theorem 6.6). Hence, if for instance X is a subspace of L_p $(2 < p < \infty)$ failing A.P., then the spaces $K(X, X^*)$ and $L(X, X^*)$ behave "as if" X or X^* has the A.P. although it is not so, but factorization does the job in this case.

Along these lines, we recall that it is not known whether there exists a reflexive Pisier space, that is, a reflexive space X such that $X\hat{\otimes}X = X\check{\otimes}X$, both algebraically and topologically, in other words such that J is one-to-one and onto. Such a space X, if it exists, fails the approximation property ([14]) but satisfies of course the conclusions of Theorem 2.2. Actually, dualizing the equation $X\hat{\otimes}X = X\check{\otimes}X$ shows that every bounded operator from X to X^* is nuclear.

3. Isometric properties of the space of compact operators

Our next observation is an extension of ([8], Remark 5.8) and shows that the reflexivity of $K(X)$ is irrelevant to such isometric considerations (see Question 5.9 in [8]). We state it in the case of reflexive spaces, but the argument can be extended to any Asplund space X with the Radon-Nikodym property.

Proposition 3.1. *Let X be a reflexive Banach space, and let Z be a subspace of $L(X)$ which strictly contains $K(X)$. Then $K(X)$ is not 1-complemented in Z.*

Proof. It suffices to show that if $T \in L(X)$ is a non-compact operator, then the space $K(X)$ is not 1-complemented in the space $E_T = K(X) \oplus \mathbf{R}T$.

Let x (resp. x^*) be strongly exposed in the unit ball of X (resp. X^*) by $x_0^* \in B_{X^*}$ (resp. by $x_0 \in B_X$). Then $x \otimes x^*$ is strongly exposed in the unit ball of $X\hat{\otimes}X^*$ by $x_0^* \otimes x_0$ ([16] or [8], Lemma 5.1). It follows that $x_0^* \otimes x_0 \in L(X)$ is a point of Fréchet smoothness of the norm of $L(X)$, with tangent linear form $x \otimes x^*$.

Assume that $\pi : E_T \to K(X)$ is a linear projection with norm 1, and let $T_0 = \pi(T)$. For any $S \in K(X)$ and any scalar λ, one has $\|S + \lambda(T - T_0)\| \geq \|S\|$. When $S = x_0^* \otimes x_0$, this inequality and smoothness of the norm shows the equation $\langle x \otimes x^*, T - T_0\rangle = 0$ and thus $\langle x^*, T(x) - T_0(x)\rangle = 0$. Since this last equation holds for all strongly exposed points x and x^* and the unit ball of any reflexive space is the norm closed convex hull of its strongly exposed points, it follows that $T = T_0$, but this contradicts $T \notin K(X)$. □

In view of the above statement, it is natural to ask when $K(X)$ is *locally* 1-complemented in Z. We recall that Y is locally 1-complemented in Z if for every finite-dimensional subspace F of Z and any $\epsilon > 0$, there is a linear operator $L : F \to Y$ with $\|L\| \leq 1 + \epsilon$ and $L(y) = y$ for all $y \in Y \cap F$, and this holds if and only if $Y^\perp$ is the kernel of a contractive projection on Z^* ([11]). In the notation of [7], this means that Y is an *ideal* in Z, in the Banach space sense since this terminology does not request the presence of an algebraic structure. The following lemma, where we use the notation of Proposition 3.1, answers our question. Its proof is relevant to the ball topology which is defined and studied in [6].

Lemma 3.2. *Let X be a reflexive Banach space, and let Z be a subspace of $L(X)$ which contains $K(X)$. Then the space $K(X)$ is locally 1-complemented in Z if and only if Z is contained in the closure of $K(X)$ with respect to the uniform convergence on compact subsets of X.*

Proof. For $T \in L(X)$, we denote as before $E_T = K(X) \oplus \mathbf{R}T$.

By Corollary 1.2 in [8], the space $K(X)^{**}$ is canonically isometric to the closure of $K(X)$ in the space $L(X)$ equipped with the topology of uniform convergence on compact subsets of X. If Z is a subspace of $K(X)^{**}$, it follows from the local reflexivity principle that $K(X)$ is locally 1-complemented in Z. We note in passing that the local reflexivity principle provides operators L which are close to being isometries, and also that if $K(X)$ is locally complemented in E_T and E_S, it is locally complemented in $\overline{\text{span}}[E_T \cup E_S]$.

Conversely, let us assume that $K(X)$ is locally 1-complemented in E_T. For any finite-dimensional subspace $F \subset E_T$ with $T \in F$ and any $\epsilon > 0$, there is a linear operator $L_{F,\epsilon} : F \to K(X)$ with $\|L_{F,\epsilon}\| \leq 1+\epsilon$ and $L_{F,\epsilon} = Id$ on $F \cap K(X)$. We set $L_{F,\epsilon}(T) = S_{F,\epsilon}$. The natural order $(F,\epsilon) \leq (G,\delta)$ when $F \subset G$ and $\delta \leq \epsilon$ defines a filter $\mathcal{F}$ on the set of pairs (F,ϵ).

We use the notation of the proof of Proposition 3.1. The operator $x_0^* \otimes x_0 \in K(X)$ is a point of Fréchet smoothness of the norm of $L(X)$, with tangent linear form $x \otimes x^*$. For any operator $S \in E_T$, we have therefore that

$$\langle x^*, S(x)\rangle = \lim_{t\to 0} t^{-1}[\|x_0^* \otimes x_0 + tS\| - 1].$$

Fréchet smoothness means that this limit is uniform on bounded subsets of E_T. Pick now $S \in K(X)$ and $t > 0$. It follows from the properties of $L_{F,\epsilon}$ that

$$\limsup_{\mathcal{F}} t^{-1}[\|x_0^* \otimes x_0 + tS_{F,\epsilon}\| - 1] \leq t^{-1}[\|x_0^* \otimes x_0 + tT\| - 1].$$

Since the set $(S_{F,\epsilon})$ is uniformly bounded, we can take the limit when $t \to 0^+$ and we get

$$\limsup_{\mathcal{F}} \langle x^*, S_{F,\epsilon}(x)\rangle \leq \langle x^*, T(x)\rangle$$

but since we can reproduce the argument with $-(x^* \otimes x)$, it follows that

$$\lim_{\mathcal{F}} \langle x^*, S_{F,\epsilon}(x)\rangle = \langle x^*, T(x)\rangle.$$

This limit holds if x^* and x are strongly exposed in their respective unit balls. Since the unit ball of any reflexive Banach space is the closed convex hull of its strongly exposed points, it follows that $T = \lim_{\mathcal{F}} S_{F,\epsilon}$ in the weak operator topology. Since the weak operator topology and the topology of uniform convergence on compact subsets of X have the same closed convex bounded sets, the result follows. □

Note that by the proof of Lemma 3.2, the metric properties of the net which is obtained from T by local complementation (e.g. by the local reflexivity principle) actually imply weak* convergence to T. This behaviour reflects the smoothness of the Banach space $K(X)$.

In what follows, we identify $K(X)^{**}$ with the closure of $K(X)$ in $L(X)$ equipped with the topology of uniform convergence on compact subsets of X (see [8], Cor. 1.2).

Lemma 3.3. *Let X be a reflexive Banach space. The space $K(X)^{**}$ is a two-sided ideal of the algebra $L(X)$, which coincide with $L(X)$ if and only if X has the compact approximation property.*

Proof. The product of $L(X)$ is separately continuous for the compact convergence, and it follows that $K(X)^{**}$ is a two-sided ideal in $L(X)$. This ideal coincide with $L(X)$ if and only if it contains I_X, and this exactly means that X has the compact approximation property. □

Let us gather in a single statement what we know about this ideal.

Theorem 3.4. *Let X be a reflexive Banach space. The following statements are equivalent:*

(1) *X has the compact approximation property.*
(2) $K(X)^{**} = L(X)$.
(3) *$K(X)$ is locally 1-complemented in $L(X)$.*
(4) *$K(X)$ is locally 1-complemented in $K(X) \oplus \mathbf{R}\, I_X$.*
(5) *$K(X)^{**}$ contains an invertible operator.*

Proof. It follows from Lemma 3.2 that (4) implies (1), and from Lemma 3.3 that (5) implies (2). The other implications are clear. □

We can summarize our observations as follows: the collection of spaces Z such that $K(X) \subset Z \subset L(X)$ and $K(X)$ is locally 1-complemented in Z (in other words, such that $K(X)$ is an ideal in Z in the Banach space sense) admits a maximal space, namely $K(X)^{**}$, and this maximal space is a two-sided ideal in the algebra $L(X)$. Note that by Lemma 3.3, if $K(X)$ is a *maximal* closed two-sided ideal in the algebra $L(X)$, exactly one of the following properties holds: X has the compact approximation property, or $K(X)$ is reflexive. And by Lemma 3.2, there is no subspace $Z \subset L(X)$ strictly containing $K(X)$ in which $K(X)$ is an ideal in the Banach space sense (i.e., in which $K(X)$ is locally 1-complemented) if and only if $K(X)$ is reflexive.

It is not known whether there exists an infinite-dimensional Banach space such that $K(X)$ is reflexive. Note that a reflexive space X with the Argyros–Haydon property $L(X) = K(X) \oplus \mathbf{R}\, I_X$ and failing the compact approximation property would be such that $K(X)$ is reflexive. It is not known at present whether a reflexive space can have the Argyros–Haydon property, however reflexive spaces which enjoy the invariant subspace property hereditarily have been constructed [1].

References

[1] S. Argyros and P. Motakis, *A reflexive space with the hereditary invariant subspace property*, to appear.

[2] P. Bandyopadhyay and G. Godefroy, *Linear structures in the set of norm-attaining functionals on a Banach space*, J. of Convex Anal. 13 (2006), 489–497.

[3] M. Feder and P.D. Saphar, *Spaces of compact operators and their dual spaces*, Israel J. Math. 21 (1975), 38–49.

[4] G. Godefroy, *The use of norm attainment*, Bulletin of the Belgian Math. Society, to appear.

[5] G. Godefroy, *Boundaries of a convex set and interpolation sets*, Math. Annalen 277 (1987), 173–184.

[6] G. Godefroy and N.J. Kalton, *The ball topology and its applications*, Contemp. Math. 85, Amer. Math. Soc. (1989), 195–237.

[7] G. Godefroy, N.J. Kalton and P.D. Saphar, *Unconditional ideals in Banach spaces*, Studia Math. 104, 1 (1993), 13–59.

[8] G. Godefroy and P.D. Saphar, *Duality in spaces of operators and smooth norms on Banach spaces*, Illinois J. of Math. 32,4 (1988), 672–695.

[9] J.R. Holub, *Reflexivity of* $L(E, F)$, Proceedings of the Amer. Math. Soc. 39 (1973), 175–177.

[10] R.C. James, *Weakly compact sets*, Trans. Amer. Math. Soc. 113 (1964), 129–140.

[11] N.J. Kalton, *Locally complemented spaces and* $\mathcal{L}^p$*-spaces for* $0 < p < 1$, Math. Nachr. 115 (1984), 71–97.

[12] G. Pisier, *Factorization of linear operators and geometry of Banach spaces*, CBMS regional conference series in mathematics 60 (1986).

[13] G. Pisier, *The operator Hilbert space* OH*, complex interpolation and tensor norms*, Memoir of the AMS. 122 (1996), no. 585.

[14] G. Pisier, *Un théorème sur les opérateurs entre espaces de Banach qui se factorisent par un espace de Hilbert*, Annales Scient. Ecole Norm. Sup. 13 (1980), 23–43.

[15] Y. I, Petunin and A.N. Plichko, *Some properties of the set of functionals that attain a supremum on the unit sphere*, Ukrain. Mat. Z. 26 (1974), 102–106.

[16] W. Ruess and C. Stegall, *Exposed and denting points in duals of operator spaces*, Israel J. Math. 53, 2 (1986), 163–190.

[17] S. Simons, *A convergence theorem with boundary*, Pacific J. Math. 40 (1972), 703–708.

[18] W. Werner, *The type of a factor with separable predual is determined by its geometry*, in Interaction between functional analysis, harmonic analysis and probability, edited by N.J. Kalton, S. Montgomery-Smith and E. Saab, M. Dekker Lecture notes 175 (1996).

G. Godefroy
CNRS-Université Paris 6
Institut de Mathématiques de Jussieu-Paris Rive Gauche
Case 247. 4, Place Jussieu
F-75252 Paris Cedex 05, France
e-mail: `godefroy@math.jussieu.fr`

Operator Theory:
Advances and Applications, Vol. 240, 195–220

Harmonic Analysis and Stochastic Partial Differential Equations: The Stochastic Functional Calculus

Brian Jefferies

Abstract. It has been recognised recently that there is a close connection between existence and regularity results for stochastic partial differential equations and functional calculus techniques in harmonic analysis. The connection is made more explicit in this paper with the notion of a *stochastic functional calculus.*

In the deterministic setting, suppose that A_1, A_2 are bounded linear operators acting on a Banach space E. A pair (μ_1, μ_2) of continuous probability measures on $[0,1]$ determines a functional calculus $f \longmapsto f_{\mu_1,\mu_2}(A_1, A_2)$ for analytic functions f by weighting all possible orderings of operator products of A_1 and A_2 via the probability measures μ_1 and μ_2. For example, $f \longmapsto f_{\mu,\mu}(A_1, A_2)$ is the Weyl functional calculus with equally weighted operator products.

Replacing μ_1 by Lebesgue measure λ on $[0,t]$ and μ_2 by stochastic integration with respect to a Wiener process W, we show that there exists a functional calculus $f \longmapsto f_{\lambda,W;t}(A+B)$ for bounded holomorphic functions f if A is a densely defined Hilbert space operator with a bounded holomorphic functional calculus and B is small compared to A relative to a square function norm. By this means, the solution of the stochastic evolution equation $dX_t = AX_t dt + BX_t dW_t$, $X_0 = x$, is represented as $t \longmapsto e^{A+B}_{\lambda,W;t}x$, $t \geq 0$. We show how to extend some of our results to L^p-spaces, $2 \leq p < \infty$ and apply them to the regularity of solutions of the Zakai equation.

Mathematics Subject Classification (2010). Primary 47A60; Secondary 47D06, 60H15.

Keywords. functional calculus, stochastic evolution equation, H^∞-functional calculus, square function, Feynman's operational calculus.

The author would like to thank Z. Brzeźniak for invaluable discussions.

1. Introduction

In a recent series of papers by J. van Neerven, M. Veraar and L. Weis, harmonic analysis techniques have been used to establish stochastic maximal regularity in L^p-spaces [22] for $p \geq 2$. Combined with fixed point arguments, the existence, uniqueness and regularity results for solutions to general nonlinear stochastic PDEs are obtained from stochastic maximal regularity estimates.

It is already apparent from the papers of F. Flandoli [6] and Z. Brzeźniak [1] that *square function estimates* facilitate the use of fixed point arguments for the solution of stochastic PDEs in Hilbert space. On the other hand, in harmonic analysis, the work of A. McIntosh and A. Yagi [19] showed that square function estimates in Hilbert space determine the existence of an H^∞-functional calculus for sectorial operators, which can be used to solve irregular boundary value problems in Hilbert space.

In the semigroup approach to linear evolution problems, the solution of the equation

$$\frac{du(t)}{dt} = Au(t) + Bu(t), \quad u(0) = u_0,$$

for a function $u : [0, \infty) \to X$ with values in a Banach space X is sought in the form

$$u(t) = e^{t(A+B)}u_0, \quad t \geq 0,$$

where the linear operator B is distinguished as a lower-order perturbation of A. In the case that X is Hilbert space and $A + B$ is selfadjoint with a spectral measure P_{A+B}, then

$$e^{t(A+B)} = \int_{\mathbb{R}} e^{t\lambda}\, dP_{A+B}(\lambda)$$

and the operator $A+B$ has a rich functional calculus associated with the spectral measure P_{A+B}. Similarly, in the stochastic setting, we find that the solution $t \longmapsto e^{A+B}_{dt,dW_t;t}x$ of the stochastic evolution equation

$$dX_t = AX_t\, dt + BX_t\, dW_t, \quad X_0 = x, \tag{1.1}$$

is defined and there is an H^∞-functional calculus

$$f \longmapsto f_{\lambda,W;t}(A + B) \tag{1.2}$$

for $A + B$. The relevant properties are that A should have an H^∞-functional calculus and B should be small compared to A relative to a "square function norm".

The notation $f_{\lambda,W;t}(A + B)$ is inspired by *Feynman's operational calculus* in which operator ordering in a functional calculus is indexed by continuous measures [11]. For example, if μ_1 and μ_2 are continuous Borel probability measures on $[0, 1]$, A_1, A_2 are bounded linear operators and $P^{1,1}(x_1, x_2) = x_1x_2$ for $x_1, x_2 \in \mathbb{R}$, then

$$P^{1,1}_{\mu_1,\mu_2}(A_1, A_2) = (\mu_1 \otimes \mu_2)(\{t_2 < t_1\})A_1A_2 + (\mu_1 \otimes \mu_2)(\{t_1 < t_2\})A_2A_1. \tag{1.3}$$

Feynman's idea was to attach time indices to each operator so that in operator expressions, operators with smaller time indices act *before* those with larger time indices, as in formula (1.3) above. If $\mu_1 = \mu_2$ is the Lebesgue measure λ on the interval $[0,t]$, $t > 0$, then

$$e^{A_1+A_2}_{\lambda,\lambda;t} = e^{t(A_1+A_2)}.$$

The expression $e^{A_1+A_2}_{\lambda,\lambda;t}$ is shorthand for $f_{\lambda,\lambda;t}(A_1, A_2)$, where f is the exponential function $f(x_1, x_2) = e^{x_1+x_2}$, $x_1, x_2 \in \mathbb{R}$, in two real variables. More generally, if $\mu = \mu_1 = \mu_2$ is *any* continuous Radon measure on $\mathbb{R}$, then

$$e^{A_1+A_2}_{\mu,\mu;t} = e^{\mu([0,t])(A_1+A_2)}$$

for every $t > 0$, because the linear operators A_1, A_2 are *equally weighted* by the measure μ in the time-ordering of operator products [12, Proposition 5.5]. The relation of the mapping (1.2) with Feynman's operational calculus is explained more fully in [10]. In the case that A and B are bounded linear operators acting on an arbitrary Banach space E, there exists an operator-valued *random evolution* $t \longmapsto e^{A+B}_{\lambda,W;t}$, $t \geq 0$, such that $X_t = e^{A+B}_{\lambda,W;t}x$, $t \geq 0$, is a strong solution of (1.1) [10, Corollary 4.8]. For a Hilbert space $E = \mathcal{H}$, the Itô isometry gives the bound [10, Theorem 5.1]

$$\|e^{A+B}_{\lambda,W;t}x\|_{L^2(\mathbb{P},H)} \leq \|x\|.\|e^{tA}\| \sum_{n=0}^{\infty} \frac{(t^{\frac{1}{2}}\|B\|)^n}{\sqrt{n!}}, \quad t \geq 0. \tag{1.4}$$

In the setting of equation (1.2) the operators A and B are generally closed and unbounded operators on the Banach space in which we expect solutions to lie, so expressions like formula (1.3) are problematic. Nevertheless, in the notation $f_{dt,dW_t;t}(A+B)$, we take the time-ordering measure associated with A to be Lebesgue measure dt and the time-ordering with respect to B is given by stochastic integration with respect to Brownian motion W_s, $s \geq 0$ on the interval $[0,t]$ for $t > 0$ – more (or least!) colourfully, we have time-ordering of B with respect to *white noise* dW_t.

A systematic study of the existence, uniqueness and regularity of solutions of parabolic stochastic evolution equations in UMD Banach spaces that includes equation (1.1) as a special case is given in [2, 21]. The emphasis here is on making sense of expressions like $f_{dt,dW_t;t}(A+B)$ for bounded holomorphic functions f and studying the joint functional calculus properties of A and B in the stochastic setting related to Feynman's operational calculus.

The *Zakai equation* arising in filtering theory is a typical evolution equation that possesses a stochastic functional calculus, where in equation (1.1), we have

$$Au(x) = \sum_{i,j=1}^{d} a_{ij}(x)\frac{\partial^2 u}{\partial x_i \partial x_j} + \sum_{i=1}^{d} q_i(x)\frac{\partial u}{\partial x_i} + r(x)u(x), \quad x \in \mathbb{R}^d,$$

$$Bu(x) = \sum_{i=1}^{d} b_i(x)\frac{\partial u}{\partial x_i} + c(x)u(x), \quad x \in \mathbb{R}^d.$$

The existence, uniqueness and regularity of solutions of the Zakai equation in UMD Banach spaces is treated in [2].

The present paper builds on [10]. Multiple stochastic integration of Hilbert space-valued functions is reviewed in Section 2. Stochastic equations in Banach spaces are discussed in Section 3 with emphasis on the van Neerven–Veraar–Weis analysis. The construction of the exponential stochastic exponential $e^{A+B}_{\lambda,W;t}$ requires norm estimates for the multiple stochastic integrals of Banach space-valued functions and we discuss such estimates for the case of M-type 2 Banach spaces in Section 4. The case of L^p-space for $p \geq 2$ is covered. The main result, Theorem 5.4 gives the construction of the stochastic functional calculus in M-type 2 Banach spaces. Applications to the Zakai equation are given in Hilbert space in Theorem 5.7.

2. Multiple stochastic integrals

The perturbation series expansion for $e^{A+B}_{dt,dW_t;t}$ is written in terms of multiple stochastic integrals with respect to the Brownian motion process, which we now define.

Let W denote Brownian motion in $\mathbb{R}$ with respect to the probability measure space $(\Omega, \mathcal{S}, \mathbb{P})$ such that $W_0 = 0$ almost surely. In the case that Ω is taken to be the set of all continuous functions $\omega : [0, \infty) \to \mathbb{R}$, the σ-algebra $\mathcal{S}$ is the Borel σ-algebra of Ω for the compact-open topology and $W_t(\omega) = \omega(t)$ for every $\omega \in \Omega$ and $t \geq 0$. There exists a unique Borel probability measure $\mathbb{P}$ on Ω – the *Wiener measure*, such that for every $0 < t_1 < \cdots < t_k$, Borel subsets $B_1, \ldots, B_k$ of $\mathbb{R}$ and $k = 1, 2, \ldots$, the measure of the elementary event

$$E = \{\omega \in \Omega : \omega(t_1) \in B_1, \ldots, \omega(t_k) \in B_k\}$$

is given by

$$\mathbb{P}(E) = \int_{B_k} \cdots \int_{B_1} p_{t_k - t_{k-1}}(x_k - x_{k-1}) \cdots p_{t_2 - t_1}(x_2 - x_1) p_{t_1}(x_1)\, dx_1 \ldots dx_k,$$

where $p_t(x) = (2\pi t)^{-\frac{1}{2}} e^{-x^2/(2t)}$, $t > 0$, $x \in \mathbb{R}$, is the associated transition function. Then Wiener measure $\mathbb{P}$ has the property that W_t, $t \geq 0$, is a process with stationary and independent increments such that W_t is a Gaussian random variable with mean zero and variance t for $t > 0$, properties which define a Brownian motion W_t, $t \geq 0$, with $W_0 = 0$ $\mathbb{P}$-a.e. over a general probability measure space $(\Omega, \mathcal{S}, \mathbb{P})$.

For a Banach space E and $1 \leq p < \infty$, the space of E-valued pth-Bochner integrable functions with respect to $\mathbb{P}$ is denoted by $L^p(\mathbb{P}, E) = L^p(\Omega, \mathcal{S}, \mathbb{P}, E)$. The linear space $L^0(\mathbb{P}, E) = L^0(\Omega, \mathcal{S}, \mathbb{P}, E)$ of equivalence classes of strongly measurable E-valued functions has the (metrisable) topology of convergence in probability.

For the purpose of expanding solutions of linear stochastic equations like (1.1) as a "stochastic Dyson series", we need to consider multiple Wiener–Itô integrals of deterministic functions. We follow the account in [17, Section 10.3] with suitable modifications for vector-valued functions. Wiener–Itô chaos in Banach spaces is treated in [18, Section 4].

Let H be a Hilbert space with inner product $\langle\cdot,\cdot\rangle_H$. Let $T>0$ and $k=1,2,\dots$. The case $k=1$ corresponds to the Wiener integral. Let $D_1=(0,T]$ and

$$D_k=\{(t_1,\dots,t_k)\in(0,T]^k : \exists i,j=1,\dots,k, i\neq j, \text{ such that } t_i=t_j\},$$

$k=2,3,\dots$. Let $A_1,\dots,A_n$ be a partition of $(0,T]$ into disjoint intervals of the form $(s,t]$ for $0\le s<t\le T$ and suppose that

$$f=\sum_{1\le j_1,\dots,j_k\le n}\alpha_{j_1,\dots,j_k}\chi_{A_{j_1}\times\dots\times A_{j_k}} \tag{2.1}$$

is an H-valued function such that $\alpha_{j_1,\dots,j_k}=0$ whenever two indices $j_1,\dots,j_k$ are equal and f vanishes on D_k. Then

$$I_k(f)=\int_{[0,T]^k} f(t_1,\dots,t_k)\,dW_{t_1}\dots dW_{t_k}$$

is defined by

$$I_k(f)=\sum_{1\le j_1,\dots,j_k\le n}\alpha_{j_1,\dots,j_k}W(A_{j_1})\cdots W(A_{j_k}).$$

Here $W((s,t])$ denotes the random variable W_t-W_s for $0\le s<t\le T$. Let $\mathcal{D}((0,T]^k,H)$ denote the linear space of H-valued step functions f of the above form. Then I_k is well defined and $I_k:\mathcal{D}((0,T]^k,H)\to L^0(\Omega,\mathcal{S},\mathbb{P},H)$ is a linear map. Moreover, the maps I_k, $k=1,2,\dots$, enjoy the following properties.

1) The integral $I_k(f)$ is invariant under the symmetrisation of the function f, that is, if $\tilde f\in\mathcal{D}((0,T]^k,H)$ is the symmetrisation

$$\tilde f(t_1,\dots,t_k)=\frac{1}{k!}\sum_{\sigma\in S_k}f(t_{\sigma(1)},\dots,t_{\sigma(k)}),\quad t_1,\dots,t_k\in(0,T]$$

of $f\in\mathcal{D}((0,T]^k,H)$ over the set S_k of all permutations of $(1,\dots,k)$, then $I_k(f)=I_k(\tilde f)$.
2) If k and k' are positive integers such that $k\neq k'$ and $f\in\mathcal{D}((0,T]^k,H)$, $g\in\mathcal{D}((0,T]^{k'},H)$, then $\mathbb{E}(\langle I_k(f),I_{k'}(g)\rangle_H)=0$.
3) If $f\in\mathcal{D}((0,T]^k,H)$ and $g\in\mathcal{D}((0,T]^k,H)$, then

$$\mathbb{E}(\langle I_k(f),I_k(g)\rangle_H)=k!\langle\tilde f,\tilde g\rangle_{L^2((0,T]^k,H)}.$$

The inner product on the right-hand side is taken in the Hilbert space $L^2((0,T]^k,H)$.

By property 3), we have a version of the Itô isometry

$$\mathbb{E}(\|I_k(f)\|_H^2)=\mathbb{E}(\|I_k(\tilde f)\|_H^2)=k!\|\tilde f\|^2_{L^2((0,T]^k,H)}\le k!\|f\|^2_{L^2((0,T]^k,H)}, \tag{2.2}$$

so that the mapping I_k can be extended to a bounded linear operator

$$I_k:L^2((0,T]^k,H)\to L^2(\Omega,\mathcal{S},\mathbb{P},H).$$

We also write $I_k(f)$ as $\int_{[0,T]^k} f(s)\,W^k(ds)$. In the case that $0 \le s < t \le T$ and $f \in L^2((0,T]^k, H)$ is zero off the simplex

$$\Delta_k(s,t) = \{(s_1,\dots,s_k) \in [s,t]^k : s < s_1 < \cdots < s_k < t\},$$

then

$$I_k(f) = \int_s^t \int_s^{t_k} \cdots \int_s^{t_2} f(t_1,\dots,t_k)\,dW_{t_1}\cdots dW_{t_k}, \tag{2.3}$$

where the right-hand side is interpreted as an iterated stochastic integral [17, pp. 299–300]. The equality is easily seen to be valid for all $f \in \mathcal{D}((0,T]^k, H)$ vanishing off $\Delta_k(s,t)$ and the linear subspace of all such functions is dense in the closed subspace of $L^2((0,T]^k, H)$ consisting of all H-valued functions belonging to $L^2((0,T]^k, H)$ which are zero almost everywhere outside $\Delta_k(s,t) \subset (0,T]^k$. The Itô isometry (2.2) for the integral (2.3) takes the form

$$\mathbb{E}(\|I_k(f)\|_H^2) = \int_s^t \int_s^{t_k} \cdots \int_s^{t_2} \|f(t_1,\dots,t_k)\|_H^2\,dt_1\cdots dt_k. \tag{2.4}$$

To check that the identity (2.4) is valid, we write

$$f_\sigma(t_1,\dots,t_k) = f(t_{\sigma(1)},\dots,t_{\sigma(k)}), \quad \text{for } t_1,\dots,t_k \in (0,T] \text{ and } \sigma \in S_k.$$

Then for $\sigma, \sigma' \in S_k$, $\sigma \ne \sigma'$, the functions f_σ and $f_{\sigma'}$ are supported by disjoint open simplexes in $(0,T]^k$, so $\langle f_\sigma, f_{\sigma'}\rangle_{L^2((0,T]^k,H)} = 0$ and we have

$$\begin{aligned}
\mathbb{E}(\|I_k(f)\|_H^2) &= \mathbb{E}(\|I_k(\tilde f)\|_H^2) \quad \text{[by property 1)]},\\
&= k!\|\tilde f\|^2_{L^2((0,T]^k,H)}, \quad \text{[by property 3)]}\\
&= \frac{1}{k!}\Big\|\sum_{\sigma\in S_k} f_\sigma\Big\|^2_{L^2((0,T]^k,H)}\\
&= \|f\|^2_{L^2((0,T]^k,H)},\\
&\qquad \text{[because } \|f_\sigma\|_{L^2((0,T]^k,H)} = \|f\|_{L^2((0,T]^k,H)} \text{ for } \sigma \in S_k].
\end{aligned}$$

The calculation for a constant function supported by $\Delta_k(0,t)$ is instructive. Using Itô's formula to compute $\int_{\Delta_k(t)} W^k(ds_1,\dots,ds_k)$ for $k = 1,2,\dots$, we have

$$\int_0^t W(ds_1) = W_t, \qquad [k=1]$$

$$\begin{aligned}
\int_0^t\int_0^{s_2} W(ds_1)W(ds_2) &= \int_0^t W_{s_2}\,W(ds_2)\\
&= \frac{1}{2}W_t^2 - \frac{1}{2}t, \qquad [k=2]\\
\int_0^t\int_0^{s_3}\int_0^{s_2} W(ds_1)W(ds_2)W(ds_3) &= \int_0^t \left(\frac{1}{2}W_{s_3}^2 - \frac{1}{2}s_3\right) W(ds_3)\\
&= \frac{1}{3!}W_t^3 - \frac{1}{2}tW_t, \qquad [k=3]
\end{aligned}$$

$$\vdots$$

$$\int_{\Delta_k(t)} W^k(ds_1,\dots,ds_k) = \frac{1}{k!}h_k(W_t/\sqrt{t})t^{k/2},$$

where $h_n(x) = (-1)^n e^{x^2/2}d^n/dx^n e^{-x^2/2}$, $x \in \mathbb{R}$, is Hermite polynomial of degree $n = 0, 1, 2, \dots$, see [17, Theorem 10.3.2].

Note that by symmetry, the equality

$$\int_{\Delta_k(\sigma;t)} W^k(ds_1,\dots,ds_k) = \int_{\Delta_k(t)} W^k(ds_1,\dots,ds_k)$$

holds for each $\sigma \in S_k$ with $\chi_{\Delta_k(\sigma;t)} := \left(\chi_{\Delta_k(t)}\right)_\sigma$, so by equation (2.2) the equality

$$\left\| \int_{\Delta_k(t)} W^k(ds_1,\dots,ds_k) \right\|^2_{L^2(\mathbb{P})} = \frac{t^k}{k!}.$$

holds for each $k = 1, 2 \dots$. This may also be obtained by applying the Itô isometry consecutively to the representation (2.3) or observing that

$$\|h_k(W_t/\sqrt{t})\|^2_{L^2(\mathbb{P})} = \frac{1}{\sqrt{2\pi}}\int_{\mathbb{R}} h_k(x)^2 e^{-x^2/2}\,dx = k!.$$

Let $T > 0$. Every element F of $L^2(\mathbb{P}, H)$ has a unique expansion (Wiener polynomial chaos) as the sum of $\mathbb{E}(F)$ and multiple stochastic integrals $I_k(f_k)$ of symmetric functions $f_k : (0,T]^k \to H$, $k = 1, 2, \dots$ [17, Theorem 10.3.3].

3. Stochastic equations in Banach spaces

A comprehensive treatment of stochastic integration of Banach space-valued deterministic functions appears in [20]. Muliple Wiener–Itô integrals for Banach space-valued functions are treated in [18, Section 3]. A full treatment requires a discussion of γ-radonifying operators and their tensor products. In some situations it is possible to get by with simpler arguments which we now describe.

3.1. Stochastic integration of vector-valued functions

We first mention some terminology related to stochastic integration. Let $\mathbb{R}_+ = [0,\infty)$.

Let $(\Omega, \mathcal{F}, \mathbb{P})$ be a probability measure space. A *filtration* is a family $\{\mathcal{F}_t : t \in \mathbb{R}_+\}$ of sub σ-algebras of $\mathcal{F}$ such that $\mathcal{F}_s \subseteq \mathcal{F}_t$, $\forall s < t$. A filtration $\{\mathcal{F}_t : t \in \mathbb{R}_+\}$ is called a *standard filtration* if

(1) $\mathcal{F}_t = \mathcal{F}_{t+} := \cap_{s>t}\mathcal{F}_s \quad \forall t$ (*right continuity*)
(2) $\mathcal{F}_0$ contains all the $\mathbb{P}$-null sets   (*completeness*)

Given an increasing family $\{\mathcal{F}_t : t \in \mathbb{R}_+\}$ of σ-algebras, a process $X : \mathbb{R} \times \Omega \to \mathbb{C}$ is *adapted* to $\mathcal{F}_t$ or *progressively measurable* if X_t is $\mathcal{F}_t$ measurable for all $t \in \mathbb{R}_+$.

Let W_t, $t \geq 0$, be a Brownian motion process on the probability space $(\Omega, \mathcal{F}, \mathbb{P})$.

Definition 3.1. Let E be a Banach space. An E-valued random process Φ_t, $t \geq 0$, is said to be *stochastically integrable* in E, if for each $\xi \in E'$, the scalar-valued process $\langle \Phi_t, \xi \rangle$, $t \geq 0$ is stochastically integrable with respect to W_t, $t \geq 0$, and there exists an E-valued random process Ψ_t, $t \geq 0$, such that

$$\langle \Psi_t, \xi \rangle = \int_0^t \langle \Phi_s, \xi \rangle \, dW_s \quad \text{a.e.} \tag{3.1}$$

for every $\xi \in E'$ and $t \geq 0$. We sometimes write

$$\Psi = \Phi.W$$

in accordance with the notion that the integral $\Phi.W$ of a vector-valued process Φ with espect to W ought to be another vector-valued process. If M is an semimartingale, then $\Phi.M$ should be a weak semimartingale, that is, $\langle \Phi, \xi \rangle.M$ is a semimartingale for each $\xi \in E'$.

Remark 3.2. It can happen that a Pettis integrable vector-valued random variable does not possess a conditional expectation with respect to a sub-σ-algebra [8], so there is a distinction between weak and strong semimartingales for vector-valued processes, even in infinite-dimensional Hilbert space, see [9] for a discussion of conditional expectation of Pettis integrable vector-valued random variables.

Let $T > 0$ and $k = 1, 2, \ldots$. An E-valued function $s \longmapsto \Phi_s$, $s \in [0,T]^k$, is said to be *k-stochastically integrable* or *W^k-integrable* in E if for each $\xi \in E'$, the scalar-valued function $t \longmapsto \langle \Phi_s, \xi \rangle$, $s \in [0,T]^k$ belongs to $L^2([0,T]^k)$, and there exists an E-valued random process Ψ_t, $t \in [0,T]$, such that

$$\langle \Psi_t, \xi \rangle = \int_{[0,t]^k} \langle \Phi_s, \xi \rangle \, W^k(ds_1, \ldots, ds_k) \quad \text{a.e.} \tag{3.2}$$

for every $\xi \in E'$ and $t \geq 0$. We shall mainly be concerned with E-valued functions of the form $\Phi_s = \chi_{\Delta_k(T)}(s) f(s)$ for $s \in [0,T]^k$.

If a deterministic function $\phi : (0,T) \to E$ is *stochastically integrable* in E and it is weakly L^2, it follows that for every Borel subset A of $(0,T)$, there exists an E-valued Gaussian random variable X_A such that

$$\langle X_A, \xi \rangle = \int_0^T \chi_A(t) \langle \phi(t), \xi \rangle \, dW_t$$

for every $\xi \in E'$ [20]: it suffices that an E-valued random variable $X_{(0,T)}$ exists.

3.2. The van Neerven–Veraar–Weis approach to the stochastic integration of vector-valued functions

For a given Banach space E, we want to find conditions to integrate an E-valued process Φ with respect to a semimartingale or just a Brownian motion process W. Bilinear stochastic equations in Banach spaces with respect to bounded linear operators A and B are treated in [10, Corollary 4.8] by a projective tensor product. For Hilbert spaces, the Itô isometry is used, see [10, Theorem 5.1].

In order to solve and find estimates for the solution of a stochastic PDE, we need to treat *unbounded linear operators* on a Banach space, in which case the argument of [10, Theorem 5.1] fails. In [21], van Neerven, Weis and Veraar show how this is done in UMD Banach spaces, such as L^p spaces for $1 < p < \infty$. A variety of bilinear stochastic PDE are solved in [2].

Let E be a real Banach space and let H be a separable Hilbert space. A continuous real linear map $T : H \to E$ is said to be *γ-radonifying* if the standard Gaussian cylindrical measure γ is mapped by T into the restriction to cylinder sets of a regular Borel probabilty measure (Radon probability) on E. If H has an orthonormal basis $\langle e_n \rangle_{n=1}^{\infty}$ and $\varphi_J(h) = \langle (h, e_n) \rangle_{n \in J}$ for any finite subset J of positive integers, then $\gamma \circ \varphi_J^{-1}$ is the standard Gaussian probability measure on $\varphi_J(H)$.

Many continuous linear maps are γ-radonifying. As mentioned in [10, Remark 4.4], nuclear maps are γ-radonifying, as are absolutely summing maps [28]. If E is a Hilbert space, then T is γ-radonifying iff T is a Hilbert–Schmidt operator iff T is absolutely summing. The mapping T is γ-radonifying iff $h \longmapsto \|Th\|_E$, $h \in H$, is a *measurable seminorm* on H in the sense of L. Gross [15]. The standard example is the map

$$T : h \longmapsto \int h\, dt, \quad h \in L^2([0,T])$$

with values in $C_0([0,T])$. Here the absolutely continuous function $\int h\, dt \in C_0([0,T])$ is the indefinite integral of h vanishing at $t = 0$ and $\gamma \circ T^{-1}$ is Wiener measure on all continuous sample paths $\omega : [0,T] \to \mathbb{R}$ satisfying $\omega(0) = 0$. As is well known, $\gamma \circ T^{-1}$ is concentrated on all Hölder continuous paths of order $0 < \alpha < \frac{1}{2}$.

Now suppose that E is an arbitrary Banach space. We say that the E-valued function $\Phi : [0,T] \to E$ is *stochastically integrable* (in the sense of [21]) if the linear map

$$T : h \longmapsto \int_0^T \Phi(t)h(t)\, dt, \quad h \in L^2([0,T]),$$

is γ-radonifying in E. The vector-valued integral here is a Pettis integral. Then $\Phi.W$ is the $L^2(\mathbb{P}, E)$-valued process defined by

$$\langle (\Phi.W)(t), \xi \rangle = \int_0^t \langle \Phi(s), \xi \rangle\, dW_s, \quad \xi \in E^*,\ t \geq 0.$$

If T is γ-radonifying, and $\langle f_n \rangle_{n=1}^{\infty}$ is an orthonormal basis of $L^2([0,T])$ then

$$\int_A \langle \Phi(s), \xi \rangle\, dW_s = \int_0^T \sum_{n=1}^{\infty} \left(\int_A \langle \Phi(s), \xi \rangle \overline{f_n(s)}\, ds \right) f_n(t)\, dW_t$$

converges in $L^2(\mathbb{P})$ for each $\xi \in E^*$ and Borel set $A \subseteq [0,T]$. The sum

$$\sum_{n=1}^{\infty} \left(\int_A \langle \Phi(s), \xi \rangle . \overline{f}_n(s)\, ds \right) \int_0^T f_n(t)\, dW_t$$

is an E-valued Gaussian series, see [20, Theorem 2.3], where the consistency with Definition 3.1 is also shown. For each $t > 0$, let $\mathcal{F}_t$ be the σ-algebra generated by the random variables $\{W_s : 0 \le s \le t\}$.

A function $\phi : \mathbb{R}_+ \to L^2(\mathbb{P}) \otimes E$ is said to be an *elementary progressively measurable function* if there exist times $0 < t_1 < \cdots < t_N$, vectors $x_{mn} \in E$ and sets $A_{mn} \in \mathcal{F}_{t_{n-1}}$, $n = 1, \ldots, N$, $m = 1, \ldots, M$ such that

$$\phi(t) = \sum_{n=1}^{N} \sum_{m=1}^{M} x_{mn} \chi_{A_{mn}} . \chi_{(t_{n-1}, t_n]}(t), \quad t \in \mathbb{R}_+.$$

Then ϕ has values in every space $L^p(\mathbb{P}) \otimes E$ for $1 \le p \le \infty$, ϕ is W-integrable in $L^p(\mathbb{P}) \otimes E \otimes L^p(\mathbb{P})$ for every $1 \le p < \infty$ and we have

$$\int_{\mathbb{R}_+} \phi \otimes dW = \sum_{n=1}^{N} \sum_{m=1}^{M} (x_{mn} \chi_{A_{mn}}) \otimes (W_{t_n} - W_{t_{n-1}}). \tag{3.3}$$

Let $\mathcal{E}$ denote the linear subspace of $L^\infty(\mathbb{P}) \otimes E \otimes L^p(\mathbb{P})$ consisting of all vectors $\int_{\mathbb{R}_+} \phi \otimes dW$ with $\phi : \mathbb{R}_+ \to L^\infty(\mathbb{P}) \otimes E$ an elementary progressively measurable function. For each $1 \le p < \infty$, let $J : L^\infty(\mathbb{P}) \otimes E \otimes L^p(\mathbb{P}) \to L^p(\mathbb{P}, E)$ be the linear map defined by $J(g \otimes x \otimes f)(\omega) = xg(\omega).f(\omega)$ for almost all $\omega \in \Omega$. The map J multiplies the L^p function f by the bounded function g and leaves $x \in E$ unchanged, that is, J is a bilinear multiplication operator.

Definition 3.3. A Banach space E is called a *UMD space* (or, E has the *unconditional martingale difference property*) if for any $1 < p < \infty$, there exists $C_p > 0$ such that for any E-valued martingale difference $\{\xi_j\}_{j=1}^n$ and $n = 1, 2, \ldots$, the inequality

$$\mathbb{E} \left\| \sum_{j=1}^{n} \epsilon_j \xi_j \right\|_E^p \le C_p \mathbb{E} \left\| \sum_{j=1}^{n} \xi_j \right\|_E^p$$

holds for every $\epsilon_j \in \{\pm 1\}$, $j = 1, \ldots, n$. By a martingale difference sequence $\{\xi_j\}_{j=1}^n$, we mean that the sum

$$\xi_1 + \sum_{j=1}^{k} \xi_j, \quad k = 1, \ldots, n,$$

is an E-valued martingale.

The following result is from [7, Theorems 2 and 2'].

Theorem 3.4. *Let E be a UMD space and $1 < p < \infty$. The multiplication map J is continuous from $\mathcal{E}$ into $L^p(\mathbb{P}, E)$ for the relative topology of $L^p(\mathbb{P} \otimes \mathbb{P}, E)$ on $\mathcal{E}$.*

By this means we can prove that elements of a wide class of E-valued processes are W-integrable in the sense of Definition 3.1, provided that E has the UMD property, see [21]. In many examples, such as [10, Theorem 4.7], the UMD property is not needed.

4. The stochastic Dyson series in M-type 2 Banach spaces

Although the theory of stochastic integration is well developed in UMD Banach spaces, we shall employ a one-sided Itô inequality valid in M-type 2 Banach spaces such as L^p-spaces with $p \geq 2$, in which stochastic maximal regularity is valid [22].

Let $1 \leq p \leq 2$. A Banach space E is said to be of *type* p if there exists $C > 0$ such that

$$\mathbb{E}\left\|\sum_{j=1}^{n} \epsilon_j x_j\right\|_E^p \leq C^p \sum_{j=1}^{n} \|x_j\|_E^p$$

for any symmetric identical independently distributed random variables $\epsilon_1, \ldots, \epsilon_n$ with values ± 1, vectors $x_1, \ldots, x_n \in E$ and $n = 1, 2, \ldots$. The smallest C with this property is denoted by $K_p(E)$.

In a type 2 Banach space E, the inequality

$$\mathbb{E}\left\|\int_0^T f(s)\, dW_s\right\|_E^2 \leq T_2(E)^2 \int_0^T \|f(t)\|_E^2\, dt$$

holds for all E-valued Borel simple functions $f : [0, T] \to E$ [25, Proposition 5.2]. Moreover, if all uniformly bounded strongly measurable functions $f : [0, T] \to E$ are stochastically integrable, then E necessarily has type 2 [25, Proposition 6.1].

For multiple stochastic integrals, we require a stronger property. Let $1 \leq p \leq 2$. A Banach space E is said to be of *M-type* p if there exists $C > 0$ such that for any E-valued martingale $\{M_j\}_{j=1}^n$, $n = 1, 2, \ldots$, the inequality

$$\sup_j \mathbb{E}\,\|M_j\|^p \leq C^p \sum_{j=1}^{n} \mathbb{E}\,\|M_j - M_{j-1}\|^p$$

holds with $M_{-1} = 0$. The smallest C with this property is denoted by $L_p(E)$.

According to [24, pp. 221–222], if a Banach space E is of M-type p, then E is of type p and reflexive, but there is a Banach space of type 2 which is of M-type p for *no* $p > 1$. If a UMD Banach space is of type p, then it is of M-type p.

For an M-type 2 Banach space E, the Itô isometry (2.2) becomes the one-sided inequality

$$\mathbb{E}(\|I_k(f)\|_E^2) \leq k! T_2(E)^{2k} \|f\|_{L^2((0,T]^k,E)}^2, \tag{4.1}$$

with $T_2(E) = L_2(E)T_2(E)$, see [1, Corollary 3.4].

We are interested mainly in L^p-spaces with $2 \leq p < \infty$, which are both M-type 2 and UMD Banach spaces. Because we are only integrating deterministic E-valued functions, we could get by with only assuming that E is a Banach space of type 2 by appealing to a deep multilinear decoupling inequality of S. Kwapien [16] analogous to Theorem 3.4, which is valid for UMD Banach spaces, see [26, Proposition 1] and [17, Theorem 6.4.1].

4.1. Sectorial operators

Let $0 < \omega < \pi/2$. The sectors $S_{\omega\pm}$ are defined by

$$S_{\omega-} = \{-z : z \in \mathbb{C},\ |\arg z| \le \omega\} \cup \{0\}, \quad S_{\omega+} = \{z : z \in \mathbb{C},\ |\arg z| \le \omega\} \cup \{0\}.$$

Suppose that $A : \mathcal{D}(A) \longrightarrow E$ is a closed densely defined linear operator acting in the Banach space E. The spectrum of A is denoted by $\sigma(A)$. If $0 \le \omega < \pi/2$, then A is said to be of *type* $\omega-$, if $\sigma(A) \subset S_{\omega-}$ and for each $\nu > \omega$, there exists $C_\nu > 0$ such that

$$\|(zI - A)^{-1}\| \le C_\nu |z|^{-1}, \quad z \notin S_{\nu-}. \tag{4.2}$$

An operator A is of *type* $\omega-$ if and only if it is the generator of an analytic semigroup e^{zA} in the region $|\arg z| < \pi/2 - \omega$ so that for each $\nu > \omega$, there exists $C_\nu > 0$ such that $\|e^{zA}\| \le C_\nu$ for all $z \in \mathbb{C}$ with $|\arg z| < \pi/2 - \nu$ [23, §2.5]. An operator A is of type $\omega+$ if and only if $-A$ is the generator of an analytic semigroup in the region $|\arg z| < \pi/2 - \omega$.

Let $T > 0$. Let E be a Banach space, A an operator of type $\omega-$, $\omega < \pi/2$ and let V be a separable Banach space with norm $\|\cdot\|_V$ such that $\mathcal{D}(A) \subset V \subset E$ with continuous inclusions and $B : V \to E$ is bounded. Suppose that there exists $c_1 > 0$ such that

$$\int_0^T \|e^{tA}x\|_V^2\, dt \le c_1^2 \|x\|_E^2 \tag{4.3}$$

for all $x \in \mathcal{D}(A)$.

Lemma 4.1. *Let $c_1 > 0$. The inequality* (4.3) *holds if and only if*

$$\int_0^T \int_0^t \|e^{(t-s)A} g(s)\|_V^2\, dsdt \le c_1^2 \int_0^T \|g(t)\|_E^2\, dt \tag{4.4}$$

for all E-valued simple functions g.

Proof. The inequality (4.3) holds for all $x \in E$ because there exists $c > 0$ such that $\|Ae^{tA}x\| \le c\|x\|/t$ for all $t > 0$. Moreover, if the bound (4.3) holds, then

$$\begin{aligned}\int_0^T \int_0^t \|e^{(t-s)A} g(s)\|_V^2\, dsdt &= \int_0^T \int_s^T \|e^{(t-s)A} g(s)\|_V^2\, dtds \\ &= \int_0^T \int_0^{T-s} \|e^{tA} g(s)\|_V^2\, dtds \\ &\le \int_0^T \int_0^T \|e^{tA} g(s)\|_V^2\, dtds \\ &\le c_1^2 \int_0^T \|g(s)\|_E^2\, ds, \qquad \text{by (4.3).}\end{aligned}$$

Now suppose that (4.4) holds. By taking $g = \chi_R.x$, $x \in E$, we obtain

$$\frac{\int_R \int_0^{T-s} \|e^{tA}x\|_V^2\, dtds}{|R|} \le c_1^2 \|x\|_E^2$$

for all finite unions R of intervals. Because $s \mapsto \int_0^{T-s} \|e^{tA}x\|_V^2\, dt$ is continuous, this is only possible if equation (4.3) holds. □

Theorem 4.2. *Let E be a Banach space of M-type 2. Suppose that the estimate (4.3) holds for all $x \in E$ and $\|Bx\|_E \le c_2\|x\|_V$ for all $x \in V$. If $c_1c_2T_2(E) < 1$, then the stochastic Dyson series*

$$e^{tA}u_0 + \sum_{k=1}^{\infty} \int_0^t \int_0^{s_k} \cdots \int_0^{s_2} \left[e^{(t-s_k)A}Be^{(s_k-s_{k-1})A}\cdots Be^{s_1A}u_0\right] dW_{s_1}\dots dW_{s_k} \tag{4.5}$$

converges absolutely in $L^2(\mathbb{P};E)$ for every $0 < t \le T$ and every $u_0 \in E$.

Proof. Suppose that the estimate (4.3) holds for all $x \in E$ and $\|Bx\|_E \le c_2\|x\|_V$ for all $x \in V$. The estimate (4.4) in Lemma 4.1 is also valid for all square integrable E-valued functions g by continuity. Then by the Itô bound (4.1), we have

$$\begin{aligned}
&\mathbb{E}\left\|\int_0^t \int_0^{s_k} \cdots \int_0^{s_2} \left[e^{(t-s_k)A}Be^{(s_k-s_{k-1})A}\cdots Be^{s_1A}u_0\right] dW_{s_1}\dots dW_{s_k}\right\|_E^2 \\
&\quad\le T_2(E)^{2k} \int_0^t \int_0^{s_k} \cdots \int_0^{s_2} \left\|e^{(t-s_k)A}Be^{(s_k-s_{k-1})A}\cdots Be^{s_1A}u_0\right\|_E^2 ds_1\dots ds_k \\
&\quad\le C^2T_2(E)^{2k} \int_0^t \int_0^{s_k} \cdots \int_0^{s_2} \left\|Be^{(s_k-s_{k-1})A}\cdots Be^{s_1A}u_0\right\|_E^2 ds_1\dots ds_k \\
&\quad\le C^2T_2(E)^{2k}c_2^2 \\
&\qquad \int_0^t \int_0^{s_k} \cdots \int_0^{s_2} \left\|e^{(s_k-s_{k-1})A}Be^{(s_{k-1}-s_{k-2})A}\cdots Be^{s_1A}u_0\right\|_V^2 ds_1\dots ds_k \\
&\quad\le C^2T_2(E)^{2k} \\
&\qquad (c_1c_2)^2\int_0^t \int_0^{s_{k-1}} \cdots \int_0^{s_2} \left\|Be^{(s_{k-1}-s_{k-2})A}\cdots Be^{s_1A}u_0\right\|_E^2 ds_1\dots ds_{k-1} \\
&\quad\vdots \\
&\quad\le C^2T_2(E)^{2k}(c_1c_2)^{2(k-1)} \int_0^t \left\|Be^{s_1A}u_0\right\|_E^2 ds_1 \\
&\quad\le C^2(c_1c_2T_2(E))^{2k}\|u_0\|_E^2.
\end{aligned}$$

Here we have used the bound $\|e^{sA}\| \le C$ for all $s \ge 0$. If $c_1c_2T_2(E) < 1$, then the sum (4.5) converges in $L^2(\mathbb{P};E)$ for every $0 < t \le T$ and every $u_0 \in E$. □

Suppose that the conditions of Theorem 4.2 hold. For each $u_0 \in E$ and $0 < t \le T$, the E-valued random variable defined by the series (4.5) is denoted by $e^{A+B}_{\lambda,W;t}u_0$. We define $e^{A+B}_{\lambda,W;0}u_0 = u_0$. The mapping $u_0 \longmapsto e^{A+B}_{\lambda,W;t}u_0$ is an element of the space $\mathcal{L}(E, L^2(\mathbb{P},E))$ of *random linear operators* [29] which we denote by

$e^{A+B}_{\lambda,W;t}$. It is easy to see that

$$t \longmapsto e^{A+B}_{\lambda,W;t}, \quad 0 \le t \le T,$$

is a continuous map from the closed interval $[0,T]$ into $\mathcal{L}(E, L^2(\mathbb{P}, E))$.

The following corollary follows from the observation that the stochastic Dyson series (4.5) is the solution obtained from the contraction mapping principle for the stochastic equation (4.6) below, see [6, Lemma 2.2]. By a *mild solution*, we mean an E-valued solution X_t, $t \ge 0$, of the stochastic equation

$$X_t = e^{tA}x + \int_0^t e^{(t-s)A} B X_s \, dW_s.$$

A general treatment of stochastic equations in Hilbert space is given in [4].

The following consequence of Theorem 4.2 may be compared with [1, Theorem 4.6], which is written in terms of the interpolation space

$$V = D_A(1/2,2) = \left\{ x \in E : \|x\|^2_{D_A(1/2,2)} = \int_0^\infty \|Ae^{tA}x\|^2_E \, dt < \infty \right\}.$$

When E is a Hilbert space and $-A$ is a positive selfadjoint operator, then

$$D_A(1/2,2) = \mathcal{D}((-A)^{\frac{1}{2}}).$$

Our point of departure is to obtain an explicit representation of the fixed point solution of the linear stochastic equation (4.6) as a "stochastic Dyson series".

Corollary 4.3. *Let E be a Banach space of* M*-type* 2*. Suppose that the conditions of Theorem* 4.2 *hold. Then for each $x \in E$, the E-valued process*

$$t \longmapsto e^{A+B}_{\lambda,W;t}x, \quad 0 \le t \le T,$$

is the unique mild solution of the stochastic equation

$$dX_t = AX_t \, dt + BX_t \, dW_t, \quad X_0 = x. \tag{4.6}$$

We can check that $t \longmapsto e^{A+B}_{\lambda,W;t}x$, $t \ge 0$, is a strong solution of the stochastic equation (4.6) for $x \in V$ if $e^{tA}V \subseteq V$ for $t \ge 0$, see [1, Proposition 4.5]. The possibility of different choices of the space V are studied in [6, §3.1] in the Hilbert space case.

For the definition of fractional powers of operators used in the next result, see [23], [14, Appendix]. The differential operator B is usually half the order of the elliptic operator A.

Corollary 4.4. *Let E be a Banach space of M-type* 2*. Suppose that there exists $c_1 > 0$ such that*

$$\int_0^\infty \|e^{tA}x\|^2_V \, dt \le c_1^2 \|x\|^2_E \tag{4.7}$$

for all $x \in E$ and $\|Bx\|_E \le c_2\|x\|_V$ for all $x \in V$. If $c_1c_2T_2(E) < 1$, then there exists $M > 0$ such that $\|e^{A+B}_{\lambda,W;t}x\|_{L^2(\mathbb{P},E)} \le M\|x\|_E$ for all $t \ge 0$.

Furthermore, suppose that A is a one-to-one operator of type $\omega-$ and the norm $\|\cdot\|_V$ is defined by $\|x\|_V = \|(-A)^{\frac{1}{2}}x\|_E$. Then for every $t > 0$, there exists

$L_t > 0$ such that $\|(-A)^{\frac{1}{2}} e^{A+B}_{\lambda,W;t}x\|_{L^2(\mathbb{P},E)} \le L_t\|x\|_E$ for all $x \in E$ and $t \longmapsto e^{A+B}_{\lambda,W;t}x$ is a predictable continuous process with values in $L^2(\mathbb{P}, \mathcal{D}((-A)^{\frac{1}{2}}))$ for $t > 0$.

Proof. Under condition (4.7), the bound giving the convergence of (4.5) is uniform in $T > 0$, from which the uniform bound for $t \longmapsto e^{A+B}_{\lambda,W;t}x$, $t > 0$, is obtained.

For the last statement, it suffices to apply Lemma 4.1 to note that

$$\mathbb{E}\left\|\int_0^t\int_0^{s_k}\cdots\int_0^{s_2}\left[(-A)^{\frac{1}{2}}e^{(t-s_k)A}Be^{(s_k-s_{k-1})A}\cdots Be^{s_1A}u_0\right]dW_{s_1}\ldots dW_{s_k}\right\|_E^2$$

$$\le T_2(E)^{2k}$$

$$\int_0^t\int_0^{s_k}\cdots\int_0^{s_2}\left\|(-A)^{\frac{1}{2}}e^{(t-s_k)A}Be^{(s_k-s_{k-1})A}\cdots Be^{s_1A}u_0\right\|_E^2 ds_1\ldots ds_k$$

$$= T_2(E)^{2k}\int_0^t\int_0^{s_k}\cdots\int_0^{s_2}\left\|e^{(t-s_k)A}Be^{(s_k-s_{k-1})A}\cdots Be^{s_1A}u_0\right\|_V^2 ds_1\ldots ds_k$$

$$\le T_2(E)^{2k}c_1^2\int_0^t\int_0^{s_k}\cdots\int_0^{s_2}\left\|Be^{(s_k-s_{k-1})A}\cdots Be^{s_1A}u_0\right\|_E^2 ds_1\ldots ds_k,$$

and then continue as in the proof of Theorem 4.2. The first term of (4.5) is treated by noting that $e^{tA}x \in \mathcal{D}(A)$ for every $x \in E$ and $t > 0$ [23, §2.5]. □

The condition $c_1c_2T_2(E) < 1$ can be relaxed if we only require the sum (4.5) to converge absolutely for small times [6]. The solution of (4.6) is then obtained by piecing together the solutions obtained from the stochastic Dyson series (4.5), so that $t \longmapsto e^{A+B}_{\lambda,W;t}x$, $t \ge 0$, has an exponential growth estimate.

5. Stochastic functional calculus

The significance of Corollary 4.4 above is that the bound (4.7) required for the existence of the solution $t \longmapsto e^{A+B}_{\lambda,W;t}x$, $t \ge 0$, of the stochastic equation (4.6) is a type of *square function estimate* for the operator A. It has been known since the work of A. McIntosh [19] that such estimates are associated with the existence of an H^∞-functional calculus for A. Furthermore, it has been shown in [5, Theorem 6.5] that the regularity of solutions of simple stochastic equations involving the operator A in Hilbert space implies that A has an H^∞-functional calculus.

A good reference for many of the results we need for an operator acting in Hilbert space is [14, Chap. 2]. We now set down the basic definitions.

5.1. H^∞ functional calculus

Let $0 < \omega < \pi/2$ and suppose that $T : \mathcal{D}(T) \to E$ is an operator of type $\omega-$ acting in the Banach space E as defined at the beginning of Section 4.

Then the bounded linear operator $f(T)$ is defined by the Riesz–Dunford formula

$$f(T) = \frac{1}{2\pi i}\int_C (zI - T)^{-1} f(z)\, dz. \tag{5.1}$$

for any function f satisfying the bounds

$$|f(z)| \le K_\nu \frac{|z|^s}{1+|z|^{2s}}, \quad z \in S_\nu^\circ.$$

The contour C can be taken to be $\{z \in \mathbb{C} : \Re(z) \le 0,\ |\Im(z)| = -\tan\theta.\Re(z)\ \}$, with $\omega < \theta < \nu$. The integral (5.1) converges as a Bochner integral in the uniform norm due to the estimate (4.2) for the resolvent $z \longmapsto (zI - T)^{-1}$ of T.

The operator T of type $\omega-$ is said to have a *bounded H^∞-functional calculus* if for each $\omega < \nu < \pi/2$, there exists an algebra homomorphism $f \longmapsto f(T)$ from $H^\infty(S_{\nu-}^\circ)$ to $\mathcal{L}(H)$ agreeing with (5.1) and a positive number C_ν such that $\|f(T)\| \le C_\nu \|f\|_\infty$ for all $f \in H^\infty(S_\nu^\circ)$. The following Hilbert space result is from [19], see also [14, Theorem 11.9].

Theorem 5.1. *Suppose that T is a one-to-one operator of type $\omega-$ in a Hilbert space H. Then T has a bounded H^∞-functional calculus if and only if for every $\omega < \nu < \pi/2$, there exists $c_\nu > 0$ such that T and its adjoint T^* satisfy the square function estimates*

$$\int_0^\infty \|\psi_t(T)u\|^2 \frac{dt}{t} \le c_\nu \|u\|^2, \quad u \in H, \tag{5.2}$$

$$\int_0^\infty \|\psi_t(T^*)u\|^2 \frac{dt}{t} \le c_\nu \|u\|^2, \quad u \in H, \tag{5.3}$$

for some function (every function) $\psi \in H^\infty(S_{\nu-}^\circ)$, which satisfies

$$\int_0^\infty \psi^2(-t) \frac{dt}{t} = 1, \text{ and} \tag{5.4}$$

$$|\psi(z)| \le K_\nu \frac{|z|^s}{1+|z|^{2s}}, \quad z \in S_{\nu-}^\circ, \tag{5.5}$$

for some $s > 0$. Here $\psi_t(z) = \psi(tz)$ for $z \in S_{\nu-}^\circ$.

For the function $\psi(z) = Cz^{\frac{1}{2}}e^z$ with $C > 0$ chosen such that (5.4) holds,

$$\begin{aligned}\int_0^\infty \|\psi_t(T)u\|^2 \frac{dt}{t} &= C^2 \int_0^\infty \|(-tT)^{\frac{1}{2}} e^{tT} u\|^2 \frac{dt}{t} \\ &= C^2 \int_0^\infty \|(-T)^{\frac{1}{2}} e^{tT} u\|^2 dt.\end{aligned}$$

With this choice for ψ, the bound (5.2) is equivalent to the bound (4.7) with $\|x\|_V = \|(-T)^{\frac{1}{2}} x\|$ for $x \in \mathcal{D}((-T)^{\frac{1}{2}})$.

5.2. Random resolvents

Suppose that $T : \mathcal{D}(T) \to E$ is a closed linear map defined in the Banach space E. Then the *resolvent* $R(\zeta)$, $\zeta \in \rho(T)$, of T is the bounded linear map defined by

$R(\zeta) = (\zeta I - T)^{-1}$ for all $\zeta \in \mathbb{C}$ belonging to the set $\rho(T)$ for which the inverse is defined. If T is the generator of a C_0-semigroup e^{tT}, $t \geq 0$, then we also have

$$(\zeta I - T)^{-1} = \int_0^\infty e^{-\zeta t} e^{tT}\, dt \tag{5.6}$$

for all $\zeta \in \mathbb{C}$ in some right half-plane. We adopt the right-hand side of equation (5.6) as the definition of a resolvent in the setting of *stochastic disentangling.*

Let E be an M-type 2 Banach space.

1) A is an operator of type $\omega-$ for $0 < \omega < \pi/2$.
2) There exists a real separable Banach space V with norm $\|\cdot\|_V$ such that $\mathcal{D}(A) \subset V \subset E$ and $B : V \to E$ is a bounded linear operator with $\|Bx\|_E \leq c_B\|x\|_V$ for all $x \in V$.
3) Let $A_\theta = e^{i\theta}A$ for $0 \leq |\theta| < \pi/2 - \omega$. For each $0 \leq |\theta| < \pi/2 - \omega$, there exists $m_\theta > 0$ such that

$$\int_0^\infty \|e^{tA_\theta}x\|_V^2\, dt \leq m_\theta^2 \|x\|_E^2 \tag{5.7}$$

for all $x \in E$.
4) There exists $0 < \delta < \pi/2 - \omega$ such that $\sup_{|\theta|\leq\delta} m_\theta c_B T_2(E) < 1$.

According to Corollary 4.4, the random process $t \longmapsto e^{A+B}_{\lambda,W;t}$, $t \geq 0$, is uniformly bounded in $\mathcal{L}(E, L^2(\mathbb{P}, E))$ by a constant K. If the pair (A, B) of linear operators satisfies conditions 1)–4) above, then so does the pair $(\beta A, \sqrt{\beta}B)$ for any $\beta > 0$, so the mapping $(\beta, t) \longmapsto e^{\beta A+\sqrt{\beta}B}_{\lambda,W;t}$, $\beta, t \geq 0$, is also uniformly bounded in $\mathcal{L}(E, L^2(\mathbb{P}, E))$ by K. Consequently, the following definition makes sense.

Definition 5.2. Let E be a Banach space of M-type 2 and suppose that the conditions 1)–4) above hold. The *stochastic resolvent* $R_{\lambda,W;t}(z; A + B)$, $t \geq 0$, of the process $t \longmapsto e^{A+B}_{\lambda,W;t}$, $t \geq 0$, is the $\mathcal{L}(E, L^2(\mathbb{P}, E))$-valued mapping $t \longmapsto R_{\lambda,W;t}(z; A + B)$, $t \geq 0$, given by

$$R_{\lambda,W;t}(z; A + B)x = \int_0^\infty e^{-z\beta} e^{A+B}_{\lambda,W;\beta t} x\, d\beta \tag{5.8}$$

for all $x \in E$, $t \geq 0$ and $\Re z > 0$.

We denote by the same symbol $R_{\lambda,W;t}(z; A + B)$ the analytic continuation of (5.8) as an element of $\mathcal{L}(E, L^2(\mathbb{P}, E))$ to the left half-plane. We obtain an $\mathcal{L}(E, L^2(\mathbb{P}, E))$-valued function of time t because we are considering disentangling over an interval $[0, t]$ as in Section 2.

Appealing to the orthogonality property 2) of multiple stochastic integrals, we see that (4.5) is a weakly orthogonal expansion in E-valued random variables. According to formula (5.8), the stochastic resolvent $R_{\lambda,W;t}(z; A + B)x$ also has a weakly orthogonal expansion in E-valued random variables. We use this expansion in order to establish the following bound.

Lemma 5.3. *Let E be a Banach space of M-type 2. Suppose that conditions 1)–4) above hold. Then the $L^2(\mathbb{P},E)$-valued function $z \longmapsto R_{\lambda,W;t}(z;A+B)x$ is holomorphic in $\mathbb{C}\setminus S_{\delta-}$ for all $t>0$ and $x\in E$ and for each $\pi/2-\delta<\mu<\pi/2$ there exists $C_\mu>0$ such that*

$$\|R_{\lambda,W;t}(z;A+B)x\|_{L^2(\mathbb{P},E)} \le \frac{C_\mu}{|z|}\|x\|, \quad z\in\mathbb{C}\setminus S_{\mu-} \tag{5.9}$$

for all $x\in E$ and $t>0$.

Proof. Let $\sqrt{z}$ denote the square root of z with positive real part. Under conditions 1)–4), replacing A by zA and B by $\sqrt{z}B$ in the expansion (4.5), we obtain a uniformly bounded $L^2(\mathbb{P},E)$-valued holomorphic function $z\longmapsto e^{zA+\sqrt{z}B}_{\lambda,W;t}x$ in $S^\circ_{\delta+}$ for each $t>0$ and $x\in E$.

For each $0<\mu<\pi/2$, let $\Xi_{\pm\mu}=\{se^{\pm i\mu}: s\ge 0\}$. Then for $0<\nu<\delta$, by the vector version of Cauchy's Theorem we have

$$R_{\lambda,W;t}(z,A+B)=\int_{\Xi_{-\nu}} e^{-z\zeta}e^{\zeta A+\sqrt{\zeta}B}_{\lambda,W;t}x\,d\zeta \tag{5.10}$$

if $\Re(ze^{-i\nu})>0$ and

$$R_{\lambda,W;t}(z,A+B)=\int_{\Xi_{\nu}} e^{-z\zeta}e^{\zeta A+\sqrt{\zeta}B}_{\lambda,W;t}x\,d\zeta \tag{5.11}$$

if $\Re(ze^{i\nu})>0$. Because $\pi/2-\delta<\mu<\pi/2$, we can choose $0<\nu<\delta$ such that $\pi/2-\nu<\mu<\pi/2$. Then the bound (5.9) follows for all $z\in\mathbb{C}\setminus S_{\mu-}$ with $\Im z\ge 0$ from the representation (5.10) and the uniform boundedness of $z\longmapsto e^{zA+\sqrt{z}B}_{\lambda,W;t}x$ in $S^\circ_{\delta+}$. For $\Im z<0$, the representation (5.11) is used. □

For any holomorphic function φ in a sector $S^\circ_{\nu-}$ with $\pi/2-\delta<\nu<\pi/2$ and satisfying the bound

$$|\varphi(z)| \;\le\; M_\nu\frac{|z|^s}{1+|z|^{2s}}, \quad z\in S^\circ_{\nu-}, \tag{5.12}$$

for some $M_\nu, s>0$, we may define the integral

$$\varphi_{\lambda,W;t}(A+B)x=\frac{1}{2\pi i}\int_C \varphi(z)R_{\lambda,W;t}(z;A+B)x\,dz, \quad x\in E, \tag{5.13}$$

in $L^2(\mathbb{P},E)$ for the contour $C=\{z\in\mathbb{C}: |\Im(z)|=-\tan\mu.\Re(z),\ \Re(z)\le 0\}$ taken anticlockwise around $S_{\delta-}$ for $\pi/2-\delta<\mu<\nu$. By Lemma 5.3 and the estimate (5.12), the contour integral converges as a Bochner integral in $L^2(\mathbb{P},E)$ and $\varphi_{\lambda,W;t}(A+B)x$ admits a weakly orthogonal expansion in E-valued random variables. In the case that $B=0$, we obtain the Riesz–Dunford formula (5.1).

The following result says that the random part $\varphi_{\lambda,W;t}(A+B)-\varphi(A)$ of $\varphi_{\lambda,W;t}(A+B)$ has an H^∞-bound under the assumptions 1)–4) above.

Theorem 5.4. *Let E be a Banach space of* M*-type* 2 *and that conditions* 1)–4) *above hold. Then for every $\pi/2 - \delta < \nu < \pi/2$, there exists $C_\nu > 0$ such that*

$$(\mathbb{E}\|\varphi_{\lambda,W;t}(A+B)x - \varphi(A)x\|^2)^{\frac{1}{2}} \le C_\nu \|\varphi\|_\infty \|x\|$$

for every holomorphic function φ on $S^\circ_{\nu-}$ satisfying the bound (5.12) *and every $t > 0$.*

Proof. For each $0 < \mu < \pi/2$, let $\Xi_{\pm\mu} = \{se^{\pm i\mu} : s \ge 0\}$ and

$$\Gamma_{\mu,1} = \{se^{i\mu} : -\infty \le s \le 0\}, \quad \Gamma_{\mu,2} = \{-se^{-i\mu} : 0 \le s < \infty\}.$$

Then for $0 < \nu < \delta$, by the vector version of Cauchy's Theorem $R_{\lambda,W;t}(z, A+B)$ is given by equation (5.10) if $\Re(ze^{-i\nu}) > 0$ and equation (5.11) if $\Re(ze^{-i\nu}) > 0$. Let φ be a uniformly bounded holomorphic function in a sector $S^\circ_{\nu-}$ with $\pi/2-\delta < \nu < \pi/2$. Let $\pi/2 - \delta < \mu < \nu$. Then

$$\begin{aligned} 2\pi i\varphi_{\lambda,W;t}(A+B)x = &\int_{\Gamma_{\mu,1}} \varphi(z)R_{\lambda,W;t}(z, A+B)x\,dz \\ &+ \int_{\Gamma_{\mu,2}} \varphi(z)R_{\lambda,W;t}(z, A+B)x\,dz, \end{aligned}$$

if the integrals converge. The Laplace transform

$$\mathcal{L}\varphi(\zeta) = \begin{cases} -\int_{\Gamma_{\mu,1}} e^{-z\zeta}\varphi(z)\,dz, & \Re(\zeta e^{i\mu}) < 0 \\ \int_{\Gamma_{\mu,2}} e^{-z\zeta}\varphi(z)\,dz, & \Re(\zeta e^{-i\mu}) < 0 \end{cases}$$

of φ is defined for $\pi/2 - \nu < |\arg\zeta| < \pi$.

From equation (4.5), the random part of $R_{\lambda,W;t}(\zeta, A+B)x$ is given by

$$\tilde{R}_{\lambda,W;t}(\zeta, A+B)x = R_{\lambda,W;t}(\zeta, A+B)x - (\zeta I - A)^{-1}x$$

In order to estimate

$$\mathbb{E}\left\|\int_{\Gamma_{\mu,2}} \varphi(\zeta)\tilde{R}_{\lambda,W;t}(\zeta, A+B)x\,d\zeta\right\|^2, \tag{5.14}$$

we apply the Itô bound (4.1) and consider the sum

$$\begin{aligned} \sum_{n=1}^{\infty} T_2(E)^{2n} \int_0^t \int_0^{t_n} \cdots \int_0^{t_2} \bigg\| \int_{\Xi_{-\theta}} &\mathcal{L}\varphi(\zeta)e^{\zeta A(t-t_n)}(\zeta^{\frac{1}{2}}B)e^{\zeta A(t_n-t_{n-1})} \\ &\cdots \zeta^{\frac{1}{2}}B)e^{\zeta A t_1}x d\zeta\bigg\|^2 dt_1 \dots dt_n \end{aligned} \tag{5.15}$$

for $\pi/2 - \mu < \theta < \pi/2 - \omega$. For each such θ, there exists $K_\theta > 0$ such that

$$|\mathcal{L}\varphi(\zeta)| \le \frac{K_\theta}{|\zeta|}\|\varphi\|_\infty, \quad \zeta \in \Xi_{-\theta},$$

for every uniformly bounded holomorphic function in a sector $S^{\circ}_{\nu-}$. It suffices to show that the sum

$$\Bigg(\sum_{n=1}^{\infty} T_2(E)^{2n}\int_0^t\int_0^{t_n}\cdots\int_0^{t_2}\Bigg(\int_{\Xi_{-\theta}}\frac{\|\varphi\|_\infty}{|\zeta|}\|e^{\zeta A(t-t_n)}(\zeta^{\frac{1}{2}}B)e^{\zeta A(t_n-t_{n-1})} \cdots\zeta^{\frac{1}{2}}B)e^{\zeta At_1}x\||d\zeta|\Bigg)^2 dt_1\dots dt_n\Bigg)^{\frac{1}{2}} \tag{5.16}$$

converges. The notation $|d\zeta|$ means arclength measure. Then an application of the Fubini–Tonelli Theorem shows that (5.14) is equal to (5.15) and is estimated by the expression (5.16). Here we don't actually appeal to the bound (5.12) which is only needed to make sense of $\varphi(A)$.

Applying Minkowski's inequality, (5.16) is estimated by

$$\begin{aligned}
&\|\varphi\|_\infty\int_{\Xi_{-\theta}}\Bigg(\sum_{n=1}^{\infty} T_2(E)^{2n}\int_0^t\int_0^{t_n}\cdots\int_0^{t_2}\Big(\|e^{\zeta A(t-t_n)}(\zeta^{\frac{1}{2}}B)e^{\zeta A(t_n-t_{n-1})}\\
&\qquad\cdots\zeta^{\frac{1}{2}}B)e^{\zeta At_1}x\|^2dt_1\dots dt_n\Bigg)^{\frac{1}{2}}\frac{|d\zeta|}{|\zeta|}\\
&=\|\varphi\|_\infty\int_0^{\infty}\Bigg(\sum_{n=1}^{\infty} T_2(E)^{2n}\int_0^t\int_0^{t_n}\cdots\int_0^{t_2}\Big(\|e^{sA_{-\theta}(t-t_n)}(s^{\frac{1}{2}}B)e^{sA_{-\theta}(t_n-t_{n-1})}\\
&\qquad\cdots s^{\frac{1}{2}}B)e^{sA_{-\theta}t_1}x\|^2dt_1\dots dt_n\Bigg)^{\frac{1}{2}}\frac{ds}{s}\\
&=\|\varphi\|_\infty\int_0^{\infty}\Bigg(\sum_{n=1}^{\infty} T_2(E)^{2n}\int_0^{st}\int_0^{s_n}\cdots\int_0^{s_2}\Big(\|e^{A_{-\theta}(st-s_n)}Be^{A_{-\theta}(s_n-s_{n-1})}\\
&\qquad\cdots Be^{A_{-\theta}s_1}x\|^2ds_1\dots ds_n\Bigg)^{\frac{1}{2}}\frac{ds}{s},\qquad [s_j=st_j \text{ for } j=1,\dots,n]\\
&=\|\varphi\|_\infty\int_0^{\infty}\Bigg(\sum_{n=1}^{\infty} T_2(E)^{2n}\int_0^{r}\int_0^{s_n}\cdots\int_0^{s_2}\Big(\|e^{A_{-\theta}(r-s_n)}Be^{A_{-\theta}(s_n-s_{n-1})}\\
&\qquad\cdots Be^{A_{-\theta}s_1}x\|^2ds_1\dots ds_n\Bigg)^{\frac{1}{2}}\frac{dr}{r},\qquad [r=st]
\end{aligned}$$

We would like to know that this integral is finite. Split it into $r\geq 1$ and $r<1$. Applying the Cauchy–Schwarz inequality for $r\geq 1$, we obtain

$$\|\varphi\|_\infty\Bigg(\int_1^{\infty}\sum_{n=1}^{\infty} T_2(E)^{2n}\int_0^{r}\int_0^{s_n}\cdots\int_0^{s_1}\Big(\|e^{A_{-\theta}(r-s_n)}Be^{A_{-\theta}(s_n-s_{n-1})} \cdots Be^{A_{-\theta}s_1}x\|^2ds_1\dots ds_n dr\Bigg)^{\frac{1}{2}}.$$

Each term

$$\int_1^\infty \int_0^r \int_0^{s_n} \cdots \int_0^{s_1} \|e^{A_{-\theta}(r-s_n)} B e^{A_{-\theta}(s_n-s_{n-1})} \cdots B e^{A_{-\theta}s_1} x\|^2 ds_1 \ldots ds_n dr$$

in the sum is bounded by

$$\int_0^\infty \int_0^r \int_0^{s_n} \cdots \int_0^{s_1} \|e^{A_{-\theta}(r-s_n)} B e^{A_{-\theta}(s_n-s_{n-1})} \cdots B e^{A_{-\theta}s_1} x\|^2 ds_1 \ldots ds_n dr. \tag{5.17}$$

For every $t > 0$ and $y \in E$, the vector $e^{tA_{-\theta}}y$ is an element of $\mathcal{D}(A)$. But $\mathcal{D}(A) \subset V \subset E$ with continuous embeddings, so there exists $C > 0$ such that (5.17) is bounded by

$$C^2 \int_0^\infty \int_0^r \int_0^{s_n} \cdots \int_0^{s_1} \|e^{A_{-\theta}(r-s_n)} B e^{A_{-\theta}(s_n-s_{n-1})} \cdots B e^{A_{-\theta}s_1} x\|_V^2 ds_1 \ldots ds_n dr. \tag{5.18}$$

Applying the inequality (5.7) and Lemma 4.1, the integral (5.17) is bounded by

$$\begin{aligned} &C^2 m_{-\theta}^2 \int_0^\infty \int_0^{s_n} \cdots \int_0^{s_1} \|B e^{A_{-\theta}(s_n-s_{n-1})} \cdots B e^{A_{-\theta}s_1} x\|^2 ds_1 \ldots ds_n \\ &\quad \le C^2 m_{-\theta}^2 c_B^2 \int_0^\infty \int_0^{s_n} \cdots \int_0^{s_1} \|e^{A_{-\theta}(s_n-s_{n-1})} B \cdots B e^{A_{-\theta}s_1} x\|_V^2 ds_1 \ldots ds_n. \end{aligned}$$

Repeating the process, we obtain the bound

$$C^2 (m_{-\theta} c_B)^{2n} \int_0^\infty \|e^{A_{-\theta}s_1} x\|_V^2 \, ds_1 \le C^2 (m_{-\theta} c_B)^{2n} m_{-\theta} \|x\|^2.$$

By condition 4), $m_{-\theta} c_B T_2(E) < 1$ and so the integral over $r \ge 1$ converges.

For $r < 1$, we can similarly estimate

$$\int_0^r \int_0^{s_n} \cdots \int_0^{s_1} \|e^{A_{-\theta}(r-s_n)} B e^{A_{-\theta}(s_n-s_{n-1})} \cdots B e^{A_{-\theta}s_1} x\|^2 ds_1 \ldots ds_n$$

to get a bound

$$C' \|\varphi\|_\infty \int_0^1 \left(\sum_{n=1}^\infty (m_{-\theta} c_B T_2(E))^{2n-2} \int_0^r \|x\|^2 ds_n \right)^{\frac{1}{2}} \frac{dr}{r}$$

which is finite. Combining the estimates for $r \ge 1$ and $r < 1$, we obtain the required bound for (5.16) and together with a similar argument for the integral over $\Gamma_{\mu,1}$, this finishes the proof of the theorem. □

Remark 5.5. The above result also holds if we replace 4) by the condition

4′) $$\sup_{\|x\| \le 1, |\theta| \le \delta} \int_0^\infty \|B e^{tA_\theta} x\|_E^2 \, dt < 1/T_2(E)^2.$$

In order to apply square function estimates for L^p-spaces for $1 < p < \infty$ given in [3] for the existence of an H^∞ functional calculus, we need to establish the convergence of (4.5) using multilinear square function estimates, which we leave to a later paper.

Combined with the characterisation of operators acting in Hilbert space with an H^∞-functional calculus [19], we have the following result establishing the existence of a *stochastic functional calculus* for "$A+B$" in Hilbert space H, where $T_2(H)=1$.

Theorem 5.6. *Suppose that A is a one-to-one operator of type $\omega-$ in a Hilbert space H such that A has an H^∞-functional calculus on $S_{\omega-}$. Let $V=\mathcal{D}((-A)^{\frac{1}{2}})$ with $\|x\|_V=\|(-A)^{\frac{1}{2}}x\|$ for $x\in V$.*

Then for every $\omega<\nu<\pi/2$, there exists $b_\nu>0$ such that for every bounded linear map $B:V\to H$ with operator norm $\|B\|_{\mathcal{L}(V,H)}<b_\nu$, there exists a linear map

$$\varphi\longmapsto\varphi_{\lambda,W;t}(A+B)$$

from $H^\infty(S_{\nu-})$ with values in the linear space $\mathcal{L}(H,L^2(\mathbb{P},H))$ such that

$$(\mathbb{E}\|\varphi_{\lambda,W;t}(A+B)x\|^2)^{\frac{1}{2}}\le C_\nu\|\varphi\|_\infty\|x\|,\ t>0,$$

for every uniformly bounded holomorphic function φ on $S^\circ_{\nu-}$.

The element $\varphi_{\lambda,W;t}(A+B)$ of $\mathcal{L}(H,L^2(\mathbb{P},H))$ is given by equation (5.13) *for every uniformly bounded holomorphic function φ on $S^\circ_{\nu-}$ satisfying the bound* (5.12)*. Furthermore, the number b_ν is given by*

$$b_\nu=\left(\sup_{\|x\|\le1,|\theta|\le\frac{\pi}{2}-\nu}\int_0^\infty\|(-A)^{\frac{1}{2}}e^{te^{i\theta}A}x\|^2\,dt\right)^{-\frac{1}{2}}. \tag{5.19}$$

Proof. Let $\omega<\nu<\pi/2$ and $\psi(z)=(-z)^{\frac{1}{2}}e^z$, for all $z\in\mathbb{C}\setminus[0,\infty)$. Then for each $0\le\theta<\pi/2-\nu$, the function $z\longmapsto\psi(e^{i\theta}z)$, $z\in S_{\nu-}$, satisfies the bound (5.5). Because A has an H^∞-functional calculus on $S_{\omega-}$, the square function estimate (5.2) holds and there exists $c_{\nu,\theta}>0$ such that

$$\begin{aligned}\int_0^\infty\|\psi_t(A)u\|^2\frac{dt}{t}&=\int_0^\infty\|(-A)^{\frac{1}{2}}e^{te^{i\theta}A}x\|^2\,dt\\&\le c_{\nu,\theta}\|x\|^2\end{aligned}$$

for all $x\in H$. Because A has an H^∞-functional calculus, the square function norms (5.2) and (5.3) are equivalent to the Hilbert space norm [19], [14, Theorem 11.9] and depend continuously on functions ψ uniformly satisfying the bound (5.5). It follows that

$$(x,\theta)\longmapsto\int_0^\infty\|(-A)^{\frac{1}{2}}e^{te^{i\theta}A}x\|^2\,dt,\quad 0\le\theta<\pi/2-\omega,\ x\in H$$

is a continuous function. By the uniform boundedness principle,

$$\sup_{\|x\|\le1,|\theta|\le\frac{\pi}{2}-\nu}\int_0^\infty\|(-A)^{\frac{1}{2}}e^{te^{i\theta}A}x\|^2$$

is finite for each $\omega<\nu<\pi/2$ and conditions 1)–4) above are satisfied with $\delta=\nu$ and the given value b_ν.

The random linear operator $\varphi_{\lambda,W;t}(A+B) \in \mathcal{L}(H, L^2(\mathbb{P}, H))$ is defined by continuous extension from functions satisfying the bound (5.5). The nonrandom part of $\varphi_{\lambda,W;t}(A+B)$ has a limit by the convergence lemma of [19] and for the random part of $\varphi_{\lambda,W;t}(A+B)$, from the proof of Theorem 5.4 it is clear that we can appeal to dominated convergence. □

Finally, we state the relevance to the space-time regularity of strong solutions of the Zakai equation

$$D_tU(t,x) = A(x,D)U(t,x) + B(x,D)U(t,x)D_tW(t), \quad t \in [0,T], x \in \mathbb{R}^d,$$

$U(0,x) = u_0(x)$ for $x \in R^d$. Here

$$A(x,D) = \sum_{i,j=1}^{d} a_{ij}(x)D_iD_j, \quad B(x,D) = \sum_{i=1}^{d} b_i(x)D_i + c(x).$$

This equation arises in filtering theory, and has been studied by many authors, see [2] and the references therein. It can be written as an abstract stochastic evolution equation of the form

$$dX_t = AX_t\,dt + BX_t\,dW_t, \qquad t \in [0,T],\ X_0 = u_0. \tag{5.20}$$

Here the linear operator A is closed and densely defined on $L^2(\mathbb{R}^d)$, the operator B is a generator of a C_0-group on $L^2(\mathbb{R}^d)$, and W is a real-valued Brownian motion on some probability space $(\Omega, \mathcal{F}, \mathbb{P})$.

Theorem 5.7. *Let A be the operator*

$$\sum_{i,j=1}^{d} \frac{\partial}{\partial x_j}\left(a_{ij}(x)\frac{\partial}{\partial x_j}\right).$$

with domain $H^2(\mathbb{R}^d)$. The operator B is given by

$$Bu(x) = \sum_{i=1}^{d} b_i(x)\frac{\partial u}{\partial x_i}(x), \quad u \in H^1(\mathbb{R}^d).$$

If the coefficients $a_{ij}(x)$ are real valued and belong to C^γ for some $\gamma \in (0,1)$ and satisfy the joint ellipticity condition

$$\sum_{i,j=1}^{d}\left(\cos\nu.a_{ij}(x) - \frac{1}{2}b_i(x)b_j(x)\right)\xi_i\xi_j \geq \rho|\xi|^2, \quad \xi \in \mathbb{R}^d,\ x \in \mathbb{R}^d,$$

for some $0 < \nu < \pi/2$, then by [14, Theorem 13.14], *the operators A and B satisfy the conditions of Theorem* 5.6, *so that (A,B) has a stochastic $H^\infty(S_{\mu-})$-functional calculus $\varphi \longmapsto \varphi_{\lambda,W;t}(A+B)$ on $L^2(\mathbb{R}^d)$ and on the sector $S_{\mu-}$ for every $\pi/2 - \nu < \mu < \pi/2$.*

Moreover, the $L^2(\mathbb{R}^d)$-valued process

$$t \longmapsto e^{A+B}_{\lambda,W;t}u_0, \quad t \in [0,T],$$

is the solution of equation (5.20) *and has paths in*

$$C([0,T];L^2(\mathbb{R}^d)) \cap C((0,T];H^2(\mathbb{R}^d)).$$

If $u_0 \in H^2(\mathbb{R}^d)$*, then the solution has paths in* $H^2(\mathbb{R}^d)$.

5.3. Further developments

In L^p-spaces with $1 < p < \infty$, if the bounds (5.2) and (5.3) are replaced by the square function estimates

$$\left\| \left(\int_0^\infty |\psi_t(T)u|^2 \frac{dt}{t} \right)^{\frac{1}{2}} \right\|_p \leq c_\nu \|u\|_p, \quad u \in L^p, \tag{5.21}$$

$$\left\| \left(\int_0^\infty |\psi_t(T')u|^2 \frac{dt}{t} \right)^{\frac{1}{2}} \right\|_{p'} \leq c_\nu \|u\|_{p'}, \quad u \in L^{p'}, \tag{5.22}$$

then we obtain conditions equivalent to the existence of an H^∞-functional calculus for T [3, Theorem 6.1, Corollary 4.5]. As mentioned in [3, p. 87], the bounds (5.2) and (5.3) may fail for the Laplacian $T = \Delta$ on $L^p(\mathbb{R}^n)$ for $p > 2$, where they are associated with Besov spaces, see for example [13, §1.2]

In order to utilise the bound (5.21) to obtain a stochastic functional calculus in an L^p-space for $1 < p < \infty$ and, say, the Laplacian operator $A = \Delta$, we would need to obtain the multilinear estimate

$$\mathbb{E}\left\| \int_0^t \int_0^{s_k} \cdots \int_0^{s_2} \left[e^{(t-s_k)A} B e^{(s_k - s_{k-1})A} \ldots B e^{s_1 A} u_0 \right] dW_{s_1} \ldots dW_{s_k} \right\|_p^2$$
$$\leq C_p^{2k} \left\| \left(\int_0^t \int_0^{s_k} \cdots \int_0^{s_2} \left| e^{(t-s_k)A} B e^{(s_k - s_{k-1})A} \ldots B e^{s_1 A} u_0 \right|^2 ds_1 \ldots ds_k \right)^{\frac{1}{2}} \right\|_p^2,$$

analogous to the one-sided Itô inequality used in the proof of Theorem 4.2. Furthermore, L^p-spaces for $1 < p < 2$ are not of type 2, and so, not of M-type 2 and the one-sided Itô inequality fails to hold. Here we would hope to apply the theory of γ-radonifying maps mentioned in Section 3.2 and a multilinear version of the embedding results of [13].

References

[1] Z. Brzeźniak, *Stochastic partial differential equations in M-type 2 Banach spaces,*. Potential Anal. **4** (1995), 1–45.

[2] Z. Brzeźniak, J. van Neerven, M.C. Veraar and L. Weis, *Itô's formula in UMD Banach spaces and regularity of solutions of the Zakai equation.* J. Differential Equations **245** (2008), 30–58

[3] M. Cowling, I. Doust, A. McIntosh and A. Yagi, *Banach space operators with a bounded* H^∞ *functional calculus.* J. Austral. Math. Soc. Ser. A **60** (1996), 51–89.

[4] G. Da Prato and J. Zabczyk, *Stochastic equations in infinite dimensions*, Encyclopedia of Mathematics and its Applications **44**, Cambridge University Press, Cambridge, 1992.

[5] J. Dettweiler, J. van Neerven and L. Weis, *Space-time regularity of solutions of the parabolic stochastic Cauchy problem.* Stoch. Anal. Appl. **24** (2006), 843–869.

[6] F. Flandoli, *On the semigroup approach to stochastic evolution equations.* Stochastic Analysis and Appl. **10** (1992), 181–203.

[7] D.J.H. Garling, *Brownian motion and UMD-spaces*, in: "Probability and Banach Spaces" (Zaragoza, 1985), 36–49, Lecture Notes in Math. **1221**, Springer-Verlag, Berlin, 1986.

[8] H. Heinich, *Esperance conditionelle pour les fonctions vectorielles.* C.R. Acad. Sci. Paris Ser. A **276** (1973), 935–938.

[9] B. Jefferies, *Conditional expectation for operator-valued measures and functions.* Bull. Austral. Math. Soc. **30** (1984), 421–429.

[10] ———, *Feynman's operational calculus and the stochastic functional calculus in Hilbert space*, in "The AMSI-ANU Workshop on Spectral Theory and Harmonic Analysis", Proc. Centre Math. Appl. Austral. Nat. Univ. **44**, Austral. Nat. Univ., Canberra, 2010, 183–210.

[11] B. Jefferies and G.W. Johnson, *Feynman's operational calculi for noncommuting operators: Definitions and elementary properties.* Russ. J. Math. Phys. **8** (2001), 153–171.

[12] ———, *Feynman's operational calculi for noncommuting systems of operators: tensors, ordered supports and disentangling an exponential factor.* Math. Notes **70** (2001), 815–838.

[13] N.J. Kalton, J.M.A.M. van Neerven, M.C. Veraar, and L.W. Weis, *Embedding vector-valued Besov spaces into spaces of γ-radonifying operators.* Math. Nachr. **281** (2008), 238–252.

[14] P. Kunstmann and L. Weis, *L_p-regularity for parabolic equations, Fourier multiplier theorems and H^∞-functional calculus.* Functional analytic methods for evolution equations, 65–311, Lecture Notes in Math. **1855**, Springer, Berlin, 2004.

[15] H.H. Kuo, *Gaussian measures in Banach spaces.* Lecture Notes in Math. **463**, Springer, Berlin, 1975.

[16] S. Kwapień, *Decoupling inequalities for polynomial chaos.* Ann. Probab. **15** (1987), 1062–1071.

[17] S. Kwapień and W. Woyczyński, *Random series and stochastic integrals: single and multiple.* Birkhäuser Boston, Inc., Boston, MA, 1992.

[18] J. Maas, *Malliavin calculus and decoupling inequalities in Banach spaces.* J. Math. Anal. Appl. **363** (2010), 383–398.

[19] A. McIntosh, *Operators which have an H_∞-functional calculus*, in: Miniconference on Operator Theory and Partial Differential Equations 1986, 212–222. Proc. Centre for Mathematical Analysis **14**, ANU, Canberra, 1986.

[20] J. van Neerven and L. Weis, *Stochastic integration of functions with values in a Banach space.* Studia Math. **166** (2005), 131–170.

[21] J. van Neerven, M.C. Veraar and L. Weis, *Stochastic evolution equations in UMD Banach spaces.* J. Funct. Anal. **255** (2008), 940–993.

[22] ———, *Stochastic maximal L^p-regularity.* Ann. Probab. **40** (2012), 788–812.

[23] A. Pazy, *Semigroups of Linear Operators and Applications to Partial Differential Equations.* Springer-Verlag, Applied Mathematical Sciences, Vol. 44, New York/Berlin/Heidelberg/Tokyo, 1983.

[24] G. Pisier, *Probabilistic methods in the geometry of Banach spaces.* Probability and analysis (Varenna, 1985), 167–241, Lecture Notes in Math. **1206**, Springer, Berlin, 1986.

[25] J. Rosiński and Z. Suchanecki, *On the space of vector-valued functions integrable with respect to the white noise. Colloq. Math.* **43** (1980), 183–201.

[26] G. Samorodnitsky and M. Taqqu, *Multiple stable integrals of Banach-valued functions.* J. Theoret. Probab. **3** (1990), 267–287

[27] H. Schaefer, *Topological Vector Spaces*, Springer-Verlag, Berlin/Heidelberg/New York, 1980.

[28] L. Schwartz, *Radon Measures in Arbitrary Topological Spaces and Cylindrical Measures*, Tata Inst. of Fundamental Research, Oxford Univ. Press, Bombay, 1973.

[29] A.V. Skorohod, *Random Linear Operators*, Riedel, 1984.

Brian Jefferies
School of Mathematics
The University of New South Wales
NSW 2052 Australia
e-mail: b.jefferies@unsw.edu.au

Operator Theory:
Advances and Applications, Vol. 240, 221–234

Subideals of Operators – A Survey and Introduction to Subideal-Traces

Sasmita Patnaik and Gary Weiss

Dedicated to the memory of Mihaly Bakonyi

Abstract. Operator ideals in $B(H)$ are well understood and exploited but ideals inside them have only recently been studied starting with the 1983 seminal work of Fong and Radjavi and continuing with two recent articles by the authors of this survey. This article surveys this study embodied in these three articles. A subideal is a two-sided ideal of J (for specificity also called a J-ideal) for J an arbitrary ideal of $B(H)$. In this terminology we alternatively call J a $B(H)$-ideal.

This surveys [5], [13] and [14] in which we developed a complete characterization of all J-ideals generated by sets of cardinality strictly less than the cardinality of the continuum. So a central theme is the impact of generating sets for subideals on their algebraic structure. This characterization includes in particular finitely and countably generated J-ideals. It was obtained by first generalizing to arbitrary principal J-ideals the 1983 work of Fong–Radjavi who determined which principal $K(H)$-ideals are also $B(H)$-ideals. A key property in our investigation turned out to be J-softness of a $B(H)$-ideal I inside J, that is, $IJ = I$, a generalization of a recent notion of $K(H)$-softness of $B(H)$-ideals introduced by Kaftal–Weiss and earlier exploited for Banach spaces by Mityagin and Pietsch. This study of subideals and the study of elementary operators with coefficient constraints are closely related. Here we also introduce and study a notion of subideal-traces where classical traces (unitarily invariant linear functionals) need not make sense for subideals that are not $B(H)$-ideals.

Mathematics Subject Classification (2010). Primary: 47L20, 47B10, 47B07; Secondary: 47B47, 47B37, 47-02, 13C05, 13C12.

Keywords. Ideals, operator ideals, principal ideals, subideals, lattices, traces, subideal-traces.

The first author was partially supported by various The Taft Foundation awards including a Charles Phelps Taft Dissertation Fellowship.
The second author was partially supported by Simons Foundation Collaboration Grant 245014, The Taft Foundation and CIRM.

1. Introduction

For general rings, an ideal (all ideals herein are two-sided ideals) is a commutative additive subgroup of a ring that is closed under left and right multiplication by elements of the ring. Herein H denotes a separable infinite-dimensional complex Hilbert space and $B(H)$ denotes the C^*-algebra of all bounded linear operators on H. Ideals of $B(H)$, with the latter regarded as a ring, have become ubiquitous throughout operator theory since their celebrated characterization by Calkin and Schatten [1], [15], in terms of "characteristic sets" of singular number sequences $s(T)$ of the operators T in the ideal. Herein these ideals, alternatively and for specificity, are called $B(H)$-ideals as one class among the classes of J-ideals defined below (next paragraph and expanded upon in Definition 2.1). This Calkin–Schatten characterization of $B(H)$-ideals has had and continues to have substantial impact in operator theory. As commutative objects in analysis, characteristic sets make more accessible the subtler properties of $B(H)$-ideals, particularly illuminating and expanding the knowledge of some of their noncommutative features. Some well-known $B(H)$-ideals are the ideal of compact operators $K(H)$, the finite rank operators $F(H)$, principal ideals (S) (i.e., singly generated $B(H)$-ideals), Banach ideals, the Hilbert–Schmidt class C_2, the trace class C_1, Orlicz ideals, Marcinkiewicz ideals and Lorentz ideals, to name a few. Definitions and properties of these ideals among others may be found in [4].

A subideal of operators is an ideal of J, for J an arbitrary $B(H)$-ideal. (For specificity we called these J-ideals.) That is, a subideal is an ideal of a $B(H)$-ideal. "Subideal" is a name coined by Gary Weiss motivated from the 1983 seminal work of Fong–Radjavi and by the new perspectives on operator ideals from work of Dykema, Figiel, Weiss and Wodzicki [4]. It is clear that every $B(H)$-ideal is a subideal, but the converse is less clear, i.e., whether or not every subideal is also a $B(H)$-ideal. Fong–Radjavi constructed the first example of a principal $K(H)$-ideal that is not a $B(H)$-ideal (Example 2.4). This shows that the class of subideals is strictly larger than the class of $B(H)$-ideals.

The main and most general results in this survey are Theorem 3.5 and Theorem 3.7 (Structure Theorem for Subideals $(\mathcal{S})_J$ for $|\mathcal{S}| < \mathfrak{c}$) in which we characterize, in terms of a new notion called softness, when a subideal generated by strictly less than $\mathfrak{c}$ elements is also a $B(H)$-ideal ($\mathfrak{c}$ denotes the cardinality of the continuum); and then we characterize its algebraic structure. Softness was first noticed by Kaftal and Weiss in [9], [11]–[12] and further exploited in [13]–[14]. Section 4 compares $B(H)$-ideals to subideals via some of their differences and similarities. And Section 5 is new research that begins the investigation of subideal-traces, an attempt at a useful analog to traces on $B(H)$-ideals which traces are themselves ubiquitous in operator theory.

2. Preliminaries

Every $B(H)$-ideal J is linear because for each $\alpha \in \mathbb{C}$, $\alpha\mathbf{1} \in B(H)$, so then for each $A \in J$, $\alpha A = (\alpha\mathbf{1})A \in J$. But surprisingly a subideal (i.e., a J-ideal) may not be linear (Section 4-Example 4.1, see also [13, Example 3.5]). A reason this proof fails for J-ideals when $J \neq B(H)$ is that $\alpha 1$ ($\alpha \neq 0$) is never contained in J.

In *Subideals of Operators* [13] we found three types of principal and finitely generated subideals: linear, real-linear and classical subideals (i.e., ideals not assumed to be linear inside $B(H)$-ideals). Indeed both the latter two types are sometimes nonlinear. Such differences in types also carry over to non-finitely generated J-ideals. The linear $K(H)$-ideals, meaning traditionally the linear ones, were studied in 1983 by Fong–Radjavi [5]. They found principal linear $K(H)$-ideals that are not $B(H)$-ideals. Herein we take all J-ideals to be linear, but as shown in [13], we expect here also that most of the results and methods apply to the two other types of subideals (the real-linear and the sometimes nonlinear classical ones).

Noting the obvious fact that intersections of ideals in any ring are themselves ideals, we begin with the following definition.

Definition 2.1.

(i) The principal $B(H)$-ideal generated by the single operator S is defined by

$$(S) := \bigcap\{I \mid I \text{ is a } B(H)\text{-ideal containing } S\}.$$

(ii) The principal J-ideal generated by S is defined by

$$(S)_J := \bigcap\{I \mid I \text{ is a } J\text{-ideal containing } S\}.$$

(iii) As above for principal J-ideals, likewise for an arbitrary subset $\mathcal{S} \subset J$, $(\mathcal{S})$ and $(\mathcal{S})_J$ denote respectively, via intersections, the smallest $B(H)$-ideal and the smallest J-ideal generated by the set $\mathcal{S}$.

(iv) Since herein all J-ideals are taken to be "linear," (iii) characterizes all of them if you set $\mathcal{S} = J$. But in [13]–[14] where J-ideals are not necessarily defined as linear, (i)–(iii) define possibly nonlinear J-ideals and are discussed below in Section 4.

Definition 2.2. For $B(H)$-ideals I, J, ideal I is called "J-soft" if $IJ = I$. (Clearly this applies only when $I \subset J$.) Equivalently in the language of s-numbers (see Remark 2.3(i), (ii), (v) below):

For every $A \in I$, $s_n(A) = O(s_n(B)s_n(C))$ for some $B \in I, C \in J$.

($s(A) := \langle s_n(A)\rangle$ is the singular number (s-number) sequence of operator A, counting multiplicities of course.)

Remark 2.3 (Standard facts and tools for operator ideals).

(i) If I, J are $B(H)$-ideals, then the traditional ideal product IJ is the $B(H)$-ideal which is alternatively described via its characteristic set

$$\Sigma(IJ) = \{\xi \in c_o^* \mid \xi \leq \eta\rho \text{ for some } \eta \in \Sigma(I) \text{ and } \rho \in \Sigma(J)\}$$

[4, Sections 2.8, 4.3] (see also [9, Section 4]). (See also Historical Background below-first paragraph.) This product operation on the lattice of $B(H)$-ideals is both associative and commutative.

(ii) If I and J are $B(H)$-ideals for which $A \in IJ$, then $A = XY$ for some $X \in I, Y \in J$ [4, Lemma 6.3].

(iii) For $T \in B(H)$, $A \in (T)$ if and only if $s(A) = O(D_m(s(T)))$ for some $m \in \mathbb{N}$. $D_m\xi$ is the m-fold sequence ampliation recalled just below in Historical Background.

Moreover, for $B(H)$-ideals I, as is well known from the polar decomposition, the inclusions $A \in I, A^* \in I, |A| \in I$ and $\operatorname{diag} s(A) \in I$ are equivalent.

(iv) The lattice of $B(H)$-ideals forms a commutative semiring with multiplicative identity $B(H)$. That is, this lattice is commutative and associative under ideal addition and multiplication (see [4, Section 2.8]) and it is distributive. Distributivity with multiplier $K(H)$ is stated without proof in [9, Lemma 5.6 – preceding comments].

One important feature of principal ideals in a general ring R is that they are building blocks for all ideals I that contain them in that:

$$I = \bigcup_{r_1,\ldots,r_n \in I,\, n \in \mathbb{N}} (r_1) + \cdots + (r_n).$$

Note also $(r) = r + Rr + rR + \sum_{\text{finite sum}} RrR$, and if R is unital, this reduces to $(r) = Rr + rR + \sum_{\text{finite sum}} RrR$.

When $R = J$ is a $B(H)$-ideal, $\sum_{\text{finite sum}} RrR = RrR$ [4, Lemma 6.3], in which case (r) collapses to $(r) = r + Rr + rR + RrR$.

(v) When $T = \sum_{i=1}^{n} A_iTB_i$ with each A_i or $B_i \in J$, one has the important s-number relation: $s(T) = O(D_m(s(T))s(C))$ for some $C \in J$ (since then $T \in (T)J$, see [9, Section 1, p. 6] and Remark 2.3(i)).

Historical Background. Calkin–Schatten completely characterized $B(H)$-ideals via the lattice preserving isomorphism between $B(H)$-ideals and characteristic sets $\Sigma \subseteq c_0^*$ where c_0^* denotes the cone of nonnegative sequences decreasing to zero; characteristic sets Σ are those subsets of c_0^* that are additive, hereditary (solid) and ampliation invariant (invariant under each m-fold ampliation $D_m\xi := \langle \xi_1, \ldots, \xi_1, \xi_2, \ldots, \xi_2, \cdots \rangle$ with each entry ξ_i repeated m times); the characteristic set $\Sigma(I) := \{\eta \in c_0^* \mid \operatorname{diag} \eta \in I\}$, so $\Sigma(K(H)) = c_0^*$.

In 1983 Fong–Radjavi [5] investigated principal $K(H)$-ideals. They found principal $K(H)$-ideals that are not $B(H)$-ideals (Example 2.4 below) by determining necessary and sufficient conditions for a principal $K(H)$-ideal to be also a $B(H)$-ideal [5, Theorem 2]. And in doing so, at least for the authors of this paper, they initiated the study of subideals. The main results of Fong–Radjavi [5] are summarized in the following theorem.

Theorem ([5, Theorems 1–2]). *For T a compact operator of infinite rank, $P := (T^*T)^{\frac{1}{2}}$, $\mathcal{I}$ the ideal in $K(H)$ generated by T, and $\mathcal{P}$ the ideal of $K(H)$ generated by P, the following are equivalent.*

(i) *$\mathcal{I}$ is an ideal in $B(H)$.*
(ii) *$\mathcal{P}$ is an ideal in $B(H)$.*
(iii) *$\mathcal{P}$ is a Lie ideal in $B(H)$.*
(iv) *$T = A_1TB_1 + \cdots + A_kTB_k$ for some k, $A_i \in K(H)$, $B_i \in B(H)$.*
(v) *$T = A_1TB_1 + \cdots + A_kTB_k$ for some k, $A_i, B_i \in K(H)$.*
(vi) *For some integer $k > 1$, $s_{nk}(P) = o(s_n(P))$ as $n \to \infty$.*

Fong–Radjavi proved this via the positive case employing Lie ideal condition (iii), but our approach below avoids considering separately the positive case and any Lie ideal considerations. Notably also, conditions (iv)–(v) above indicate the relevance of elementary operators with coefficient constraints.

Example 2.4. Condition (vi) of the above theorem shows that if the singular number sequence of the operator P is given by $s(P) = \left\langle \frac{1}{2^n} \right\rangle$, then the principal $K(H)$-ideal generated by P is a $B(H)$-ideal. But if $s(P) = \left\langle \frac{1}{n} \right\rangle$, then the principal $K(H)$-ideal generated by P is not a $B(H)$-ideal.

Indeed, $\frac{\frac{1}{2^{nk}}}{\frac{1}{2^n}} = \frac{1}{2^{n(k-1)}} \to 0$ but $\frac{\frac{1}{nk}}{\frac{1}{n}} = \frac{1}{k} \nrightarrow 0$ as $n \to \infty$.

3. Subideals of operators

Motivated by the Calkin–Schatten characterization and the seminal work of Fong–Radjavi, a natural question to ask is:

What can be said about subideals? Can they be characterized in some way?

A conventional approach to attack the characterization problem for J-ideals is to begin at the elementary level as did Fong–Radjavi, albeit they did not consider characterizations except implicitly for principal $K(H)$-ideals in one of their proofs. So we first investigate principal J-ideals, then finitely generated J-ideals and then J-ideals $\mathcal{I} = (\mathcal{S})_J$ generated by sets $\mathcal{S}$ of higher cardinalities including the countable case. We fully generalize Fong–Radjavi's result [5, Theorem 2] from principal $K(H)$-ideals to arbitrary principal J-ideals and then to finitely generated J-ideals. The reason to consider the finitely generated case separate from the principal case is that, unlike $B(H)$-ideals where every finitely generated $B(H)$-ideal is always a principal $B(H)$-ideal, a finitely generated J-ideal need not be a principal J-ideal (see Section 4 – Example 4.2 for the case $J = K(H)$). Consequently, we characterize all J-ideals generated by sets of cardinality strictly less than the cardinality of the continuum, including finitely and countably generated J-ideals. A key property in this characterization turned out to be J-softness of a $B(H)$-ideal I inside J, that is, $IJ = I$ (Definition 2.2) a generalization of a recent notion of $K(H)$-softness of $B(H)$-ideals introduced by Kaftal–Weiss [9] and earlier exploited for Banach and Hilbert spaces by Mityagin and Pietsch.

We first begin with the following algebraic description of the principal J-ideal generated by $S \in J$ (see Remark 2.3(iv)).

Proposition 3.1. *For $S \in J$, an algebraic description of principal J-ideal $(S)_J$ is given by*

$$(S)_J = \left\{ \alpha S + AS + SB + \sum_{i=1}^{m} A_i S B_i \mid A,\ B,\ A_i,\ B_i \in J,\ \alpha \in \mathbb{C},\ m \in \mathbb{N} \right\}$$

That is, $(S)_J = \mathbb{C}S + JS + SJ + J(S)J$.

The following theorem generalizes Fong–Radjavi's result from principal $K(H)$-ideals to principal J-ideals by determining necessary and sufficient conditions for a principal J-ideal to be also a $B(H)$-ideal. Here is where J-softness first played a prominent role.

For compact operators S, T, the product $s(S)s(T)$ denotes the pointwise product of their s-number sequences.

Theorem 3.2. *For $S \in J$ and $(S)_J$, the principal J-ideal generated by S, the following are equivalent.*

(i) *$(S)_J$ is a $B(H)$-ideal.*
(ii) *The principal $B(H)$-ideal (S) is J-soft, i.e., $(S) = J(S)$; (equivalently, $(S) = (S)J$).*
(iii) *$S = AS + SB + \sum_{i=1}^{m} A_i S B_i$ for some $A,\ B,\ A_i,\ B_i \in J,\ m \in \mathbb{N}$.*
(iv) *$s(S) = O(D_k(s(S))s(T))$ for some $T \in J$ and $k \in \mathbb{N}$.*

Proof of (i) $\Rightarrow$ (ii) *only.* This is the main part of the proof so we provide here an outline. For every unitary map $\phi : H \to H \oplus H$, $S \mapsto \phi S \phi^{-1}$ preserves s-number sequences and hence also ideals via Calkin–Schatten's representation. Since $(S)_J$ is a $B(H)$-ideal containing S, $\phi^{-1}(S \oplus 0)\phi$, $\phi^{-1}(0 \oplus S)\phi \in (S)_J$ since they possess the same s-numbers as S. Then by Proposition 3.1 for principal J-ideal $(S)_J$,

$$\phi^{-1}(S \oplus 0)\phi = \alpha S + X \quad \text{and} \quad \phi^{-1}(0 \oplus S)\phi = \beta S + Y$$

for some $X, Y \in JS + SJ + J(S)J$, $\alpha, \beta \in \mathbb{C}$.

If $\alpha = 0$ or $\beta = 0$, then $\phi^{-1}(S \oplus 0)\phi$ or $\phi^{-1}(0 \oplus S)\phi \in J(S)$. Then, in either case, $S \in J(S)$, hence $(S) \subseteq J(S)$ and since the other inclusion is automatic, one has $(S) = J(S)$. If $\alpha, \beta \neq 0$, multiplying the first equation by $-\beta$ and the second equation by α and adding obtains $\phi^{-1}(-\beta S \oplus \alpha S)\phi = -\beta X + \alpha Y \in J(S)$. Multiplying $-\beta S \oplus \alpha S$ in $B(H \oplus H)$ by a suitable diagonal projection one obtains $\phi^{-1}(S \oplus 0)\phi \in J(S)$. Hence, also $S \in J(S)$, again equivalent to (ii). □

Remark 3.3. Using basic linear algebra techniques, we extended Theorem 3.2 from principal J-ideals to finitely generated J-ideals by solving a large system of linear equations which we then project into a finite-dimensional quotient space [14, Theorem 4.5].

The techniques for finitely generated subideals do not work for countably generated subideals because the latter case involves an intractable infinite system of equations, so a more sophisticated approach was needed. Based on the Hamel dimension of a related quotient space (Proposition 3.4 next), a necessary and sufficient softness condition is found for a subideal with a generating set of cardinality strictly less than $\mathfrak{c}$, to be also a $B(H)$-ideal. In particular, this softness condition applies to all countably generated subideals (Theorem 3.5, see also [14, Theorem 4.1]). We then use this condition to characterize the structure of these subideals (Theorem 3.7, see also [14, Theorem 4.4]). To investigate this in [14], we began with the following proposition.

Proposition 3.4 ([14, Proposition 3.1]). *For the J-ideal $(\mathcal{S})_J$ generated by a set $\mathcal{S}$ and defining*

$$(\mathcal{S})_J^0 := \operatorname{span}\{\mathcal{S}J + J\mathcal{S}\} + J(\mathcal{S})J,$$

the Hamel dimension of the quotient space $(\mathcal{S})_J/(\mathcal{S})_J^0$ is at most the cardinality of the generating set $\mathcal{S}$.

The main softness theorem for when a J-ideal is also a $B(H)$-ideal [14]:

Theorem 3.5 ([14, Theorem 4.1]). *A J-ideal $(\mathcal{S})_J$ generated by a set $\mathcal{S}$ of cardinality strictly less than $\mathfrak{c}$ is a $B(H)$-ideal if and only if the $B(H)$-ideal $(\mathcal{S})$ is J-soft.*

Sketch of proof. Here we sketch only the proof of the first implication, that is, that $(\mathcal{S})_J$ is a $B(H)$-ideal implies $(\mathcal{S})$ is J-soft. The reverse implication is somewhat routine. The algebraic structure of $(\mathcal{S})_J$ is given by $(\mathcal{S})_J = \operatorname{span}\{\mathcal{S}\} + (\mathcal{S})_J^0$ and so the quotient space $(\mathcal{S})_J/(\mathcal{S})_J^0 = \operatorname{span}\{[S_\alpha]\}$ where S_α ranges over $\mathcal{S}$. Hence the Hamel dimension of $(\mathcal{S})_J/(\mathcal{S})_J^0$ is strictly less than $\mathfrak{c}$. And by minimality $(\mathcal{S})_J = (\mathcal{S})$, since $(\mathcal{S})_J$ is also a $B(H)$-ideal.

The assumption that $(\mathcal{S})J \subsetneq (\mathcal{S})$ provides an operator in their difference which we use to construct an imbedding of ℓ^p into $(\mathcal{S})_J/(\mathcal{S})_J^0$. But the Hamel dimension of ℓ^p is $\mathfrak{c}$ [7, Lemma 3.4] and the Hamel dimension of $(\mathcal{S})_J/(\mathcal{S})_J^0$ is strictly less than $\mathfrak{c}$, a contradiction. Therefore, the condition $(\mathcal{S})_J$ is a $B(H)$-ideal implies that $(\mathcal{S})J = (\mathcal{S})$, that is, $(\mathcal{S})$ is J-soft. □

Remark 3.6. Theorem 3.5 on the equivalence of a J-ideal $(\mathcal{S})_J$ being a $B(H)$-ideal and $(\mathcal{S})$, the $B(H)$-ideal it generates, being J-soft motivates the question on whether this is always true independent of its various classes of generators. The answer is no from the following example. And Theorem 3.5 yields new information about the possible cardinality of any class of its generators.

The $K(H)$-ideal $(\operatorname{diag}\langle\frac{1}{n}\rangle)$ is also a principal $B(H)$-ideal but is not $K(H)$-soft [14, Section 4, Example 4.5]. Thus $\mathcal{I}$ being a $B(H)$-ideal is not equivalent to J-softness of the $B(H)$-ideal $(\mathcal{I})$, for $\mathcal{I}$ a J-ideal and $(\mathcal{I})$ the $B(H)$-ideal generated by $\mathcal{I}$. Moreover, by Theorem 3.5, $(\operatorname{diag}\langle\frac{1}{n}\rangle)$ which is also a $K(H)$-ideal, cannot be generated in $K(H)$ by less than $\mathfrak{c}$ generators.

As a consequence of Theorem 3.5 we obtain a characterization of all J-ideals generated by sets of cardinality strictly less than the cardinality of the continuum. These are the countably generated J-ideals when assuming the continuum hypothesis, and otherwise these include more J-ideals than the countably generated ones.

Theorem 3.7 (Structure Theorem for $(\mathcal{S})_J$ when $|\mathcal{S}| < \mathfrak{c}$). *The algebraic structure of the J-ideal $(\mathcal{S})_J$ generated by a set $\mathcal{S}$ of cardinality strictly less than $\mathfrak{c}$ is given by*

$$(\mathcal{S})_J = \operatorname{span}\{\mathcal{S} + J\mathcal{S} + \mathcal{S}J\} + J(\mathcal{S})J,$$

$J(\mathcal{S})J$ is a $B(H)$-ideal, $\operatorname{span}\{J\mathcal{S} + \mathcal{S}J\} + J(\mathcal{S})J$ *is a J-ideal, and*

$$J(\mathcal{S})J \subset \operatorname{span}\{J\mathcal{S} + \mathcal{S}J\} + J(\mathcal{S})J \subset (\mathcal{S})_J$$

This inclusion collapses to

$$J(\mathcal{S})J = (\mathcal{S})_J$$

if and only if

$$(\mathcal{S}) \text{ is } J\text{-soft (i.e., } (\mathcal{S})J = (\mathcal{S})).$$

4. Comparison of subideals to $B(H)$-ideals

As mentioned in Preliminaries Section 2, a subideal may not be linear. This led the authors of this paper to introduce three kinds of J-ideals, namely, linear, real-linear and classical J-ideals ([13, Definition 2.1])(the latter two are nonlinear). The term "classical" is meant in the sense of abstract rings, for instance, ideals where scalar multiplication may not make sense. The classical principal J-ideal generated by S is defined by $\langle S\rangle_J := \bigcap\{I \mid I \text{ is a classical } J\text{-ideal containing } S\}$. From Remark 2.3(iv) one deduces that

$$\langle S\rangle_J = \left\{nS + AS + SB + \sum_{i=1}^{m} A_iSB_i \mid A,\ B,\ A_i,\ B_i \in J,\ n \in \mathbb{Z},\ m \in \mathbb{N}\right\}.$$

Example 4.1 (A concrete nonlinear principal ideal is: $\langle \operatorname{diag}\langle 1/n\rangle\rangle_{K(H)}$). Indeed, if it were linear, then the principal $B(H)$-ideal $(\operatorname{diag}\langle \frac{1}{n}\rangle)$ would be $K(H)$-soft, which is not the case. (Combine Example 2.4 and Theorem 3.2.)

The explicit description of the principal J-ideal generated by S given in Proposition 3.1 implies that every principal J-ideal contains $J(S)J$. It is well known that every proper $B(H)$-ideal contains $F(H)$, the $B(H)$-ideal of all finite rank operators [6, Chapter III, Section 1, Theorem 1.1]. Likewise one sees here that every nonzero principal J-ideal contains $F(H)$ (since $S \neq 0$ implies $(S)_J \supset J(S)J \neq \{0\}$ and so the $B(H)$-ideal $J(S)J \supset F(H)$) and hence so also for every nonzero J-ideal because each is algebraically spanned by its principal ones. The intersection of all $B(H)$-ideals properly containing $F(H)$ is precisely $F(H)$ [12, Corollary 3.8(ii)], and since every $B(H)$-ideal is a J-ideal, it is clear then that the intersection of all J-ideals properly containing $F(H)$ is also precisely $F(H)$.

Some striking differences between J-ideals and $B(H)$-ideals are described next for the case $J = K(H)$ in Examples 4.2–4.5. Every finitely generated $B(H)$-ideal is always a principal $B(H)$-ideal because, as is straightforward to see, the $B(H)$-ideal generated by $\mathcal{S} = \{S_1, \dots, S_n\} \subset B(H)$, namely $(\mathcal{S})$, is precisely the principal ideal $(|S_1| + \cdots + |S_n|)$ where $|S| := (S^*S)^{1/2}$. But finitely generated J-ideals (classical, linear or real-linear) may not be principal as seen in the following example.

Example 4.2 (A doubly generated J-ideal of any of the three types that is not principal). For $J = K(H)$,

$$S_1 = \operatorname{diag}\left(1, 0, \frac{1}{2}, 0, \frac{1}{3}, \cdots\right) \quad \text{and} \quad S_2 = \operatorname{diag}\left(0, 1, 0, \frac{1}{2}, 0, \frac{1}{3}, \cdots\right),$$

$(\{S_1, S_2\})_{K(H)}$ is not a principal linear $K(H)$-ideal, and likewise for the classical and real-linear cases $\langle\{S_1, S_2\}\rangle_J$ and $(\{S_1, S_2\})_J^{\mathbb{R}}$ [13, Section 4, Example 4.1].

For $T \in B(H), (T) = (|T|)$, but this need not be true for principal linear $K(H)$-ideals (Example 4.3). Moreover, all $B(H)$-ideals are selfadjoint, but this is not necessarily true for principal linear $K(H)$-ideals (Example 4.4) and unlike $B(H)$-ideals, $K(H)$-ideals need not necessarily commute under ideal product (Example 4.5).

Example 4.3. If $J = K(H)$ and operator $T = \operatorname{diag}\langle \frac{i^n}{n} \rangle$, then $(T)_{K(H)} \neq (|T|)_{K(H)}$. In fact, $(|T|)_{K(H)} \not\subseteq (T)_{K(H)}$ and $(T)_{K(H)} \not\subseteq (|T|)_{K(H)}$ [13, Section 5, Example 5.1].

Example 4.4 ($K(H)$-ideal that is not closed under the adjoint operation). $T^* \notin (T)_{K(H)}$ where $T = \operatorname{diag}\langle \frac{i^n}{n} \rangle$, [13, Section 5, Example 5.2].

Example 4.5 ($K(H)$-ideals that do not commute). For $J = K(H)$ and with respect to the standard basis take S to be the diagonal matrix

$$S := \operatorname{diag}(1, 0, 1/2, 0, 1/3, 0, \dots)$$

and T to be the weighted shift with this same weight sequence.

Then $(S)_{K(H)}(T)_{K(H)} \neq (T)_{K(H)}(S)_{K(H)}$, [14, Section 5, Example 5.4].

5. Subideal-Traces

Subideals $\mathcal{I}$ that are not $B(H)$-ideals need not be invariant under unitary equivalence, i.e., $U\mathcal{I}U^* \not\subseteq \mathcal{I}$ for some unitary operator U (Examples 5.1–5.2 below). Therefore, the definition of trace on a $B(H)$-ideal, that is, a unitarily invariant linear functional, need not make sense on a subideal. Motivated by our work in [2] on unitary operators of the form $U = \mathbf{1} + A$ for $A \in K(H)$ we observe that subideals $\mathcal{I}$ are invariant under these unitaries (i.e., $U\mathcal{I}U^* \subset \mathcal{I}$). This led the authors of this paper to introduce the notion of a subideal-trace as defined below in Definition 5.3 (see also Remark 5.9).

Example 5.1 (A $K(H)$-ideal that is *not* invariant under unitary equivalence). For $J = K(H)$ and a unitary map $\phi : H \to H \oplus H$, consider $S = \phi^{-1}(D \oplus 0)\phi$ for $D = \operatorname{diag}\langle \frac{1}{n} \rangle$. Then $(S)_{K(H)}$ the principal $K(H)$-ideal generated by S is not invariant under unitary equivalence. We prove this by constructing one unitary operator U for which $USU^* \notin (S)_{K(H)}$. Indeed, assume $(S)_{K(H)}$ is invariant under unitary equivalence. We then have the following contradiction. Since

$$\phi^{-1} \begin{pmatrix} 0 & 1 \\ 1 & 0 \end{pmatrix} \phi \text{ is a unitary operator in } B(H),$$

it follows that

$$\phi^{-1} \begin{pmatrix} 0 & 1 \\ 1 & 0 \end{pmatrix} \phi \; S \; \phi^{-1} \begin{pmatrix} 0 & 1 \\ 1 & 0 \end{pmatrix} \phi = \phi^{-1} \begin{pmatrix} 0 & 0 \\ 0 & D \end{pmatrix} \phi \in (S)_{K(H)}$$

Using the algebraic structure of $(S)_{K(H)}$ (Proposition 3.1) one obtains,

$$\phi^{-1} \begin{pmatrix} 0 & 0 \\ 0 & D \end{pmatrix} \phi = \alpha S + X,$$

where $X \in K(H)S + SK(H) + K(H)(S)K(H) \subset \left(\operatorname{diag}\langle \frac{1}{n} \rangle\right) K(H)$ (since $s(S) = s(D), (S) = \left(\operatorname{diag}\langle \frac{1}{n} \rangle\right)$). That is,

$$\phi^{-1} \begin{pmatrix} -\alpha D & 0 \\ 0 & D \end{pmatrix} \phi \in (\operatorname{diag}\left\langle \frac{1}{n} \right\rangle) K(H).$$

This implies that $D \in (\operatorname{diag}\langle \frac{1}{n} \rangle)K(H)$, a contradiction to the non-softness of $(\operatorname{diag}\langle \frac{1}{n} \rangle)$ [13, Example 3.3]. Therefore, $(S)_{K(H)}$ is not invariant under unitary equivalence.

Example 5.2 ($K(H)$-ideal that *is* invariant under unitary equivalence). Varga [16] constructed a concrete example of a $K(H)$-ideal generated by the unitary orbit of a positive compact operator that is not a $B(H)$-ideal, namely, $(\mathcal{U}(A))_{K(H)}$ where $0 \le A \in K(H)$ and $\mathcal{U}(A) = \{UAU^* \mid U^* = U^{-1}\}$. Using Remark 2.3 (iv) for an ideal written as the union of finite sums of its principal ideals, and Proposition 3.1 giving the algebraic structure of the principal $K(H)$-ideal $(UAU^*)_{K(H)}$ generated by UAU^*: for each $T \in (UAU^*)_{K(H)}$ and V a unitary operator in $B(H)$, from Proposition 3.1 one has

$$\begin{aligned} VTV^* &= V(\alpha UAU^* + BUAU^* + UAU^*C + A'XB')V^* \\ &\qquad \text{(where } B, C, A', B' \in K(H), X \in (UAU^*)\text{)} \\ &= \alpha VUAU^*V^* + VBV^*VUAU^*V^* \\ &\quad + VUAU^*V^*VCV^* + VA'V^*VXV^*VB'V^* \\ &\qquad (VXV^* \in (VUAU^*V^*) \text{ since } X \in (UAU^*)) \\ &\in (VUAU^*V^*)_{K(H)} \subset (\mathcal{U}(A))_{K(H)} \qquad \text{(since } VU \text{ is unitary).} \end{aligned}$$

Therefore the $K(H)$-ideal $(\mathcal{U}(A))_{K(H)}$ is invariant under unitary equivalence.

Denote by $\mathcal{U}(H)$ the full group of unitary operators in $B(H)$. Recall the essential feature of traces: their unitary invariance, that is, τ is a trace on a $B(H)$-ideal I when it is a linear functional for which $\tau(UTU^*) = \tau(T)$ for all $T \in I, U \in \mathcal{U}(H)$. And essential for this is that Ad_U preserves I, that is, for every $X \in I$ and $U \in \mathcal{U}(H), Ad_U(X) := UXU^* \in I$. But for J-ideals $\mathcal{I}$, Ad_U may not preserve $\mathcal{I}$ (Example 5.1 above). However some adjustments can be made to preserve much of the trace notion.

Definition 5.3. For a J-ideal $\mathcal{I}$ and the subgroup of unitary operators

$$\mathcal{U}_J(H) := \{\mathbf{1} + A \in \mathcal{U}(H) |\, A \in J\},$$

a linear functional

$$\tau : \mathcal{I} \to \mathbb{C}$$

is called a subideal-trace if $\tau(X) = \tau(UXU^*)$ for every $X \in \mathcal{I}$, $U \in \mathcal{U}_J(H)$. In other words, τ is called a subideal-trace if τ is $\mathrm{Ad}_{\mathcal{U}_J(H)}$-invariant, that is, if $\tau(X) = \tau(Ad_U(X))$ for $U \in \mathcal{U}_J(H)$ and $X \in \mathcal{I}$.

Remark 5.4. In particular, if $J = B(H)$ (so $\mathcal{U}_{B(H)}(H) = \mathcal{U}(H)$), then $\mathcal{I}$ is a $B(H)$-ideal and hence Ad_U preserves $\mathcal{I}$ for $U \in \mathcal{U}(H)$ and Definition 5.3 becomes the standard definition of a trace on a $B(H)$-ideal.

Example 5.5 (A simple example of a subideal-trace). Consider $(S)_J$, a principal linear J-ideal generated by $S \in J$ that is not a $B(H)$-ideal, and recall Proposition 3.1 on the structure of its elements. Define the map $\tau : (S)_J \to \mathbb{C}$ as

$$\tau\left(\alpha S + AS + SB + \sum_{k=1}^{m} A_k S B_k\right) := \alpha,$$

where $A, B, A_k, B_k \in J, \alpha \in \mathbb{C}, m \in \mathbb{N}$. By our methods developed earlier, it is elementary to show that τ is a well-defined linear functional on $(S)_J$ when $(S)_J$ is not a $B(H)$-ideal. Indeed, if $\alpha S + X = \beta S + Y$ for $X, Y \in SJ + JS + J(S)J$, then $(\alpha - \beta)S \in SJ + JS + J(S)J$. Since $(S)_J$ is not a $B(H)$-ideal, $\alpha = \beta$ (otherwise $S \in J(S)$ which by Theorem 3.2 implies $(S)_J$ is a $B(H)$-ideal). Therefore $\tau(\alpha S + X) = \tau(\beta S + Y)$, hence τ is a well-defined map. It is elementary to show that τ is a linear map. And since

$$(\mathbf{1}+A)\left(\alpha S + AS + SB + \sum_{k=1}^{m} A_k S B_k\right)(\mathbf{1}+A^*) = \alpha S + X \text{ for } X \in SJ + JS + J(S)J,$$

it follows that τ is $\mathrm{Ad}_{\mathcal{U}_J(H)}$-invariant. Hence τ is a subideal-trace on $(S)_J$.

The commutator space of a $B(H)$-ideal I, $[I, B(H)]$, is the linear span of single commutators $[A, B]$ for $A \in I, B \in B(H)$. Since $UXU^* - X = [UX, U^*] \in [I, B(H)]$ for every $X \in I$ and every unitary operator $U \in \mathcal{U}(H)$, and since unitary operators span $B(H)$, unitarily invariant linear functionals on I are precisely the linear functionals on I that vanish on the commutator space $[I, B(H)]$ [11, Section 2].

Because every operator is the linear combination of four unitary operators, the well-known commutator space $[I, B(H)]$ is also the linear span of the single commutators $[A, U]$ for $A \in I, U \in \mathcal{U}(H)$. That is, $[I, \mathcal{U}(H)] = [I, B(H)]$. Observing that $\mathcal{U}_{B(H)}(H) = \mathcal{U}(H)$, we make the following analog.

Definition 5.6. The $\mathcal{U}_J(H)$-commutator space of J-ideal $\mathcal{I}$ is defined as

$$[\mathcal{I}, \mathcal{U}_J(H)] := \text{linear span}\{[X, U] \mid X \in \mathcal{I}, U \in \mathcal{U}_J(H)\}.$$

Notice that if $\mathcal{I}$ is a $B(H)$-ideal, then the $\mathcal{U}_J(H)$-commutator space of $\mathcal{I}$ is precisely $[\mathcal{I}, B(H)]$, the commutator space of $\mathcal{I}$.

In the following proposition we obtain a necessary and sufficient condition for a linear functional on a subideal to be a subideal-trace. This is an analog of the trace case just described.

Proposition 5.7. *For a J-ideal $\mathcal{I}$, a linear functional $\tau : \mathcal{I} \to \mathbb{C}$ is a subideal-trace if and only if τ vanishes on the $\mathcal{U}_J(H)$-commutator space of $\mathcal{I}$, that is, τ vanishes on $[\mathcal{I}, \mathcal{U}_J(H)]$.*

Proof. Suppose τ is a subideal-trace. It suffices to show that τ vanishes on single commutators $[X, U]$ for $X \in \mathcal{I}$ and $U \in \mathcal{U}_J(H)$. For $X \in \mathcal{I}$ and $\mathbf{1} + B \in \mathcal{U}(H)$ where $B \in J$, $X(\mathbf{1} + B) = X + XB \in \mathcal{I}$. Since τ is $Ad_{\mathcal{U}_J(H)}$-invariant,

$$\begin{aligned} \tau(X(\mathbf{1} + B)) &= \tau((\mathbf{1} + B)X(\mathbf{1} + B)(\mathbf{1} + B^*)) \\ &= \tau((\mathbf{1} + B)X)), \end{aligned}$$

i.e.,

$$\tau([X, (\mathbf{1} + B)]) = 0.$$

Therefore $\tau([X, U]) = 0$ for every $U \in \mathcal{U}_J(H)$.

Next we prove the reverse implication, that is, if τ vanishes on the $\mathcal{U}_J(H)$-commutator space of $\mathcal{I}$, $[\mathcal{I}, \mathcal{U}_J(H)]$, then τ is a subideal-trace. That is, for $U \in \mathcal{U}_J(H)$, $\tau(X) = \tau(UXU^*)$.

Since τ vanishes on $[\mathcal{I}, \mathcal{U}_J(H)]$, in particular, $\tau([X, (\mathbf{1} + B)]) = 0$ implying $\tau(BX) = \tau(XB)$ for all $X \in \mathcal{I}$ and $(\mathbf{1} + B) \in \mathcal{U}_J(H)$. Since $U = \mathbf{1} + B$ is a unitary operator, $(\mathbf{1} + B)(\mathbf{1} + B^*) = \mathbf{1}$ hence $B + B^* + BB^* = 0$.

$$\begin{aligned} \tau((\mathbf{1} + B^*)X(\mathbf{1} + B)) - \tau(X) &= \tau(XB) + \tau(B^*X) + \tau(B^*XB) \\ &= \tau(BX) + \tau(B^*X) + \tau(BB^*X) \\ &\qquad \text{(since } B^*X \in \mathcal{I}) \\ &= \tau((B + B^* + BB^*)X) = \tau(0) = 0 \end{aligned}$$

Therefore linear functional τ is $Ad_{\mathcal{U}_J(H)}$-invariant, and so by Definition 5.3, τ is a subideal-trace on $\mathcal{I}$. □

Corollary 5.8. *The set of all subideal-traces on a J-ideal $\mathcal{I}$ can be identified with the elements of the linear dual of the quotient space $\frac{\mathcal{I}}{[\mathcal{I}, \mathcal{U}_J(H)]}$.*

Indeed, for a given subideal-trace τ on a subideal $\mathcal{I}$, define a functional $f_\tau : \frac{\mathcal{I}}{[\mathcal{I}, \mathcal{U}_J(H)]} \to \mathbb{C}$ as $f_\tau([X]) := \tau(X)$ where $[X]$ is the coset of the element $X \in \mathcal{I}$.

Since $[X] = [Y]$ *implies* $X - Y \in [\mathcal{I}, \mathcal{U}_J(H)]$ *and* τ *a subideal-trace,* $\tau(X - Y) = 0$ *which implies that* f_τ *is a well-defined linear functional on the quotient space. On the other hand, given a linear functional* f *on the quotient space* $\frac{\mathcal{I}}{[\mathcal{I},\mathcal{U}_J(H)]}$*, define a function* $\tau : \mathcal{I} \to \mathbb{C}$ *as* $\tau(X) := f([X])$*. Since* f *is a linear functional,* τ *is also a linear functional. And for every element* $Y \in [\mathcal{I}, \mathcal{U}_J(H)], f([Y]) = 0$ *implying* $\tau(Y) = 0$*. Hence* τ *vanishes on* $[\mathcal{I}, \mathcal{U}_J(H)]$*. Therefore by Proposition* 5.7*,* τ *is a subideal-trace on* $\mathcal{I}$*.*

Remark 5.9. A subideal $\mathcal{I}$ may be invariant under a larger class than $\mathcal{U}_J(H)$ but not invariant under the full group of unitary operators $\mathcal{U}(H)$. For instance, $U = \lambda(\mathbf{1} + B)$ for $|\lambda| = 1$ and $(\mathbf{1} + B) \in \mathcal{I}$. But there may be more less obvious unitary operators under which $\mathcal{I}$ is invariant (Example 5.10 below). This leads us to suggest the following alternative definition of a subideal-trace (Definition 5.11 below). However we will not explore it further here.

Example 5.10 (A $K(H)$-ideal invariant under a larger class of unitaries, but not invariant under the full group $\mathcal{U}(H)$). Using the principal $K(H)$-ideal $(S)_{K(H)}$ and the unitary map ϕ of Example 5.1, the unitary operator $U := \phi^{-1}(\mathbf{1} \oplus (-\mathbf{1}))\phi \in \mathcal{U}(H) \setminus \mathcal{U}_{K(H)}(H)$. That $U \notin \mathcal{U}_{K(H)}(H)$ is a simple computation. Then $(S)_{K(H)}$ is invariant under Ad_U because $USU^* = S$ (an easy verification combining the definition of U here with the definition of S in Example 5.1), but $(S)_{K(H)}$ is not invariant under Ad_U for $U \in \mathcal{U}(H)$ which again follows from Example 5.1.

Definition 5.11. For a J-ideal $\mathcal{I}$ and $\mathcal{U}^{\mathcal{I}}(H) := \{U \in \mathcal{U}(H) | \, UXU^* \in \mathcal{I} \text{ for } X \in \mathcal{I}\}$, a linear functional

$$\tau : \mathcal{I} \to \mathbb{C}$$

is called a $\mathcal{U}^{\mathcal{I}}(H)$-subideal-trace if $\tau(X) = \tau(UXU^*)$ for every $X \in \mathcal{I}$ and $U \in \mathcal{U}^{\mathcal{I}}(H)$, that is, τ is $\mathrm{Ad}_{\mathcal{U}^{\mathcal{I}}(H)}$-invariant.

The following inclusion holds for a subideal $\mathcal{I}$:

$$Ad_{\mathcal{U}^{\mathcal{I}}(H)}\text{-invar. subideal-traces of } \mathcal{I} \subset Ad_{\mathcal{U}_J(H)}\text{-invar. subideal-traces of } \mathcal{I}$$

The next natural question is whether or not these inclusions are proper. In particular, do Definition 5.3 and Definition 5.11 define different classes of functionals on a subideal that is not a $B(H)$-ideal? When $\mathcal{I}$ is a $B(H)$-ideal, Remark 5.4 tells us that they are the same class.

References

[1] Calkin, J.W., *Two-sided ideals and congruences in the ring of bounded operators in Hilbert space,* Ann. of Math. **42** (2)(1941), 839–873.

[2] Beltita, D., Patnaik, S., and Weiss, G., *On Cartan subalgebras of operator ideals,* in preparation.

[3] Dixmier, J., *Existence de traces non normales,* C. R. Acad. Sci. Paris Sér. A-B 262 (1966), A1107–A1108.

[4] Dykema, K., Figiel, T., Weiss, G., and Wodzicki, M., *The commutator structure of operator ideals*, Adv. Math., **185** (1) (2004), 1–79.

[5] Fong, C.K. and Radjavi, H., *On ideals and Lie Ideals of Compact Operators*, Math. Ann. **262**, 23–28 (1983).

[6] Gohberg, C.I. and Krein, M.G., *Introduction to the theory of nonselfadjoint operators*, Transl. Amer. Math. Soc. **18**, Providence, RI (1969).

[7] Halbeisen, Lorenz and Hungerbühler, Norbert, *The cardinality of Hamel bases of Banach spaces*, East-West J. Math., (2000) 153–159.

[8] Kaftal, V. and Weiss, G., *Traces, ideals, and arithmetic means*, Proc. Nat. Acad. Sci. U.S.A. 99(2002), 7356-7360.

[9] Kaftal, V. and Weiss, G., *Soft ideals and arithmetic mean ideals*, Integral equations and Operator Theory 58 (2007), 363–405.

[10] Kaftal, V. and Weiss, G., *A survey on the interplay between arithmetic mean ideals, traces, lattices of operator ideals, and an infinite Schur–Horn majorization theorem.* Hot topics in operator theory, Theta 2008, 101–135.

[11] Kaftal, V. and Weiss, G., *Traces on operator ideals and arithmetic means*, J. Operator Theory, 63 Issue 1, Winter 2010, 3–46.

[12] Kaftal, V. and Weiss, G., *$B(H)$ lattices, density and arithmetic mean ideals*, Houston J. Math., **37**(1)(2011), 233–283.

[13] Patnaik, S. and Weiss, G., *Subideals of Operators*, Journal of Operator Theory, 101–122 (2011).

[14] Patnaik, S. and Weiss, G., *Subideals of Operators II*, Integral Equations and Operator Theory, Volume 74, Issue 4 (2012), pp. 587–600.

[15] Schatten, R., *Norm ideals of completely continuous operators*, Ergebnisse der Mathematik und ihrer Grenzgebiete, Neue Folge, Heft, Vol. 27, Springer, Berlin (1960).

[16] Varga, J., *On unitary invariant ideals in the algebra of compact operators*, Proc. Amer. Math. Soc., Volume 108, Number 3 (1990).

Sasmita Patnaik
Department of Mathematics and Statistics
Indian Institute of Technology
Kanpur, Uttar Pradesh, India 208016
Telephone: 8127989114
e-mail: sasmita@iitk.ac.in
sasmita 19@yahoo.co.in

Gary Weiss
University of Cincinnati
Department of Mathematics
Cincinnati, OH, 45221-0025, USA
e-mail: gary.weiss@uc.edu

Operator Theory:
Advances and Applications, Vol. 240, 235–242

Multipliers and L^p-operator Semigroups

Werner J. Ricker

Abstract. Deciding whether the generator of certain semigroups of operators in $L^p(\mathbb{R})$ are unbounded scalar-type spectral operators can be reduced to deciding when $e^{i\varphi}$, for specific unbounded functions $\varphi : \mathbb{R} \to \mathbb{R}$, is a p-multiplier. We illustrate how van der Corput's lemma is an effective technique in this regard.

Mathematics Subject Classification (2010). 42A45, 47B40, 47D06.

Keywords. Operator semigroup, p-multiplier, van der Corput lemma.

When considering spectral properties of infinitesimal generators of certain semigroups of operators in L^p-spaces one quickly arrives at questions concerning p-multipliers. For ease of presentation we will consider only the real line $\mathbb{R}$. So, let $D = -i\frac{d}{dx}$ be the closed, densely defined operator of differentiation in L^p with domain $\mathcal{D}(D) = \{f \in L^p : f \in AC,\ f' \in L^p\}$, where L^p denotes $L^p(\mathbb{R})$ and AC is the space of functions on $\mathbb{R}$ which are absolutely continuous on bounded intervals. By $\mathcal{L}(L^p)$, $1 < p < \infty$, we denote the space of bounded linear operators of L^p into itself. Let $\mathcal{M}^{(p)}$ denote the Banach algebra of all Fourier multipliers for L^p relative to the group $\mathbb{R}$; briefly, p-multipliers. Then each function $\psi \in \mathcal{M}^{(p)}$ specifies an element $\psi(D)$ of $\mathcal{L}(L^p)$ via the formula $(\psi(D)f)\widehat{\ } = \psi\widehat{f}$, for $f \in L^2 \cap L^p$, where $\widehat{\cdot}$ denotes the Fourier transform. The notation $\psi(D)$ is consistent with the fact (for $1 < p \leq 2$) that $\mathcal{D}(D) = \{f \in L^p : \xi\widehat{f}(\xi) = \widehat{g}(\xi)$, for some $g \in L^p\}$ and, for $f \in \mathcal{D}(D)$, that $Df = g$ where g is the unique element of L^p satisfying $\xi\widehat{f}(\xi) = \widehat{g}(\xi)$, for $\xi \in \mathbb{R}$. The multiplier norm $\|\psi\|_p$ of $\psi \in \mathcal{M}^{(p)}$ is defined to be the operator norm $\|\psi(D)\|$ of $\psi(D) \in \mathcal{L}(L^p)$.

Let $\varphi(x) = \sum_{j=0}^{n} a_j x^j$, for $x \in \mathbb{R}$, be any polynomial of *even* degree with real coefficients a_j, for $0 \leq j \leq n$. Suppose that $a_n > 0$. Then there exists $\alpha \in \mathbb{R}$ such that $\varphi(x) \geq \alpha$, for all $x \in \mathbb{R}$. By translating, if necessary, it may be assumed that $\alpha = 0$. Let $\mathcal{H}^+ = \{z \in \mathbb{C} : \ \mathrm{Re}(z) > 0\}$. For each $z \in \mathcal{H}^+$ the function $\varphi_z(x) = e^{-z\varphi(x)}$, for $x \in \mathbb{R}$, is rapidly decreasing and so $\{\varphi_z : z \in \mathcal{H}^+\} \subseteq \mathcal{M}^{(p)}$, for every $1 < p < \infty$. The corresponding family of Fourier p-multiplier operators $\{\varphi_z(D) : z \in \mathcal{H}^+\}$ is then an analytic semigroup in $\mathcal{L}(L^p)$ with infinitesimal generator $-\varphi(D)$. In L^2 the operator D is selfadjoint and hence, so is $\varphi(D) = \sum_{j=0}^{n} a_j D^j$, where D^j is defined in the usual way for non-negative integral powers of an unbounded operator. Accordingly, $\varphi(D) = \int_{\mathbb{R}} \varphi(\lambda)\, dQ(\lambda)$ has an integral representation with respect to the resolution of the identity Q of D. Of particular interest is the Laplace operator $-\frac{d^2}{dx^2}$ corresponding to $\varphi(x) = x^2$. The question arises of whether the case $p = 2$ carries over to other values of p in $(1, \infty)$, that is, whether $\varphi(D) = \int_{\mathbb{R}} \varphi(\lambda)\, dQ(\lambda) = \int_0^{\infty} \lambda\, dP(\lambda)$ for some spectral measure P defined on the Borel subsets of $[0, \infty)$ and with values in $\mathcal{L}(L^p)$? If this were the case, then it would follow that the semigroup of Fourier p-multiplier operators $\{\varphi_z(D) : z \in \mathcal{H}^+\}$ coincides with the operators $\int_0^{\infty} e^{-z\varphi(\lambda)}\, dP(\lambda)$, for $z \in \mathcal{H}^+$, defined via the usual calculus for scalar-type spectral operators, [1, Ch. XVII], and hence, $\{\varphi_z(D) : z \in \mathcal{H}^+\}$ is uniformly bounded in $\mathcal{L}(L^p)$. Standard multiplier convergence theorems would then imply that the "boundary group" $\varphi_{it}(D) = e^{-it\varphi(D)}$, for $t \in \mathbb{R}$, exists in $\mathcal{L}(L^p)$ and also consists of Fourier p-multiplier operators. So, the question of whether or not operators of the form $\varphi(D)$ are scalar-type spectral operators in L^p is reduced to determining when functions of the type

$$x \mapsto e^{i\varphi(x)}, \qquad x \in \mathbb{R}, \tag{1}$$

belong to $\mathcal{M}^{(p)}$. This formulation has a meaning for all $\mathbb{R}$-valued polynomials φ on $\mathbb{R}$ (not just those of even degree). So, it is of interest to determine which functions (1), with $\varphi : \mathbb{R} \to \mathbb{R}$ say a polynomial, belong to $\mathcal{M}^{(p)}$? Results of L. Hörmander, [3], are closely related to this question. Of course, various techniques are available in this situation. Our aim in this note is to show how a well-known result of van der Corput can be effectively used in treating this problem.

van der Corput's lemma. *Let $j \geq 1$ be an integer and $h \in C^{(j)}(a, b)$ be a $\mathbb{R}$-valued function satisfying $|h^{(j)}(w)| \geq \lambda > 0$, for all $w \in [a, b]$, where $-\infty < a < b < \infty$. In the case of $j = 1$ it is assumed that h' is monotone. Then there exists a constant $c_j > 0$ (independent of a, b and h) such that*

$$\left| \int_a^b e^{ih(w)}\, dw \right| \leq c_j \lambda^{-1/j}.$$

Remark 1. For $j = 1, 2$ we refer to [6, p.197]; see [5] for arbitrary j.

As an immediate application we have the following result.

Lemma 1. *Let $1 < p < \infty$. Let $\varphi : \mathbb{R} \to \mathbb{R}$ be a polynomial of degree at least two. Then $w \mapsto e^{i\varphi(w)}$, for $w \in \mathbb{R}$, is not an element of $\mathcal{M}^{(p)}$ unless $p = 2$.*

Proof. It suffices to consider $2 < p < \infty$. Fix $x \in \mathbb{R}$ and $t \in \mathbb{R} \setminus \{0\}$ and let $j = \deg(\varphi)$ denote the degree of φ. If $h_{x,t}(w) = xw - \varphi(tw)$, for $w \in \mathbb{R}$, then $h_{x,t}^{(j)}(w) = \alpha t^j$, for $w \in \mathbb{R}$, for some constant $\alpha \in \mathbb{R}$. Accordingly, for any bounded interval $[a, b] \subset \mathbb{R}$ it follows that

$$\left| h_{x,t}^{(j)}(w) \right| = |\alpha|\ |t|^j, \qquad w \in \mathbb{R}. \tag{2}$$

The L^1-function $f = \chi_{[a,b]}$ satisfies $\widehat{f}(v) = iv^{-1}(e^{-ibv} - e^{-iav})$, for $v \neq 0$, from which it is clear (observe $\widehat{f} \in C_0(\mathbb{R})$) that $\widehat{f} \in L^r$, for every $1 < r < \infty$. In particular, there exists $u \in L^2 \cap L^p$ such that $\widehat{u} = f$.

For $t \in \mathbb{R}$, define

$$u_t(x) = (2\pi)^{-1} \int_{\mathbb{R}} e^{i[xw - \varphi(tw)]}\, \widehat{u}(w)\, dw, \qquad x \in \mathbb{R}.$$

Then the formula $u_t(x) = \frac{1}{2\pi} \int_a^b e^{ih_{x,t}(w)}\, dw$, van der Corput's lemma and (2) imply that $\|u_t\|_\infty \leq c_j |t|^{-1}$, for $t \in \mathbb{R} \setminus \{0\}$ and some constant $c_j > 0$. Since $p > 2$, it follows that

$$\|u_t\|_p^p \leq \|u_t\|_\infty^{p-2} \|u_t\|_2^2 \;\leq\; d_j \, \|u_t\|_2^2 \, |t|^{2-p}, \qquad t \neq 0.$$

But, $\widehat{u}_t(\cdot) = \widehat{u}(\cdot) e^{i\varphi(t\cdot)}$ and so Parseval's formula implies that $\|u_t\|_2^2 = \|u\|_2^2$ from which it follows that

$$\|u_t\|_p^p \leq d_j \, \|u\|_2^2 \, |t|^{2-p}, \qquad t \neq 0.$$

Arguing as in the proof of Lemma 1.3 in [2] it follows that if $e^{i\varphi(t\cdot)} \in \mathcal{M}^{(p)}$ for some $t \neq 0$, then actually $e^{i\varphi(t\cdot)} \in \mathcal{M}^{(p)}$ for every $t \in \mathbb{R}$ and $\sup_{t \in \mathbb{R}} \|e^{i\varphi(t\cdot)}\|_p < \infty$. So, there is $\kappa_p > 0$ such that $\|u\|_p \leq \kappa_p \|u_t\|_p$ and hence, $\|u\|_p \leq r_j |t|^{(2p^{-1}-1)}$, for $t \neq 0$ and for some constant $r_j > 0$. Since $2p^{-1} < 1$ this contradicts $u \neq 0$ (let $t \to 0$). Accordingly, $e^{i\varphi(t\cdot)} \notin \mathcal{M}^{(p)}$ for every $t \neq 0$. □

Remark 2.

(a) Since $e^{i\varphi} \in \mathcal{M}^{(p)}$ whenever $\varphi(x) = \alpha x + \beta$ with $\alpha, \beta \in \mathbb{R}$ (for every $1 < p < \infty$), Lemma 1 answers the question of when $e^{i\psi}$ belongs to $\mathcal{M}^{(p)}$ for a given polynomial $\psi : \mathbb{R} \to \mathbb{R}$.

(b) It is an immediate consequence of Lemma 1 that the infinitesimal generator $-\varphi(D)$ of the analytic semigroup $\{\varphi_z(D) : z \in \mathcal{H}^+\}$ is not a scalar-type spectral operator in L^p if $p \neq 2$, where $\varphi : \mathbb{R} \to \mathbb{R}$ is any polynomial of even (and positive) degree. □

Lemma 1 suggests the question of also determining which functions of the type (1) are p-multipliers for φ a $\mathbb{R}$-valued *rational function* on $\mathbb{R}$. Given any $\mathbb{R}$-valued rational function f/g on $\mathbb{R}$, where f and g are $\mathbb{R}$-valued polynomials on $\mathbb{R}$, it will henceforth be assumed that all linear and quadratic factors (over the field $\mathbb{R}$) common to both f and g have been cancelled.

Theorem 1. *Let $1 < p < \infty$, with $p \neq 2$, and f/g be a $\mathbb{R}$-valued rational function on $\mathbb{R}$, where f and g are $\mathbb{R}$-valued polynomials on $\mathbb{R}$.*

(i) *If g has no real zero, then $e^{if/g} \in \mathcal{M}^{(p)}$ if and only if*

$$\deg(f) \leq 1 + \deg(g).$$

(ii) *If g has at least one real zero, then $e^{if/g} \notin \mathcal{M}^{(p)}$.*

This result, which incorporates Lemma 1, will be proved in a series of steps.

Lemma 2. *Let $1 < p < \infty$, with $p \neq 2$, and f/g be a $\mathbb{R}$-valued rational function on $\mathbb{R}$, where f and g are $\mathbb{R}$-valued polynomials on $\mathbb{R}$, and suppose that g has no real zero. Then $e^{if/g} \in \mathcal{M}^{(p)}$ if and only if*

$$\deg(f) \leq 1 + \deg(g).$$

Proof. We begin with the following

Claim. *If* $\deg(f) < \deg(g)$, *then $e^{if/g} \in \mathcal{M}^{(p)}$.*

The stated condition implies that f/g is of bounded variation on $\mathbb{R}$ (it is bounded and piecewise monotonic) and hence, by Stečkin's theorem, $f/g \in \mathcal{M}^{(p)}$ for every $1 < p < \infty$. Since $\mathcal{M}^{(p)}$ is a commutative Banach algebra and $z \mapsto e^{iz}$ is an entire function in $\mathbb{C}$, it follows that $e^{if/g} = \sum_{n=0}^{\infty}(if/g)^n/n!$ determines an element of $\mathcal{M}^{(p)}$; the series converges absolutely with respect to the norm $|\!|\!| \cdot |\!|\!|_p$ in $\mathcal{M}^{(p)}$ because

$$\sum_{n=0}^{\infty} \frac{|\!|\!|(if/g)^n|\!|\!|_p}{n!} \leq \sum_{n=0}^{\infty} \frac{|\!|\!|if/g|\!|\!|_p^n}{n!} = e^{|\!|\!|if/g|\!|\!|_p} < \infty.$$

Of course, $e^{if/g}$ so defined is the function $x \mapsto e^{if(x)/g(x)}$, for $x \in \mathbb{R}$. This establishes the claim.

Suppose now that $\deg(f) \leq 1 + \deg(g)$. Then there exist constants $\alpha, \beta \in \mathbb{R}$ and a real polynomial r with $\deg(r) < \deg(g)$ such that

$$\frac{f(x)}{g(x)} = \alpha + \beta x + \frac{r(x)}{g(x)}, \qquad x \in \mathbb{R}.$$

By the above claim (for the pair r, g in place of f, g) we have that $e^{ir/g} \in \mathcal{M}^{(p)}$. Since also $e^{i(\alpha+\beta x)} \in \mathcal{M}^{(p)}$, it follows that $e^{if/g} \in \mathcal{M}^{(p)}$.

Assume now that $\deg(f) > 1 + \deg(g)$. Then there exist real polynomials q, s with $\deg(s) < \deg(g)$ and $\deg(q) \geq 2$ such that

$$\frac{f(x)}{g(x)} = q(x) + \frac{s(x)}{g(x)}, \qquad x \in \mathbb{R}.$$

Since $e^{is/g} \in \mathcal{M}^{(p)}$ (via the above claim for the pair s, g in place of f, g), it would follow that $e^{iq} \in \mathcal{M}^{(p)}$ *if* it were the case that $e^{if/g} \in \mathcal{M}^{(p)}$. But, this would contradict Lemma 1 as $\deg(q) \geq 2$. Accordingly, we must have that $e^{if/g} \notin \mathcal{M}^{(p)}$. This proves that necessarily $\deg(f) \leq 1 + \deg(g)$ whenever $e^{if/g} \in \mathcal{M}^{(p)}$. □

Lemma 3. *Let $1 < p < \infty$, with $p \neq 2$, and $\varphi(x) = \alpha(\beta x - \gamma)^{-n}$, for each $x \in \mathbb{R} \setminus \{\gamma/\beta\}$, where α, β, γ are real numbers and n is a positive integer. Then $e^{i\varphi} \notin \mathcal{M}^{(p)}$.*

Proof. It suffices to consider $2 < p < \infty$. Since $\varphi = \psi \circ \rho$, where $\rho(x) = \beta x - \gamma$ is affine and $\psi(w) = \alpha w^{-n}$, it suffices to show that $e^{i\psi} \notin \mathcal{M}^{(p)}$, [2, Theorem 1.3].

Fix $x \in \mathbb{R}$ and $t \in \mathbb{R} \setminus \{0\}$. Define

$$h_{x,t}(w) = xw - \psi(tw) = xw - \alpha t^{-n} w^{-n}, \qquad w \in [1/2, 1],$$

in which case $h_{x,t}^{(2)}(w) = -\alpha n(n+1)t^{-n}w^{-(n+2)}$, for $w \in [1/2, 1]$. Hence,

$$\left|h_{x,t}^{(2)}(w)\right| \geq |\alpha| n(n+1)|t|^{-n}, \qquad w \in [1/2, 1]. \tag{3}$$

Let $u \in L^2 \cap L^p$ satisfy $\widehat{u} = \chi_{[1/2,1]}$. Define, for each $t \in \mathbb{R}$, the function

$$u_t(x) = \frac{1}{2\pi} \int_{\mathbb{R}} e^{i(xw - \psi(tw))} \,\widehat{u}(w)\, dw = \int_{1/2}^{1} e^{ih_{x,t}(w)}\, dw, \qquad x \in \mathbb{R}.$$

Via van der Corput's lemma (with $j = 2$) and (3) it follows that $\|u_t\|_\infty \leq c|t|^{n/2}$ for some constant c. Assume $e^{i\psi(t\cdot)} \in \mathcal{M}^{(p)}$ for some $t \neq 0$. Arguing as in the proof of Lemma 1 we have $\|u_t\|_\infty^{p-2} \leq c^*|t|^{n(p-2)/2}$ and hence, for some $d > 0$, that $\|u\|_p \leq d|t|^{n(1-2p^{-1})}$. Arguing again as in the proof of Lemma 1 and using the fact that $\lim_{t\to 0^+} |t|^{n(1-2p^{-1})} = 0$ (for $p > 2$) gives the desired contradiction. □

A slightly more general result is the following one.

Lemma 4. *Let $1 < p < \infty$, with $p \neq 2$, and*

$$\varphi(x) = \sum_{j=1}^{n} a_j (x - \alpha)^{-j}, \qquad x \in \mathbb{R} \setminus \{\alpha\},$$

where a_j, for $1 \leq j \leq n$, and α are real numbers with $a_n \neq 0$ and n is a positive integer. Then $e^{i\varphi} \notin \mathcal{M}^{(p)}$.

Proof. It suffices to consider $2 < p < \infty$ and (by Lemma 3) the case $n \geq 2$. Furthermore, as $\varphi = \psi \circ \rho$, where $\rho(x) = x - \alpha$ is affine and $\psi(w) = \sum_{j=1}^{n} a_j w^{-j}$, for $w \neq 0$, it suffices to show $e^{i\psi} \notin \mathcal{M}^{(p)}$. Multiplying by -1, if necessary, it may be assumed that $a_n > 0$.

Fix $x \in \mathbb{R}$ and $t \in \mathbb{R} \setminus \{0\}$. Define

$$h_{x,t}(w) = xw - \psi(tw) = xw - \sum_{j=1}^{n} a_j t^{-j} w^{-j}, \qquad w \neq 0,$$

in which case its second derivative

$$h_{x,t}^{(2)}(w) = -w^{-2-n} t^{-n} \sum_{j=1}^{n} j(j+1) a_j\, t^{n-j} w^{n-j}, \qquad w \neq 0.$$

Since $\lim_{s\to 0^+} \sum_{j=1}^{n} j(j+1)a_j s^{n-j} = 0$ there exists $0 < \delta < 1$ such that

$$\left|\sum_{j=1}^{n-1} j(j+1) a_j s^{n-j}\right| < \frac{1}{2} n(n+1) a_n, \qquad |s| < \delta.$$

Accordingly, for all t and w satisfying $0 < |tw| < \delta$ we have

$$\begin{aligned}\left|h^{(2)}_{x,t}(w)\right| &= |w|^{-2-n}|t|^{-n}\left|n(n+1)a_n - \left(-\sum_{j=1}^{n-1} j(j+1)a_j(tw)^{n-j}\right)\right| \\ &\geq |w|^{-2-n}|t|^{-n}\left|n(n+1)a_n - \left|\sum_{j=1}^{n-1} j(j+1)a_j(tw)^{n-j}\right|\right| \\ &\geq \frac{1}{2}\,|w|^{-2-n}|t|^{-n}\,n(n+1)a_n.\end{aligned}$$

Choose any $\gamma \in (0,\delta)$. Then for every $0 < t < 1$ and $w \in [\gamma,\delta]$ it is the case that $|tw| < \delta$ and $|w|^{-2-n} > 1$, from which it follows that

$$\left|h^{(2)}_{x,t}(w)\right| \geq \frac{1}{2}\,n(n+1)a_n t^{-n}, \qquad w \in [\gamma,\delta], \tag{4}$$

for every $0 < t < 1$. Assume that $e^{i\psi(s\,\cdot)} \in \mathcal{M}^{(p)}$ for some $s \neq 0$. If $u \in L^2 \cap L^p$ satisfies $\widehat{u} = \chi_{[\gamma,\delta]}$, then it is possible to argue as in the proof of Lemma 3 (with (4) replacing (2) there) that $\|u\|_p \leq ct^{n(1-2p^{-1})}$ for $0 < t < 1$ and some $c > 0$. Since $t^{n(1-2p^{-1})} \to 0$ as $t \to 0^+$ this gives the desired contradiction. □

Lemma 5. *Let $1 < p < \infty$, with $p \neq 2$, and define*

$$\varphi(x) := \sum_{r=1}^{k}\sum_{j=1}^{n(r)} a_j^{(r)}(x-\alpha_r)^{-j}, \qquad x \in \mathbb{R}\setminus\{\alpha_r\}_{r=1}^k,$$

where, for $1 \leq r \leq k$, the $a_j^{(r)}$, $1 \leq j \leq n(r)$, are real numbers with $a_{n(r)}^{(r)} \neq 0$, the α_r are distinct real numbers and the $n(r)$ are positive integers. Then $e^{i\varphi} \notin \mathcal{M}^{(p)}$.

Proof. Applying an affine transformation, if needed, we may assume that $\alpha_1 = 0$ and $a_{n(1)}^{(1)} > 0$. For ease of reading set $n := n(1)$. Assume $2 < p < \infty$. Write $\varphi = \varphi_1 + \varphi_2$, with $\varphi_1(w) = \sum_{j=1}^n a_j^{(1)} w^{-j}$ and $\varphi_2(w) = \sum_{r=2}^k \sum_{j=1}^{n(r)} a_j^{(r)}(w-\alpha_r)^{-j}$. Fix elements $x \in \mathbb{R}$ and $t \in \mathbb{R}\setminus\{0\}$. If $h_{x,t}(w) = xw - \varphi(tw)$, then $h^{(2)}_{x,t}(w) = -\varphi_1^{(2)}(tw) - \varphi_2^{(2)}(tw)$ and hence,

$$\left|h^{(2)}_{x,t}(w)\right| \geq \left|\,\left|\varphi_1^{(2)}(tw)\right| - \left|\varphi_2^{(2)}(tw)\right|\,\right|, \tag{5}$$

where the second derivatives of φ_1 and φ_2 are taken with respect to w. An examination of the proof of Lemma 4 shows that there exist γ and δ with $0 < \gamma < \delta < 1$ and

$$\left|\varphi_1^{(2)}(tw)\right| \geq \frac{1}{2}(n+1)na_n^{(1)}t^{-n}, \qquad w \in [\gamma,\delta], \tag{6}$$

for all $0 < t < 1$. It is clear that δ can be chosen so that $[0,\delta]$ does not contain any of the other singular points α_r, for $2 \leq r \leq k$, of φ. Then

$$\mu := \sup\left\{\left|\varphi_2^{(2)}(tw)\right|;\ w \in [\gamma,\delta],\ t \in [0,1]\right\} < \infty$$

and there exists $s \in (0,1)$ such that $\frac{1}{2}(n+1)na_n^{(1)} > \mu s^n$. It follows from (5) and (6) that

$$|h_{x,t}^{(2)}(w)| \geq \frac{1}{2}(n+1)na_n^{(1)}t^{-n} - \mu, \qquad w \in [\gamma, \delta],$$

for all $t \in (0,s)$. Then van der Corput's lemma implies that there is a constant $c > 0$ such that

$$\left| \int_\gamma^\delta e^{ih_{x,t}(w)}\, dw \right| \leq c\big((n+1)na_n^{(1)}2^{-1}t^{-n} - \mu\big)^{-1/2},$$

for all $t \in (0,s)$. Since the right-side of this inequality tends to zero as $t \to 0^+$ it can be argued as in the proof of Lemma 4 that $e^{i\varphi} \notin \mathcal{M}^{(p)}$. □

An examination of the proof of Lemma 5 shows that whenever $\varphi(x) = \psi(x) + \sum_{j=1}^n a_j(x-\alpha)^{-j}$, where the α, a_j, for $1 \leq j \leq n$ are real numbers (and $a_n \neq 0$) with n a positive integer, ψ is a $C^{(2)}$-function in (u,v) for some u and v (satisfying $u < \alpha$ and $\alpha + 1 < v$) and

$$\mu := \sup\left\{ \left|\partial^2\psi(t(w+\alpha))/\partial w^2\right| : t \in [0,1],\ w - \alpha \in [\rho_1, \rho_2] \right\} < \infty$$

for some interval $[\rho_1, \rho_2] \subseteq (\alpha - u, v - \alpha - 1)$ for which it is known that

$$\left| \partial^2 \left(\sum_{j=1}^n a_j t^{-j} w^{-j} \right) \Big/ \partial w^2 \right| \geq ct^{-n}, \qquad w \in [\rho_1, \rho_2],$$

for all $0 < t < 1$, then $e^{i\varphi} \notin \mathcal{M}^{(p)}$ for every $p \neq 2$. This applies, in particular, to the case of

$$\psi(x) = h(x) + \sum_{r=1}^k \sum_{j=1}^{n(r)} a_j^{(r)}(x - \alpha_r)^{-j},$$

where h is a polynomial, the $a_j^{(r)}$ are real numbers and the α_r, for $1 \leq r \leq k$, are distinct real numbers with $\alpha \notin \{\alpha_1, \dots, \alpha_k\}$.

The proof of Theorem 1 follows from these remarks, Lemmas 1, 2 and 5 and the partial fraction decomposition of rational functions.

We conclude with a few remarks. As noted previously, if $\varphi : \mathbb{R} \to \mathbb{R}$ is already an element of $\mathcal{M}^{(p)}$, then so is $e^{i\varphi}$. For φ within the class of polynomials it was observed that $e^{i\varphi} \in \mathcal{M}^{(p)}$, for $1 < p < \infty$ with $p \neq 2$, if an only if $\varphi(x) = \alpha x + \beta$ for some $\alpha, \beta \in \mathbb{R}$. There are, of course, other unbounded (non-polynomial) functions φ for which $e^{i\varphi} \in \mathcal{M}^{(p)}$, e.g., for $\varphi(x) = |x|$ and $\varphi(x) = ln|x|$ (in which case $e^{i\varphi(x)} = |x|^i$); see [4, p.96]. For certain φ it is again possible to apply van der Corput's lemma to show that $e^{i\varphi} \notin \mathcal{M}^{(p)}$ unless $p = 2$. For example, this is the case if $\varphi(x) = |x|^\alpha$, for some $\alpha \in \mathbb{R} \setminus \{0,1\}$. Indeed, if $x \in \mathbb{R}$ and $t \in \mathbb{R} \setminus \{0\}$ are fixed and

$$h_{x,t}(w) = xw - \varphi(tw) = xw - |t|^\alpha |w|^\alpha, \qquad w \neq 0,$$

then (for some $c > 0$) we have $|h_{x,t}^{(2)}(w)| \geq c\alpha|\alpha - 1|.|t|^\alpha$, for $w \in [1/2, 1]$, whenever $\alpha > 0$ (with $\alpha \neq 1$) and, for some $c^* > 0$, that $|h_{x,t}^{(2)}(w)| \geq c^*|\alpha|(1+|\alpha|)|t|^{-\alpha}$, for $w \in [1/2, 1]$, whenever $\alpha < 0$. Using these estimates and the van der Corput lemma it can be argued as before that $e^{i|x|^\alpha} \notin \mathcal{M}^{(p)}$, for every $1 < p < \infty$ with $p \neq 2$. The same is true of e^{ie^x}; in this case $|h^{(2)}_{x,t}(w)| \geq t^2$, for all $w \in \mathbb{R}$. However, these are all ad hoc cases. It is clear that the arguments given in this note are of limited use when considering the question of when $e^{i\varphi} \in \mathcal{M}^{(p)}$ for arbitrary measurable functions $\varphi : \mathbb{R} \to \mathbb{R}$.

Similar questions as considered in this note also arise in $\mathbb{R}^n$; see [3], for example, and the references therein.

References

[1] Dunford N. and Schwartz J.T., *Linear Operators III: Spectral Operators*, Wiley-Interscience, New York, 1971.

[2] Hörmander L., *Estimates for translation invariant operators in L^p spaces.* Acta Math. **104** (1960), 93–140.

[3] Ricker W.J., *Non-spectrality of generators of some classical analytic semigroups.* Indag. Math. (New Ser.) **1** (1990), 95–103.

[4] Stein E.M., *Singular Integrals and Differentiability Properties of Functions*, Princeton Math. Series No. 30, Princeton University Press, Princeton, 1970.

[5] Wainger S., *Averages and singular integrals over lower dimensional sets.* Ann. Math. Studies **112** (1986), 357–421.

[6] Zygmund A., *Trigonometric Series I* (2nd ed.), Cambridge University Press, Cambridge, 1988.

Werner J. Ricker
Math.-Geogr. Fakultät
Katholische Universität Eichstätt-Ingolstadt
D-85072 Eichstätt, Germany
e-mail: `werner.ricker@ku-eichstaett.de`

Operator Theory:
Advances and Applications, Vol. 240, 243–256

Taylor Approximations of Operator Functions

Anna Skripka

Abstract. This survey on approximations of perturbed operator functions addresses recent advances and some of the successful methods.

Mathematics Subject Classification (2010). Primary 47A55, 47B10.

Keywords. Perturbation theory, Taylor approximation.

1. Introduction

An active mathematical investigation of perturbed operator functions started in as early as 1950's, following a series of physics papers by I.M. Lifshits on the change of the free energy of a crystal due to appearance of a small defect. The latter research in physics gave birth to the Lifshits–Krein spectral shift function [32, 27, 28], which has become a fundamental object in perturbation problems of mathematical physics. Subsequent attempts to include more general perturbations than those in [27, 28] have resulted in consideration of higher-order Taylor approximations of perturbed operator functions and introduction of Koplienko's higher-order spectral shift functions [26, 37, 18, 43, 47, 45].

Approximation of operator functions also arises in problems of noncommutative geometry involving spectral flow (see, e.g., [8]) and spectral action functional (see, e.g., [16]). This investigation was initially carried out independently of the study of the spectral shift functions. However, a recent unified approach to the Lifshits–Krein spectral shift function and the spectral flow allowed to establish that these two objects essentially coincide [5]. Higher-order Taylor formulas have been derived for spectral actions in [58], with restrictions on the operators relaxed in [56] by applying more universal perturbation theory techniques.

The proof of existence of the first-order (Lifshits–Krein) spectral shift function, which is due to M.G. Krein, relied on the theory of analytic functions and was of a different nature than the proofs of the other mentioned results on the

Research supported in part by NSF grant DMS-1249186.

approximations of operator functions. An important object in higher-order Taylor approximations is the Gâteaux derivative of an operator function. When the initial operator and the perturbation do not commute, the Gâteaux derivative is a complex object, whose complexity increases with the order of differentiation. Treatment of such derivatives and subsequent derivation of Taylor approximations was based on a delicate noncommutative analysis, which had been developing for some 60 years.

To proceed to a detailed discussion of the aforementioned and further results and methods, we need to fix some notation. We work with a pair of operators defined in a separable Hilbert space $\mathcal{H}$, denoting the initial operator H_0 and its perturbation V. The perturbation is always a bounded operator and, moreover, some summability restrictions are imposed either on V or H_0. In some instances, H_0 is allowed to be unbounded, and we will consider only closed densely defined unbounded operators. For sufficiently nice scalar functions f, we consider the operator functions $f(H_0)$ and $f(H_0+V)$ given by the functional calculus. We are interested in some scalar characteristics associated with perturbations that are calculated using traces (a canonical trace Tr, a Dixmier trace Tr_ω, a normal trace on a semi-finite von Neumann algebra τ, and, more generally, any trace $\tau_{\mathcal{I}}$ on a normed ideal $\mathcal{I}$ continuous with respect to the ideal norm).

We consider the remainders of the Taylor approximations

$$R_{n,H_0,V}(f) = f(H_0+V) - \sum_{k=0}^{n-1} \frac{1}{k!} \frac{d^k}{dt^k}\bigg|_{t=0} f(H_0+tV),$$

where $n \in \mathbb{N}$ and the Gâteaux derivatives $\frac{d}{dt}\big|_{t=0} f(H_0+tV)$ are evaluated in the uniform operator topology. If the nth order Gâteaux derivative is continuous on $[0,1]$, then we have the integral representation for the remainder

$$R_{n,H_0,V}(f) = \frac{1}{(n-1)!} \int_0^1 (1-t)^{n-1} \frac{d^n}{dt^n} f(H_0+tV)\,dt, \tag{1.1}$$

which can be proved by applying functionals in the dual space $(\mathcal{B}(\mathcal{H}))^*$ of the algebra of bounded linear operators on $\mathcal{H}$ and reducing the problem to the scalar case. The questions we are interested in consist in establishing more specific properties of the remainders $R_{n,H_0,V}(f)$.

2. Schatten class perturbations

In this section, we discuss Taylor approximations in the classical setting of perturbations belonging to the Schatten–von Neumann ideals of compact operators

$$S^\alpha = \big\{A \in \mathcal{B}(\mathcal{H}) : \|A\|_\alpha := \big(\mathrm{Tr}(|A|^\alpha)\big)^{\frac{1}{\alpha}} < \infty\big\}, \quad \alpha \in [1,\infty)$$

(see, e.g., [52]). The operator functions under consideration come from either polynomials $\mathcal{P}$ or the functions with nice Fourier transforms

$$\mathcal{W}_n = \big\{f : f^{(j)}, \widehat{f^{(j)}} \in L^1(\mathbb{R}), j = 0,\dots,n\big\}.$$

The class $\mathcal{W}_n$ includes such widely used sets of functions as $C_c^{n+1}(\mathbb{R})$ and the rational functions in $C_0(\mathbb{R})$, which we denote by $\mathcal{R}$.

2.1. Spectral shift functions

As a joint finding of many investigations, we have the following representation for the Taylor remainders corresponding to self-adjoint perturbations of self-adjoint operators.

Theorem 2.1. *If $H_0 = H_0^*$ and $V = V^* \in S^n$, $n \in \mathbb{N}$, then there exists a unique real-valued function $\eta_n = \eta_{n,H_0,V} \in L^1(\mathbb{R})$ and a constant $c_n > 0$ such that*

$$\|\eta_n\|_1 \le c_n \|V\|_n^n$$

and

$$\operatorname{Tr}\left(f(H_0+V) - \sum_{k=0}^{n-1} \frac{1}{k!} \frac{d^k}{dt^k}\bigg|_{t=0} f(H_0+tV) \right) = \int_{\mathbb{R}} f^{(n)}(t)\, \eta_n(t)\, dt, \tag{2.1}$$

for $f \in \mathcal{W}_n$.

The cases $n = 1$, $n = 2$, and $n \ge 3$ are due to [27] (see also [29]), [26], and [43], respectively. The formula (2.1) has been extended from $\mathcal{W}_n$ to the Besov class $B^n_{\infty 1}(\mathbb{R})$ in [38], [39], and [3], respectively. Differentiability of operator functions in the setting most applicable to Theorem 2.1 is discussed in [40] and [6]. The results of [6, 40] can also be used to justify that the trace on the left-hand side of (2.1) is well defined.

The function η_n provided by Theorem 2.1 is called the nth order spectral shift function associated with the pair of operators $(H_0, H_0 + V)$. The name to η_1 was given by M.G. Krein and can be understood from the formula

$$\eta_1(\lambda) = \operatorname{Tr}\big(E_{H_0}((-\infty,\lambda))\big) - \operatorname{Tr}\big(E_{H_0+V}((-\infty,\lambda))\big)$$

holding for H_0 and V finite matrices, where E_H denotes the spectral measure of H. A number of remarkable connections of the first-order spectral shift function to other objects of mathematical physics can be found in the brief survey [9]. More detailed discussion of the first-order spectral shift function can be found in [12, 52, 59] and of the second-order one in [23]. When a perturbation V is in the Hilbert–Schmidt class S^2, the higher-order spectral shift functions η_n can be expressed via the lower-order ones (see [26] in case $n = 2$ and [18, 54] in case $n \ge 3$). The former are more sensitive to the displacement of the spectrum under perturbation, as demonstrated in [53, 55].

The question of validity of

$$\operatorname{Tr}\big(f(H_0+V) - f(H_0)\big) = \int_{\Omega} f'(t)\, \eta_1(t)\, dt, \tag{2.2}$$

was also investigated for non-self-adjoint operators H_0 and $H_0 + V$. Here the set $\Omega \subset \mathbb{C}$ is determined by H_0 and V. The trace formula (2.2) with $\Omega = \mathbb{T}$ (the unit circle) was proved in [28] for unitary operators H_0 and $H_0 + V$ such that $V \in S^1$. The case of arbitrary bounded operators H_0 and $H_0 + V$ differing by $V \in S^1$ is naturally harder than the case of self-adjoint operators. If H_0 and $H_0 + V$ are contractions, then (2.2) holds with $\Omega = \mathbb{T}$ for every f analytic on a disc centered at zero of radius $r > 1$ [31]. Attempts to extend (2.2) to more general functions f resulted in consideration of only selected pairs of contractions and brought to modification of (2.2) with passage to a more general type of integration. The relevant discussion (also for dissipative operators H_0 and $H_0 + V$) can be found in [1, 2, 30, 34, 35, 36, 49, 50, 51].

The higher-order version of (2.2) for pairs of bounded operators has the following formulation.

Theorem 2.2. *Let H_0 and H_0+V be contractions and assume that $V \in S^n$, $n \geq 2$. Then, there exists a function $\eta_n = \eta_{n,H_0,V}$ in $L^1(\mathbb{T})$ such that*

$$\operatorname{Tr}\left(f(H_0+V) - \sum_{k=0}^{n-1} \frac{1}{k!} \frac{d^k}{dt^k}\bigg|_{t=0} f(H_0+tV)\right) = \int_{\mathbb{T}} f^{(n)}(z)\,\eta_n(z)\,dz, \tag{2.3}$$

for $f \in \mathcal{P}$. Furthermore, there exists a constant $c_n > 0$ such that a function η_n satisfying (2.3) *can be chosen so that*

$$\|\eta_n\|_1 \leq c_n \|V\|_n^n. \tag{2.4}$$

The case $n = 2$ for H_0 and H_0+V unitaries, where the derivative is evaluated along a multipicative path of unitaries instead of the path of contractions $t \mapsto H_0 + tV$, is due to [37] (with later extension of the class of functions f in [39]) and for arbitrary pairs of contractions H_0 and $H_0 + V$ joined by the path $t \mapsto H_0 + tV$ is due to [47]. The case $n \geq 3$ is established in [45]. The spectral shift function η_n satisfying Theorem 2.2 is determined uniquely only up to an analytic term (that is, the equivalence class of η_n in the quotient space $L^1(\mathbb{T})/H^1(\mathbb{T})$ is uniquely determined). Theorem 2.2 can be extended to analytic functions f.

2.2. Proof strategy

The proofs of Theorems 2.1 and 2.2 are very subtle and technically involved, so we will give only a flavor of some basic ideas. For simplicity we assume that $\|H_0\| \leq 1$, $\|H_0 + V\| \leq 1$, $V \in S^n$, and $f \in \mathcal{P}$. Then our goal is the formula

$$\operatorname{Tr}\big(R_{n,H_0,V}(f)\big) = \int_{\mathbb{T}} f^{(n)}(z)\nu_n(dz)\,dz, \tag{2.5}$$

where ν_n is a finite measure, with total variation bounded by

$$\|\nu_n\| \leq c_n \|V\|_n^n. \tag{2.6}$$

From the integral representation for the remainder (1.1), we derive

$$\operatorname{Tr}\big(R_{n,H_0,V}(f)\big) = \frac{1}{(n-1)!} \int_0^1 (1-t)^{n-1} \operatorname{Tr}\left(\frac{d^n}{ds^n}\bigg|_{s=t} f(H_0+sV)\right) dt.$$

Thus, if we prove

$$\sup_{t\in[0,1]} \left|\frac{1}{n!} \operatorname{Tr}\left(\frac{d^n}{ds^n}\bigg|_{s=t} f(H_0+sV)\right)\right| \leq c_n \|V\|_n^n \cdot \big\|f^{(n)}\big\|_\infty, \tag{2.7}$$

then application of the Hahn–Banach theorem and the Riesz representation theorem for the dual space of $C(\mathbb{T})$ implies existence of a measure ν_n satisfying (2.5) and (2.6).

For $n = 1$, we have

$$\operatorname{Tr}\left(\frac{d}{ds}\bigg|_{s=t} f(H_0 + sV)\right) = \operatorname{Tr}\big(f'(H_0 + tV)V\big),$$

which in case of f a polynomial follows from the straightforward calculation of the derivative and some combinatorics. Applying the Hölder and von Neumann inequalities then implies (2.7) with $n = 1$ and $c_1 = 1$. This reasoning does not allow to establish the absolute continuity of ν_1 (which was established in [27, 48]), but it can be generalized to apply to the higher-order case. If, in addition, we take H_0 and V to be self-adjoint, then application on the spectral theory allows to derive an explicit formula for ν_1, as it was done in [10].

Apart from the case of commuting H_0 and $V \in S^2$, we do not have the convenient equality $\operatorname{Tr}\left(\frac{d^2}{ds^2}\big|_{s=t} f(H_0 + sV)\right) = \operatorname{Tr}\big(f''(H_0 + tV)V^2\big)$. However, since the set function $A_1 \times A_2 \mapsto \operatorname{Tr}\big(E_{H_0+tV}(A_1)VE_{H_0+tV}(A_2)V\big)$, where A_1, A_2 are Borel subsets of $\mathbb{R}$, uniquely extends to a measure on $\mathbb{R}^2$ with total variation $\|V\|_2^2$, we have

$$\operatorname{Tr}\left(\frac{d^2}{dt^2}\bigg|_{s=t} f(H_0 + sV)\right) = \int_{\mathbb{R}^2} (f')^{[1]}(\lambda_1, \lambda_2)\operatorname{Tr}\big(E_{H_0+tV}(d\lambda_1)VE_{H_0+tV}(d\lambda_2)V\big)$$

(see, e.g., [55, Theorem 3.12]), which along with the estimate for the divided difference $\|(f')^{[1]}\|_\infty \le \|f''\|_\infty$ implies (2.7) with $n = 2$ and $c_2 = \frac{1}{2}$.

When $n \ge 3$, the set function $A_1 \times \cdots \times A_n \mapsto \operatorname{Tr}\big(E_{H_0+tV}(A_1)V \dots E_{H_0+tV}(A_n)V\big)$ can fail to extend to a measure of finite variation on $\mathbb{R}^n$ (see [18, Section 4]). This is one of the reasons suggesting that the case $n \ge 3$ requires much more delicate (noncommutative) analysis of operator derivatives than the case $n < 3$.

Pioneering estimates for norms of nth order operator derivatives are attributed to Yu.L. Daleckii and S.G. Krein [17]. In [17], $H_0 = H_0^*$ and $V = V^* \in \mathcal{B}(\mathcal{H})$, a scalar function f belongs to $C^{2n}(\mathbb{R})$, and the estimates depend on the size of the spectrum of the operator H_0. Development of the Birman–Solomyak double operator integration (see, e.g., [11]) and subsequent multiple operator integration (see [40] and also [6]) resulted in significant improvement of the estimates for operator derivatives. It follows from [40] that for $H_0 = H_0^*$ and $V = V^* \in S^n$,

$$\sup_{t\in[0,1]} \left|\operatorname{Tr}\left(\frac{d^n}{ds^n}\bigg|_{s=t} f(H_0 + sV)\right)\right| \le c_n \|f\|_{B^n_{\infty 1}(\mathbb{R})} \cdot \|V\|_n^n,$$

where $f \in B^n_{\infty 1}(\mathbb{R})$; however, the norm $\|f\|_{B^n_{\infty 1}(\mathbb{R})}$ is greater than the norm $\|f^{(n)}\|_{L^\infty(\mathbb{R})}$. The powerful estimates (2.7) are established in the following theorems.

Theorem 2.3. ([45]) *If $\|H_0\| \le 1$, $\|H_0 + V\| \le 1$, and $n \in \mathbb{N}$, then there exists a constant $c_n > 0$ such that for every $f \in \mathcal{P}$ the following estimates hold.*

(i) *If $\beta > n$ and $V \in S^\beta$, then*

$$\sup_{t\in[0,1]} \left\| \frac{d^n}{ds^n}\Big|_{s=t} f(H_0 + sV) \right\|_{\frac{\beta}{n}} \le c_n \|V\|_\beta^n \cdot \|f^{(n)}\|_{L^\infty(\mathbb{T})}.$$

(ii) *If $V \in S^n$, then*

$$\sup_{t\in[0,1]} \left| \mathrm{Tr}\left(\frac{d^n}{ds^n}\Big|_{s=t} f(H_0 + sV) \right) \right| \le c_n \|V\|_n^n \cdot \|f^{(n)}\|_{L^\infty(\mathbb{T})}.$$

Theorem 2.4. ([43]) *If $H_0 = H_0^*$, $V = V^*$, and $n \in \mathbb{N}$, then there exists a constant $c_n > 0$ such that for every $f \in \mathcal{W}_n$ the following estimates hold.*

(i) *If $\beta > n$ and $V \in S^\beta$, then*

$$\sup_{t\in[0,1]} \left\| \frac{d^n}{ds^n}\Big|_{s=t} f(H_0 + sV) \right\|_{\frac{\beta}{n}} \le c_n \|V\|_\beta^n \cdot \|f^{(n)}\|_{L^\infty(\mathbb{R})}.$$

(ii) *If $V \in S^n$, then*

$$\sup_{t\in[0,1]} \left| \mathrm{Tr}\left(\frac{d^n}{ds^n}\Big|_{s=t} f(H_0 + sV) \right) \right| \le c_n \|V\|_n^n \cdot \|f^{(n)}\|_{L^\infty(\mathbb{R})}.$$

The proofs of Theorems 2.3 and 2.4 (and also analogous estimates for polylinear transformations more general than operator derivatives) include a subtle synthesis of advanced techniques from harmonic, functional, complex analysis and noncommutative L^p spaces as well as development of a novel approach to multiple operator integration. The principal two cases here are the ones of self-adjoints and unitaries, while the case of contractions reduces to the case of unitaries by applying the Sz.-Nagy–Foiaş dilation theory [57].

2.3. Operator Lipschitz functions

Derivation of the estimates of Theorems 2.3 and 2.4 was preceded by resolution of Krein's conjecture on whether every Lipschitz function on $\mathbb{R}$ is operator Lipschitz. Detailed discussion of the problem, including references to partial results, can be found in [41, 46]; here we only state the concluding result and mention some generalizations.

Theorem 2.5. ([46]) *Let f be a Lipschitz function on $\mathbb{R}$. Then, for every $\alpha \in (1,\infty)$, there is a constant $c_\alpha > 0$ such that*

$$\|f(B) - f(A)\|_\alpha \le c_\alpha \|B - A\|_\alpha \cdot \|f\|_{Lip},$$

for all $A = A^*$, $B = B^*$, defined in $\mathcal{H}$ with $B - A \in S^\alpha$.

The best constant $c_\alpha \sim \frac{\alpha^2}{\alpha-1}$ is obtained in [15]. It is known that not every Lipschitz function is operator Lipschitz in S^1 and in $\mathcal{B}(\mathcal{H})$ (i.e., when $\alpha \in \{1,\infty\}$) [20, 21, 22]. Operator Lipschitzness of functions of normal operators and of functions of several variables is discussed in [4, 24].

3. Some natural generalizations

If a perturbation V is not compact and no additional restriction on H_0 is imposed, then the canonical trace Tr of $R_{n,H_0,V}(f)$ is not defined. Depending on the problem, one can consider another trace that is defined on $R_{n,H_0,V}(f)$ for rather general H_0, V, and f, or impose extra restrictions on H_0, f, and/or V to ensure $R_{n,H_0,V}(f) \in S^1$.

3.1. Compact resolvents and similar conditions

Perturbations that arise in the study of differential operators are multiplications by functions defined on $\mathbb{R}^d$, which are not compact operators. In this case, the condition $V \in S^n$ gets replaced by a restriction on the resolvent of the initial operator H_0.

If H_0 equals the negative Laplacian $-\Delta$ and the operator V act as multiplication by a real-valued function in $L^1(\mathbb{R}^3) \cap L^\infty(\mathbb{R}^3)$, then

$$(H_0 - zI)^{-1} - (H_0 + V - zI)^{-1} \in S^1, \quad z \in \mathbb{C} \setminus \mathbb{R} \tag{3.1}$$

(see, e.g., [12]). Due to the invariance principle for the first-order spectral shift function (see, e.g., [12]), the problem for a pair of self-adjoint operators (H_0, V) satisfying (3.1) reduces to the problem for a pair of unitaries with difference in S^1, and (2.1) with $n = 1$ holds for $f \in C_c^\infty(\mathbb{R}) \cup \mathcal{R}$, as established in [28]. In this case, η_1 is an element of $L^1\big(\mathbb{R}, \frac{1}{1+t^2}dt\big)$. Existence of the first-order spectral shift function under more general resolvent conditions is discussed in [25, 60].

If $H_0 = -\Delta$ and V is a multiplication by a real-valued function in $L^2(\mathbb{R}^3) \cap L^\infty(\mathbb{R}^3)$, then instead of the condition (3.1), we have

$$(I + H_0^2)^{-1/4} V \in S^2 \tag{3.2}$$

(see, e.g., [44]). It is established in [26] that for a pair of self-adjoint operators (H_0, V) satisfying (3.2), there exists $\eta_2 \in L^1\big(\mathbb{R}, \frac{1}{1+t^2}dt\big)$ such that the trace formula (2.1) with $n = 2$ holds for $f \in \mathcal{R}$. A modified trace formula is obtained in [44] for a pair (H_0, V) satisfying $(I + H_0^2)^{-1/2} V \in S^2$. The proofs are based on multiple operator integration techniques developed to partly compensate for the lack of the invariance principle under the assumption (3.2).

In perturbation problems of noncommutative geometry, typical assumptions on the operators are that the resolvent of H_0 is compact and $V \in \mathcal{B}(\mathcal{H})$. The following result is obtained in [56], relaxing assumptions on H_0 and V made in [58].

Theorem 3.1. *Let $H_0 = H_0^*$ be defined in $\mathcal{H}$ and have compact resolvent and let $V = V^* \in \mathcal{B}(\mathcal{H})$. Let $\{\mu_k\}_{k=1}^\infty$ be a sequence of eigenvalues of H_0 counting multiplicity and let $\{\psi_k\}_{k=1}^\infty$ be an orthonormal basis of the respective eigenvectors.*

Then, for each function $f \in C_c^{n+1}(\mathbb{R})$, *with* $n \in \mathbb{N}$,

$$\begin{aligned}
&\operatorname{Tr}\big(f(H_0+V)\big) - \operatorname{Tr}\big(f(H_0)\big)\\
&= \sum_{p=1}^{n-1} \frac{1}{p} \sum_{i_1,\dots,i_p} (f')^{[p-1]}(\mu_{i_1},\dots,\mu_{i_p})\, \langle V\psi_{i_1},\psi_{i_2}\rangle \cdots \langle V\psi_{i_p},\psi_{i_1}\rangle\\
&\quad + \operatorname{Tr}\big(R_{H_0,f,n}(V)\big),
\end{aligned}$$

where

$$\operatorname{Tr}\big(R_{H_0,f,n}(V)\big) = \mathcal{O}\big(\|V\|^n\big).$$

Moreover, the trace formula (2.1) with $f \in C_c^3(\mathbb{R})$ is established in [5] for $n = 1$ (this is also discussed in the next subsection) and, under the additional assumption $(I + H_0^2)^{-1/2} \in S^2$, in [56] for $n = 2$. The respective spectral shift functions η_1 and η_2 are locally integrable.

Taylor asymptotic expansions and spectral distributions have also been considered in the study of pseudodifferential operators (see, e.g., [13]).

3.2. Operators in a semifinite von Neumann algebra

Let $\mathcal{M}$ be a semifinite von Neumann algebra of bounded linear operators defined on $\mathcal{H}$ and let τ be a semifinite normal faithful trace τ on $\mathcal{M}$. (The definitions can be found in, e.g., [33].) Note that $(\mathcal{B}(\mathcal{H}), \operatorname{Tr})$ is one of examples of $(\mathcal{M}, \tau)$. Let H_0 be either an element of $\mathcal{M}$ or an unbounded closed densely defined self-adjoint operator affiliated with $\mathcal{M}$ (that is, all the spectral projections of H_0 are elements of $\mathcal{M}$). The perturbation V is taken to be a bounded element of the noncommutative L^p-space associated with $(\mathcal{M}, \tau)$, that is,

$$V \in \mathcal{L}^n = \big\{A \in \mathcal{M} :\ \|A\|_n := \tau(|A|^n)^{\frac{1}{n}} < \infty\big\}, \quad n \in \mathbb{N}.$$

Theorem 3.2. *If* $H_0 = H_0^*$ *is affiliated with* $\mathcal{M}$ *and* $V = V^* \in \mathcal{L}^n$, $n \in \mathbb{N}$, *then there exists a unique real-valued function* $\eta_n = \eta_{n,H_0,V} \in L^1(\mathbb{R})$ *and a constant* $c_n > 0$ *such that*

$$\|\eta_n\|_1 \le c_n \|V\|_n^n$$

and

$$\tau\left(f(H_0+V) - \sum_{k=0}^{n-1} \frac{1}{k!} \frac{d^k}{dt^k}\bigg|_{t=0} f(H_0+tV) \right) = \int_{\mathbb{R}} f^{(n)}(t)\, \eta_n(t)\, dt, \qquad (3.3)$$

for $f \in \mathcal{W}_n$.

The case $n = 1$ was established first for a bounded operator H_0 in [14] and then for an unbounded operator in [7]. The case $n = 2$ is due to [18, 54] and $n \ge 3$ is due to [43]. The strategy of the proof is as described in Subsection 2.2; this strategy can be implemented because noncommutative L^p-spaces have much in common with Schatten ideals (see, e.g., [42]).

The first-order spectral shift function for a pair of τ-Fredholm operators differing by a τ-compact perturbation is known to coincide with the spectral flow [5, Theorem 3.18]. It is also established in [5] that (3.3) with $n = 1$ holds for

H_0 having τ-compact resolvent. (In the case $(\mathcal{M},\tau) = (\mathcal{B}(\mathcal{H}),\mathrm{Tr})$, a τ-compact operator is merely a compact operator.)

Theorem 3.3. ([5]) *If $H_0 = H_0^*$ is affiliated with $\mathcal{M}$ and has a τ-compact resolvent and if $V = V^* \in \mathcal{M}$, then, for $f \in C_c^3((a,b))$,*

$$\tau\big(f(H_0+V)\big) = \tau\big(f(H_0)\big) + \int_{\mathbb{R}} f'(\lambda)\tau\big(E_{H_0}((a,\lambda]) - E_{H_0+V}((a,\lambda])\big)\,d\lambda.$$

Analogs of (3.3) with $n = 1$ and $n = 2$ for pairs of arbitrary (non-self-adjoint) operators in $\mathcal{M}$ differing by a perturbation $V \in \mathcal{L}^n$ are obtained in [19]. As to the case $n \geq 3$, the results of Theorem 2.4 can be extended to pairs of operators in $\mathcal{M}$ by applying dilation of contractions in $\mathcal{M}$ to unitary operators in semi-finite von Neumann algebras constructed in [19].

3.3. General traces

The canonical trace Tr is widely used, but it is not the most "typical" trace. The distinctive feature of Tr is that it is normal, i.e, has the property of monotonicity. A continuous trace on a normed ideal of compact operators in $\mathcal{B}(\mathcal{H})$ other than S^1 has a singular component, which vanishes on finite rank operators. Detailed discussion of traces and applications of singular traces to classical and noncommutative geometry can be found in [33].

Let $\mathcal{M}$ be a semifinite (von Neumann) factor and $\mathcal{I}$ a normed ideal of $\mathcal{M}$ with norm $\|\cdot\|_{\mathcal{I}}$. (The definitions can be found, e.g., in [19, 33].) Let $\tau_{\mathcal{I}}$ be a trace on $\mathcal{I}$ bounded with respect to the ideal norm $\|\cdot\|_{\mathcal{I}}$. Examples of $(\mathcal{I},\tau_{\mathcal{I}})$ include (S^1,Tr), $(\mathcal{L}^1,\tau)$, where τ is the normal faithful semifinite trace on $\mathcal{M}$, and $(\mathcal{L}^{(1,\infty)},\mathrm{Tr}_\omega)$, where $\mathcal{L}^{(1,\infty)}$ denotes the dual Macaev ideal and Tr_ω the Dixmier trace on it corresponding to a generalized limit ω on $\ell^\infty(\mathbb{N})$.

The following results are obtained in [19].

Hypotheses 3.4. *Consider a set Ω, a closed, densely defined operator H_0 affiliated to $\mathcal{M}$, an operator $V \in \mathcal{I}$ and a space $\mathcal{F}$ of functions that satisfy one of the following assertions.*

(i) $\Omega = \mathrm{conv}\big(\sigma(H_0) \cup \sigma(H_0+V)\big)$, $H_0 = H_0^* \in \mathcal{M}$, $V = V^*$, $\mathcal{F} = C^3(\mathbb{R})$;

(ii) $\Omega = \mathbb{R}$, *H_0 and H_0+V are maximal dissipative operators (that is, closed, densely defined operators whose quadratic forms have nonnegative imaginary parts), and*

$$\mathcal{F} = \mathrm{span}\left\{\lambda \mapsto (z-\lambda)^{-k} : k \in \mathbb{N},\, \mathrm{Im}(z) < 0\right\};$$

(iii) $\Omega = \mathbb{T}$, $\|H_0\| \leq 1$, $\|H_0+V\| \leq 1$, *and $\mathcal{F}$ is the set of all functions that are analytic on discs centered at 0 and of radius strictly larger than 1.*

Theorem 3.5. *Let Ω, H_0, V and $\mathcal{F}$ satisfy Hypotheses 3.4. Then, there exists a (countably additive, complex) measure $\nu_1 = \nu_{1,H_0,V}$ on Ω such that*

$$\|\nu_1\| \leq \min\left\{\tau_{\mathcal{I}}\big(|\mathrm{Re}(V)|\big) + \tau_{\mathcal{I}}\big(|\mathrm{Im}(V)|\big),\, \|\tau_{\mathcal{I}}\|_{\mathcal{I}^*}\cdot\|V\|_{\mathcal{I}}\right\}$$

and

$$\tau_{\mathcal{I}}\big(f(H_0+V)-f(H_0)\big)=\int_\Omega f'(\lambda)\,\nu_1(d\lambda),$$

for all $f\in\mathcal{F}$. *If Hypotheses* 3.4(i) *are satisfied, then the measure* ν_1 *is real and unique.*

When $\mathcal{I}=S^1$, the measure ν_1 is absolutely continuous, but when $\mathcal{I}$ is the dual Macaev ideal (with the Dixmier trace), the measure ν_1 can be of any type [19, Theorem 4.4]. Moreover, we do not have an explicit formula for ν_1 in case of a general trace $\tau_{\mathcal{I}}$. Derivation of an explicit formula for ν_1 in case $\mathcal{I}=S^1$, $H_0=H_0^*$, and $V=V^*$ relies on the fact that $\mathrm{Tr}\big(E_{H_0}(\cdot)V\big)$ is a (countably-additive) measure, while the set function $\mathrm{Tr}_\omega\big(E_{H_0}(\cdot)V\big)$ can fail to be countably-additive (see [19, Section 3]).

As another consequence of singularity of Tr_ω (and, more generally, of every trace satisfying $\tau_{\mathcal{I}}(\mathcal{I}^2)=\{0\}$), we have the following linearization formula.

Theorem 3.6. *Assume Hypotheses* 3.4 *and assume* $\tau_{\mathcal{I}}(\mathcal{I}^2)=\{0\}$. *Then,*

$$\tau_{\mathcal{I}}\big(f(H_0+V)-f(H_0)\big)=\tau_{\mathcal{I}}\big(f'(H_0)V\big).$$

Below we consider perturbations in the normed ideal $\mathcal{I}^{1/2}=\big\{A\in\mathcal{M}:|A|^2\in\mathcal{I}\big\}$ and impose an additional natural assumption $\|AB\|_{\mathcal{I}}\le\|A\|_{\mathcal{I}^{1/2}}\|B\|_{\mathcal{I}^{1/2}}$, which, in particular, holds for the ideals S^1, $\mathcal{L}^1$, and $\mathcal{L}^{(1,\infty)}$.

Hypotheses 3.7. *Consider a set* Ω, *a closed, densely defined operator* H_0 *affiliated with* $\mathcal{M}$, $V\in\mathcal{I}^{1/2}$ *and a set* $\mathcal{F}$ *of functions that satisfy one of the following assertions:*

(i) $\Omega=\mathbb{R}$, H_0 *and* H_0+V *are maximal dissipative operators, and*

$$\mathcal{F}=\mathrm{span}\left\{\lambda\mapsto(z-\lambda)^{-k}:\,k\in\mathbb{N},\,\mathrm{Im}(z)<0\right\};$$

(ii) $\Omega=\mathbb{T}$, $\|H_0\|\le1$, $\|H_0+V\|\le1$, *and* $\mathcal{F}$ *is the set of all functions that are analytic on discs centered at* 0 *and of radius strictly larger than* 1.

Theorem 3.8. *Let* Ω, H_0, V *and* $\mathcal{F}$ *satisfy Hypotheses* 3.7. *Then, there exists a (countably additive, complex) measure* $\nu_2=\nu_{2,H_0,V}$ *on* Ω *such that*

$$\|\nu_2\|\le\frac{1}{2}\,\tau_{\mathcal{I}}(|V|^2)$$

and

$$\tau_{\mathcal{I}}\left(f(H_0+V)-f(H_0)-\frac{d}{dt}\bigg|_{t=0}f(H_0+tV)\right)=\int_\Omega f''(\lambda)\,\nu_2(d\lambda),$$

for every $f\in\mathcal{F}$.

Theorem 3.9. *Suppose* $\tau_{\mathcal{I}}(\mathcal{I}^{3/2})=\{0\}$. *Either assume Hypotheses* 3.7 *or else take* $H_0=H_0^*\in\mathcal{M}$, $V=V^*\in\mathcal{I}^{1/2}$, *and* $\mathcal{F}=C^4(\mathbb{R})$. *Then, for every* $f\in\mathcal{F}$,

$$\tau_{\mathcal{I}}\left(f(H_0+V)-f(H_0)-\frac{d}{dt}\bigg|_{t=0}f(H_0+tV)\right)=\frac{1}{2}\,\tau_{\mathcal{I}}\left(\frac{d^2}{dt^2}\bigg|_{t=0}f(H_0+tV)\right).$$

The major components in the proofs of Theorems 3.5 and 3.8 are analogs of the estimates (2.7), which hold due to the continuity of $\tau_{\mathcal{I}}$ with respect to $\|\cdot\|_{\mathcal{I}}$. However, presence of a singular component in the trace $\tau_{\mathcal{I}}$ requires more careful treatment of the operator derivatives than in the case of the normal trace Tr.

References

[1] V.M. Adamjan, H. Neidhardt, *On the summability of the spectral shift function for pair of contractions and dissipative operators*, J. Operator Theory **24** (1990), no. 1, 187–205.

[2] V.M. Adamjan, B.S. Pavlov, *Trace formula for dissipative operators*, Vestnik Leningrad. Univ. Mat. Mekh. Astronom. **1979**, no. 2, 5–9, 118 (Russian).

[3] A.B. Aleksandrov, V.V. Peller, *Trace formulae for perturbations of class S_m*, J. Spectral Theory, **1** (2011), no. 1, 1–26.

[4] A.B. Aleksandrov, V.V. Peller, D. Potapov, F.A. Sukochev, *Functions of normal operators under perturbations*, Adv. Math. **226** (2011), no. 6, 5216–5251.

[5] N.A. Azamov, A.L. Carey, F.A. Sukochev, *The spectral shift function and spectral flow*, Comm. Math. Phys. **276** (2007), no. 1, 51–91.

[6] N.A. Azamov, A.L. Carey, P.G. Dodds, F.A. Sukochev, *Operator integrals, spectral shift, and spectral flow*, Canad. J. Math. **61** (2009), no. 2, 241–263.

[7] N.A. Azamov, P.G. Dodds, F.A. Sukochev, *The Krein spectral shift function in semifinite von Neumann algebras*, Integral Equations Operaor Theory **55** (2006), 347–362.

[8] M.-T. Benameur, A.L. Carey, J. Phillips, A. Rennie, F.A. Sukochev, K.P. Wojciechowski, *An analytic approach to spectral flow in von Neumann algebras.* Analysis, geometry and topology of elliptic operators, 297–352, World Sci. Publ., Hackensack, NJ, 2006.

[9] M. Sh. Birman, A.B. Pushnitski, *Spectral shift function, amazing and multifaceted. Dedicated to the memory of Mark Grigorievich Krein* (1907–1989), Integral Equations Operator Theory **30** (1998), no. 2, 191–199.

[10] M.Sh. Birman, M.Z. Solomyak, *Remarks on the spectral shift function*, Zapiski Nauchn. Semin. LOMI **27** (1972), 33–46 (Russian). Translation: J. Soviet Math. **3** (1975), 408–419.

[11] M.Sh. Birman, M. Solomyak, *Double operator integrals in a Hilbert space*, Integral Equations Operator Theory **47** (2003), no. 2, 131–168.

[12] M. Sh. Birman, D.R. Yafaev, *The spectral shift function. The papers of M.G. Krein and their further development*, Algebra i Analiz **4** (1992), no. 5, 1–44 (Russian). Translation: St. Petersburg Math. J. **4** (1993), no. 5, 833–870.

[13] J.-M. Bouclet, *Trace formulae for relatively Hilbert-Schmidt perturbations*, Asymptot. Anal. **32** (2002), 257–291.

[14] R.W. Carey, J.D. Pincus, *Mosaics, principal functions, and mean motion in von Neumann algebras*, Acta Math. **138** (1977), 153–218.

[15] M. Caspers, S. Montgomery-Smith, D. Potapov, F. Sukochev, *The best constants for operator Lipschitz functions on Schatten classes*, preprint.

[16] A.H. Chamseddine, A. Connes, *The spectral action principle*, Comm. Math. Phys. **186** (1997), 731–750.

[17] Yu.L. Daleckii, S.G. Krein, *Integration and differentiation of functions of Hermitian operators and applications to the theory of perturbations*, (Russian) Vorone. Gos. Univ. Trudy Sem. Funkcional. Anal. **1956** (1956), no. 1, 81–105.

[18] K. Dykema, A. Skripka, *Higher order spectral shift*, J. Funct. Anal. **257** (2009), 1092–1132.

[19] K. Dykema, A. Skripka, *Perturbation formulas for traces on normed ideals*, Comm. Math. Phys. **325** (2014), no 3, 1107–1138.

[20] Y.B. Farforovskaya, *An estimate of the nearness of the spectral decompositions of self-adjoint operators in the Kantorovič–Rubinštein metric*, Vestnik Leningrad Univ. **22** (1967), no. 19, 155–156.

[21] Y.B. Farforovskaya, *The connection of the Kantorovič–Rubinštein metric for spectral resolutions of self-adjoint operators with functions of operators*, Vestnik Leningrad Univ. **23** (1968), no. 19, 94–97.

[22] Y.B. Farforovskaya, *An example of a Lipschitz function of self-adjoint operators with non-nuclear difference under a nuclear perturbation*, Zap. Nauchn. Sem. Leningrad. Otdel. Mat. Inst. Steklov. (LOMI) **30** (1972), 146–153.

[23] F. Gesztesy, A. Pushnitski, B. Simon, *On the Koplienko spectral shift function, I. Basics*, Zh. Mat. Fiz. Anal. Geom. **4** (2008), no. 1, 63–107.

[24] E. Kissin, D. Potapov, V. Shulman, F. Sukochev, *Operator smoothness in Schatten norms for functions of several variables: Lipschitz conditions, differentiability and unbounded derivations*, Proc. London Math. Soc. **105** (2012), no. 4, 661–702.

[25] L.S. Koplienko, *Local conditions for the existence of the function of spectral shift.* Investigations on linear operators and the theory of functions, VIII. Zap. Nauchn. Sem. Leningrad. Otdel. Mat. Inst. Steklov. (LOMI) **73** (1977), 102–117 (Russian). Translation: J. Soviet Math. **34** (1986), no. 6, 2080–2090.

[26] L.S. Koplienko, *Trace formula for perturbations of nonnuclear type*, Sibirsk. Mat. Zh. **25** (1984), 62–71 (Russian). Translation: Siberian Math. J. **25** (1984), 735–743.

[27] M.G. Krein, *On a trace formula in perturbation theory*, Matem. Sbornik **33** (1953), 597–626 (Russian).

[28] M.G. Krein, *On the perturbation determinant and the trace formula for unitary and self-adjoint operators*, Dokl. Akad. Nauk SSSR **144** (1962), 268–271 (Russian). Translation: Soviet Math. Dokl. **3** (1962), 707–710.

[29] M.G. Krein, *Some new studies in the theory of perturbations of self-adjoint operators*, First Math. Summer School, Part I, 1964, pp. 103–187, Izdat. "Naukova Dumka", Kiev (Russian). Translation: Topics in differential and integral equations and operator theory, Birkhäuser Verlag, Basel, 1983, pp. 107–172.

[30] M.G. Krein, *Perturbation determinants and a trace formula for some classes of pairs of operators*, J. Operator Theory **17** (1987), no. 1, 129–187 (Russian).

[31] H. Langer, *Eine Erweiterung der Spurformel der Störungstheorie*, Math. Nachr. **30**, 123–135 (1965) (German).

[32] I.M. Lifshits, *On a problem of the theory of perturbations connected with quantum statistics*, Uspehi Matem. Nauk (N.S.) **7** (1952), no. 1 (47), 171–180 (Russian).

[33] S. Lord, F. Sukochev, D. Zanin, Singular Traces, de Gruyter Studies in Mathematics, 46, Walter de Gruyter & Co., Berlin, 2012.

[34] K.A. Makarov, A. Skripka, M. Zinchenko, *On perturbation determinant for antidissipative operators*, preprint.

[35] H. Neidhardt, *Scattering matrix and spectral shift of the nuclear dissipative scattering theory.* Operators in indefinite metric spaces, scattering theory and other topics (Bucharest, 1985), 237–250, Oper. Theory Adv. Appl., **24**, Birkhäuser, Basel, 1987.

[36] H. Neidhardt, *Scattering matrix and spectral shift of the nuclear dissipative scattering theory. II.* J. Operator Theory **19** (1988), no. 1, 43–62.

[37] H. Neidhardt, *Spectral shift function and Hilbert–Schmidt perturbation: extensions of some work of L.S. Koplienko*, Math. Nachr. **138** (1988), 7–25.

[38] V.V. Peller, *Hankel operators in the perturbation theory of unbounded self-adjoint operators.* Analysis and partial differential equations, Lecture Notes in Pure and Applied Mathematics, 122, Dekker, New York, 1990, pp. 529–544.

[39] V.V. Peller, *An extension of the Koplienko–Neidhardt trace formulae*, J. Funct. Anal. **221** (2005), 456–481.

[40] V.V. Peller, *Multiple operator integrals and higher operator derivatives*, J. Funct. Anal. **223** (2006), 515–544.

[41] V.V. Peller, *The behavior of functions of operators under perturbations.* A glimpse at Hilbert space operators, 287–324, Oper. Theory Adv. Appl., **207**, Birkhäuser Verlag, Basel, 2010.

[42] G. Pisier, Q. Xu, *Noncommutative L^p-spaces.* Handbook of the Geometry of Banach spaces, 2, North-Holland, Amsterdam, 2003, pp. 1459–1517.

[43] D. Potapov, A. Skripka, F. Sukochev, *Spectral shift function of higher order*, Invent. Math., **193** (2013), no. 3, 501–538.

[44] D. Potapov, A. Skripka, F. Sukochev, *On Hilbert–Schmidt compatibility*, Oper. Matrices, **7** (2013), no. 1, 1–34.

[45] D. Potapov, A. Skripka, F. Sukochev, *Higher order spectral shift for contractions*, Proc. London Math. Soc. **108** (2014), no 3, 327–349.

[46] D. Potapov, F. Sukochev, *Operator-Lipschitz functions in Schatten–von Neumann classes*, Acta Math., **207** (2011), 375–389.

[47] D. Potapov, F. Sukochev, *Koplienko spectral shift function on the unit circle*, Comm. Math. Phys., **309** (2012), 693–702.

[48] D. Potapov, F. Sukochev, D. Zanin, *Krein's trace theorem revisited*, J. Spectral Theory, in press.

[49] A.V. Rybkin, *The spectral shift function for a dissipative and a selfadjoint operator, and trace formulas for resonances*, Mat. Sb. (N.S.) **125(167)** (1984), no. 3, 420–430 (Russian).

[50] A.V. Rybkin, *The spectral shift function, the characteristic function of a contraction and a generalized integral*, Mat. Sb. **185** (1994), no. 10, 91–144 (Russian). Translation: Russian Acad. Sci. Sb. Math. **83** (1995), no. 1, 237–281.

[51] A.V. Rybkin, *On A-integrability of the spectral shift function of unitary operators arising in the Lax–Phillips scattering theory*, Duke Math. J. **83** (1996), no. 3, 683–699.

[52] B. Simon, Trace ideals and their applications. Second edition. Mathematical Surveys and Monographs, 120. American Mathematical Society, Providence, RI, 2005.

[53] A. Skripka, *Trace inequalities and spectral shift*, Oper. Matrices **3** (2009), no. 2, 241–260.

[54] A. Skripka, *Higher order spectral shift, II. Unbounded case*, Indiana Univ. Math. J. **59** (2010), no. 2, 691–706.

[55] A. Skripka, *Multiple operator integrals and spectral shift*, Illinois J. Math., **55** (2011), no. 1, 305–324.

[56] A. Skripka, *Asymptotic expansions for trace functionals*, J. Funct. Anal. **266** (2014), no 5, 2845–2866.

[57] B. Sz.-Nagy, C. Foiaş, Harmonic analysis of operators on Hilbert space. Translated from the French and revised, North-Holland Publishing Co., Amsterdam-London, 1970.

[58] W.D. van Suijlekom, *Perturbations and operator trace functions*, J. Funct. Anal. **260** (2011), no. 8, 2483–2496.

[59] D.R. Yafaev, Mathematical scattering theory: general theory, Providence, R.I., AMS, 1992.

[60] D.R. Yafaev, *The Schrödinger operator: perturbation determinants, the spectral shift function, trace identities, and more*, Funktsional. Anal. i Prilozhen. **41** (2007), no. 3, 60–83 (Russian). Translation: Funct. Anal. Appl. **41** (2007), no. 3, 217–236.

Anna Skripka
Department of Mathematics and Statistics
University of New Mexico
400 Yale Blvd NE, MSC01 1115
Albuquerque, NM 87131, USA
e-mail: `skripka@math.unm.edu`

MIX
Papier aus verantwortungsvollen Quellen
Paper from responsible sources
FSC® C105338

If you have any concerns about our products,
you can contact us on
ProductSafety@springernature.com

In case Publisher is established outside the EU,
the EU authorized representative is:
Springer Nature Customer Service Center GmbH
Europaplatz 3, 69115 Heidelberg, Germany

Printed by Libri Plureos GmbH
in Hamburg, Germany